AMERICA'S SOLAR POTENTIAL

PRAEGER STUDIES IN ENERGY CONSERVATION AND SOLAR ENERGY

General Editor:
Avraham Shama

AMERICA'S SOLAR POTENTIAL

A National Consumer Study

Barbara Farhar-Pilgrim
Charles T. Unseld

PRAEGER

PRAEGER SPECIAL STUDIES • PRAEGER SCIENTIFIC

Library of Congress Cataloging in Publication Data

Farhar-Pilgrim, Barbara.
Consumer and residential solar energy.

(Praeger studies in energy conservation and solar energy)
Bibliography: p.
1. Solar energy—Economic aspects—United States.
2. Solar houses—United States. 3. Solar energy industries—United States. I. Unseld, Charles T.
II. Title. III. Series.
HD9681.U62F37 690′.869 82-5231
ISBN 0-03-061696-4 AACR2

Published in 1982 by Praeger Publishers
CBS Educational and Professional Publishing
a Division of CBS Inc.
521 Fifth Avenue, New York, New York 10175 U.S.A.

23456789 052 987654321

Printed in the United States of America

FOREWORD

Avraham Shama

Solar energy and conservation energy have become important sources of energy in recent years. Together, they can potentially supply 30-50 percent of all energy services in the United States in years to come. But while it is clear that demand for solar and conservation energy services will increase in the future, our knowledge of the factors which may explain and perhaps speed this transition is very limited. This is because solar and conservation energy innovations are still a novelty for most people. These innovations supply renewable energy; do not affect the environment negatively as do many fossil fuels; have positive economic impacts; and are often cheaper or more appealing than their alternatives. As a result, both interest in and use of these energy sources have been increasing in the past decade, and more such growth is expected in the future.

Praeger Studies in Energy Conservation and Solar Energy are designed to improve our understanding of these energy sources, their diffusion and market potential, and their impact on various aspects of economic and social life; and to help design more effective public and business policies regarding renewable energy.

First in this study series was Marketing Solar Energy Innovations edited by Avraham Shama and published in 1981. It described solar energy innovations, examined relevant market penetration models, reviewed marketing and consumer use of solar energy, assessed the effectiveness of relevant public and business policies, and demonstrated how the diffusion of innovations framework may be used to accelerate the adoption of solar energy innovations.

The present book, Consumers and Residential Solar Energy: Findings from the National Study by Barbara Farhar-Pilgrim and Charles T. Unseld, is designed to describe residential users of solar energy innovations; discuss their experience with these new energy systems; analyze their purchase and use motivations; and outline consumer segments likely to adopt solar energy innovations in the near future. It uses the conceptual framework of the diffusion of innovations, and is based on two very large surveys—one of homeowners and one of solar homeowners—as well as on the authors' day-to-day study and research of solar energy innovations for over three years. The net result, I think, is a very comprehensive and useful book which includes a wealth of information about residential use of solar energy innovations so far, their market potential, and the strategies to increase their market share. As such, the book is a benchmark study that should be of interest to all students of energy use and consumer behavior, including energy and solar energy business executives, national, state, and local policy makers, and utilities.

In summary, I think the authors made a significant and lasting contribution to our knowledge of residential use of solar energy innovations and their potential. They have much to be proud of, and we have much to be pleased with.

A high quality standard has been set in this study: We hope to maintain it throughout the Praeger Energy Conservation and Solar Energy Series.

FOREWORD

Everett M. Rogers

Almost all of the thousands of past investigations of the diffusion of innovations studied new ideas that were "bigger and better." Generally the innovations studied cost more than the practices they replaced, used more energy, and made life more convenient or luxurious for their adopters. Today, most American citizens realize that the past era of unlimited growth and low-cost energy has been replaced by a period of living within our means. This phase-change in our thinking implies a need to better understand the diffusion of technologies that will enable us to live more energy-efficient lives.

Thus the present book, which reports results from the largest diffusion study yet completed of solar energy, is especially significant. This investigation was conducted while the authors were at the Solar Energy Research Institute (SERI), near Denver, and was funded by the U.S. Department of Energy. The data were gathered in late 1980, a time at which the U.S. government was highly committed to solar energy diffusion as a means of coping with the U.S. energy crisis; for instance, tax-rebate incentives were paid to solar adopters by federal and many state governments. The rate of adoption of solar energy innovations (including solar flat-plate collectors and other types of active and passive systems) is still only a few percent of all U.S. households. But these hundreds of thousands of adopters represent a major behavior change, in particular when we remember that there were only an estimated 150 adopters of solar panels at the time of the first energy crunch in 1973.

Whether the rate of adoption of solar energy will continue to grow remains to be seen, now that the federal government no longer promotes this innovation so enthusiastically. The present book provides important findings about who has adopted solar energy: households of higher income, more education, and those more committed to energy savings. We also learn about the sources and channels of communication leading a householder to adopt solar energy, the effects of the tax-rebate incentives, the high degree of satisfaction with solar energy by adopters, and so forth. These research results lead us to expect a continued diffusion of solar innovations.

So do the present data obtained from a national sample of U.S. households who represent the potential audience of solar adopters. The authors show that the U.S. public has a rather high degree of awareness and some understanding of solar innovations, and a generally favorable attitude toward them. At the heart of the diffusion process is a satisfied adopter telling a friend or neighbor about his or her experiences with the innovation. A great number of these interpersonal networks now seem to be activated for solar energy in the United States, especially in the states where the rate of adoption is high due to climatic and other factors.

This book tells us a great deal about the diffusion of solar energy in the United States. It is unique in many respects and will be read by diffusion scholars, solar enthusiasts, and those individuals who wish to understand better the process through which a national social problem leads to public reaction. I find this a fascinating account of solar diffusion with implications for policy-makers, scholars of human behavior, and the educated public.

PREFACE

Most of this study was done at the Solar Energy Research Institute (SERI) in Golden, Colorado. The Institute was established under the Energy Reorganization Act of October 26, 1974. In that Act, Congress specified that a research and development program be initiated to resolve the major technical problems impeding the diffusion and adoption of solar energy in the United States. Technologies to be part of the R & D program included those suitable for residential applications: solar thermal energy conversion, photovoltaic conversion, wind power, solar heating and cooling of buildings, and bioconversion.

The Act specifies that studies should be done of incentives "to promote broader utilization and consumer acceptance of solar energy technologies." The Act further states that information on energy conservation technologies and energy sources shall be disseminated to the public "through the use of mass communications," thus necessitating research into the information channels about solar energy that various consumers use.

This study is important because it describes who supports solar energy, who buys it, why they buy it, and what they like and don't like about it. A truism about social science studies is that they "discover the obvious." Whether the results reported here are viewed as "obvious" depends upon what the reader thought in advance about the study's topic. If you already knew all of it, at least you have empirical verification for your opinion. If you learn something new and useful, then perhaps the findings were not so obvious. In either case, we will feel that we have done our job well.

ACKNOWLEDGMENTS

A project of this scope and depth could never have occurred without the assistance and support of significant numbers of people. These contributors are here acknowledged and publicly thanked for their invaluable ideas, time, and effort.

First, Tina Hobson and Sandra Bregman, then of DOE's Office of Consumer Affairs, supported the project in a variety of ways. Polly Craighill provided program management. K. Gary Moore of DOE's Active Solar Program provided funding. Without their involvement, the National Study of the Residential Solar Consumer would not have happened.

Second, project consultants provided expertise and moral support. Everett M. Rogers, Institute for Communication Research, Stanford University; Robert Gruber, University of Denver; Harry Cotugno, The Gallup Organization, Inc.; and K. L. Berry, statistical consultant, made a variety of theoretical, methodological, and practical contributions to the study's success.

Third, Rebecca Vories, formerly consumer specialist at the Solar Energy Research Institute (SERI) and currently partner in Infinite Energy, Inc., worked diligently on development of the project, with particular emphasis on its marketing aspects. She has continued to be involved, since starting her own firm, through review and suggestions.

Fourth, SERI staff also made important contributions. F. Floyd Shoemaker and Bill Belew reviewed the project's technical plans, requests for proposal (RFPs), and manuscripts for scientific acceptability, particularly with regard to theory, method, and statistics. Robin Crews provided extensive research assistance on various phases of the project for over two years. Linda Taylor provided excellent administrative and research assistance for over two years. Craig Piernot aided in contractual and conceptual development. Rick Jackson and Doug Madison performed computer analyses for the National Study data. Other SERI reviewers included David Roessner, Barbara Burns, Peter Ketels, Lita Dunham, William Hoagland, Rob Koontz, Avraham Shama, Robert Odland, and Bert Mason.

Finally, under contract to SERI were the Gallup Organization, Inc., which provided consulting, sampling, data collection, editing, and preliminary analysis for the national homeowners survey; Jane Zimmerman, University of Arizona, who conducted the pilot study of nine solar families; and Marketing Management Concepts, Inc., which handled the solar homeowners data. To all of these contributors goes credit for much of what is worthwhile in the study.

The National Study was formally approved by the Energy Information Administration and the Office of Management and Budget.

This book does not reflect the views of the U.S. Department of Energy or the Solar Energy Research Institute, who are not responsible for its content.

CONTENTS

LIST OF TABLES

Table		Page

LIST OF FIGURES

Figure		Page

INTRODUCTION AND GUIDE TO THE BOOK

The National Study of the Residential Solar Consumer was designed to identify factors affecting decisions to adopt solar energy in homes and to learn about the experiences of those already living with solar energy systems. Many analysts have identified motivations for and barriers to the adoption of solar energy systems for homes, yet little empirical data existed on factors influencing people's decisions about solar energy (Farhar et al. 1979b; Unseld and Crews 1979; Farhar et al. 1980).

In 1979, a national goal was established to derive 20 percent of all U.S. energy from the sun by the year 2000. A significant portion of this solar contribution will have to occur in 81 million existing American households, which currently consume about 25 percent of the energy used in the United States, and in new homes constructed in the next two decades. If such changes are to occur in the single-family residential sector, they will require decisions by millions of current and potential homeowners to adopt solar technologies, ranging from domestic hot water to energy-conscious passive designs.

Estimates of active residential solar energy systems in the United States in 1980 range from about 30,000 to more than 140,000 installed systems or significantly less than 1 percent of existing homes (Solar Energy Institute of North America 1979; U.S. Department of Energy 1978; Ward 1979; Ramsay and Niland 1979). The existing estimates of active solar applications do not include homes with passive solar or wind energy systems. The figures will have to increase dramatically by the year 2000 if the national goal is to be met.

The primary objectives of the National Study were to discover:

- Who are those now using solar energy in their homes
- Why they use it
- Who might use it and why
- What the experiences are of those already living with solar energy

Answers to these questions can form the basis for effective incentives, information, and marketing programs for solar energy by industry and government.

Two concurrent national surveys have been completed:

- Personal interviews with a national probability sample of 2,023 homeowners in the 48 contiguous states; the population from which homeowners were selected comprised all owner-occupied year-round housing units in the 48 contiguous states, including mobile homes but excluding town houses, condominiums, and military bases. The 1980 U.S. Census estimate for this population is 52.2 million units.

- A mail questionnaire with a national nonrandom sample of 3,809 solar homeowners; the population from which solar homeowners were selected consisted of all households in the United States using solar energy systems to heat, cool, or light the home. A nonrandom sample of residential solar energy users was compiled by SERI. This list* contained approximately 6,500 residential users of all types of residential solar energy systems. The solar owners list was formed only partly on the basis of self-selection. This is the most extensive sampling of solar owners in existence.

The Solar Energy Research Institute (SERI) developed the survey instruments for both surveys. The Gallup Organization, Inc., conducted the homeowner interviews from October 16 through November 24, 1980. SERI conducted the mail survey of solar homeowners beginning September 10, with responses accepted through December 24, 1980. Marketing Management Concepts, Inc., provided data handling services for the solar homeowners survey.

"Solar energy systems," as used in this study, include those for which individual homeowners make decisions, such as active and passive water heating, space heating and cooling, and wind energy systems. The National Study is not limited to any one solar technology, and its findings are relevant to all solar applications that can be used on homes.

A great deal is known about how technological innovations spread in society, but we still need to know whether the normal processes apply equally well to solar energy technologies or if there is something special about them that distinguishes their diffusion processes from others. Based on what we already know about diffusion of innovations generally, and solar energy specifically, we would anticipate that the following key factors are among those that would affect consumer willingness to adopt solar energy:

- Awareness and knowledge of solar technologies
- Perceived relative advantage of adoption, including both economic and noneconomic benefits

---

*The SERI list is not available to external users; it is protected under the Privacy Act by a system of records, DOE-59.

- Concern about the risk of adoption
- Perceived feasibility and availability of systems
- Opinions of neighbors, friends, and leaders
- Perception of the energy situation

Survey questions were designed to collect data on these and other relevant items, as well as on a full array of demographic variables.

The National Study provides a rich and diverse set of data on homeowner knowledge and attitudes relative to solar energy and on the positive and negative experiences of those already living with solar energy. A vast array of findings are made available here, so a guide to what follows is essential. The study was designed using a theoretical framework to answer a series of questions, such as who the next likely users of solar energy will be. That framework forms the study's "guiding ideas," which are discussed in detail in Chapter 1. The organization of variables in Chapter 1 is used for the later presentation of findings. Preparation for the two national surveys and the way in which they were conducted are described in Chapter 2, on the study's method.

Some of the findings from the national homeowners survey are presented in Chapter 3, where frequency of response to each item in the survey is presented, and the beginnings of relating the study's variables, one to another, are discussed. The reader who wishes to go straight to detailed results should begin with Chapter 3. Chapter 4, also concerning the national homeowners survey, combines both theory and findings on the relationship between demographics and attitudes toward solar energy. Chapter 5 focuses on homeowner attitudes toward utility company involvement in promoting residential solar use—a key question in these troubled energy times.

In Chapter 6, we introduce solar homeowner findings, presenting the frequency of response to each question in the national solar homeowners survey and some key analyses, particularly on solar homeowners' satisfaction with their solar energy systems.

A major multivariate analysis using the study's data has been completed: the "next tiers" analysis. In Chapter 7, the results of this analysis are presented, answering the question of who the next likely residential users of solar energy are. The findings from the National Study have important marketing and policy implications. These are discussed in Chapter 8.

1

GUIDING IDEAS OF THE NATIONAL STUDY

Two major kinds of variables were studied: (1) knowledge, attitudes, and practices regarding solar energy, and (2) respondent characteristics. The first kind of variable, which can be termed "social-psychological," is discussed in this chapter on the study's guiding ideas. The second kind is discussed in Chapter 4, where the description of the study's guiding ideas is continued. Social-psychological variables are the knowledge, attitude, and action sets relative to solar energy, specifically. Respondent characteristics include demographics (such as age and income) and values and life-style preferences that describe respondents, irrespective of their attitudes toward solar energy.

The National Study has been conducted in the diffusion-of-innovations research tradition. The first part of this chapter summarizes that tradition and presents the model of the solar adoption decision process used to guide the research. How the model is used to answer the question of who the next solar users will be is described in the section on "next tiers." Finally, the National Study's dependent variables—what we are explaining—and its independent variables—factors affecting consumer decisions to use solar energy—are presented and discussed.

When we were preparing for the two national surveys, we conducted pilot field interviews on solar energy with solar owners, consumers who nearly became solar owners, and others who had not yet considered adopting solar energy. Quotes from these pilot interviews are included in this discussion to illustrate the concepts that formed the study's guiding ideas.

THE DIFFUSION OF INNOVATIONS

A long-established research tradition views the adoption of innovations as a process occurring over time. Much previous research

FIGURE 1.1. Categorization and Distribution of Adopters of Innovations

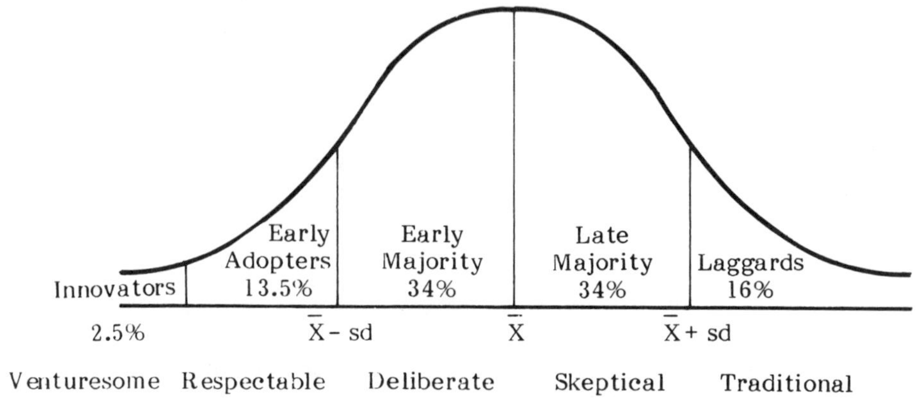

Source: Adapted with permission of Macmillan Publishing Co., Inc. from Everett M. Rogers and F. Floyd Shoemaker, Communication of Innovations, Second Edition. New York: Free Press, 1971. Copyright © 1971 by the Free Press, a Division of Macmillan Publishing Co., Inc.

has focused on characteristics of innovations that affect their rate of adoption—the speed with which the process occurs. Other work has focused on characteristics of innovation adopters. People have been categorized into five types by how quickly they adopt innovations (Figure 1.1). A small proportion (the first 2.5 percent of the population to adopt) are "innovators," the leading edge of adopters. Next, a group of about 13.5 percent is defined as "early adopters," those who benefit from the experience of innovators, maximizing their advantages in adopting the innovation, while minimizing their risks. Early adopters, it turns out, are also frequently "opinion leaders" and, thus, serve as an important social catalyst to shift the innovation's penetration from the select few to the "early majority" (34 percent). Gradually, the "late majority" (34 percent) adopts the innovation, for not to do so would leave them in a worse position relative to everyone else. Finally "the laggards" (16 percent) get around to it after it is unlikely that they will derive much benefit from it, or they never adopt at all. The payoff in adopting any innovation is the advantage (in money, prestige, or satisfaction) that it provides, and the risk in adopting is whether or not the adopter turns out to be right. When most people have adopted an innovation, the market is said to be "saturated."

The diffusion process varies in length for each innovation (Figure 1.2). Besides the characteristics of the innovation, the level of innovation decision making also affects the speed with which innovations are adopted, termed the adoption rate. Innovations that can be adopted by individuals, such as the birth control pill, can reach saturation in as quickly as five years. Innovations requiring organizational and community decisions, such as kindergarten, can take as long as 50 years. Diffusion theory would thus predict that decentralized residential energy conservation and solar applications will diffuse more rapidly than centralized solar use by utility companies, adoption of community solar energy systems, and use of solar technologies for commercial and multifamily buildings. Centralized solar use requires a more complex, time-consuming decision process than does individual solar adoption.

However, it cannot be expected that residential solar energy systems will diffuse in only five years. The solar decision process is complex enough, as will be seen, to require a considerable length of time to complete. Action, such as advertising and consumer education programs, can be taken to speed up the process, yet solar as an energy form is sufficiently different from conventional forms to

FIGURE 1.2. Examples of Diffusion Curves

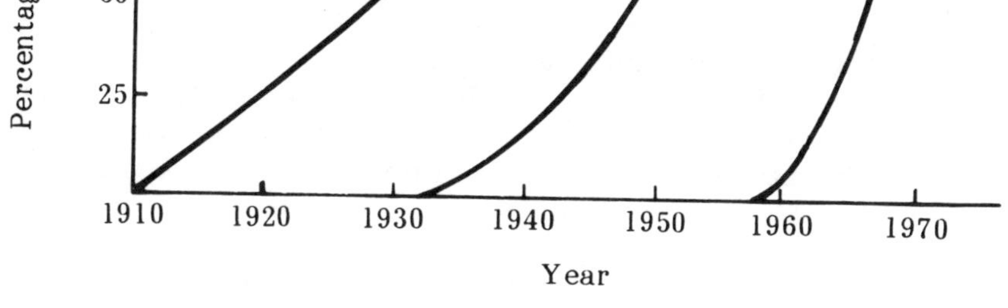

Source: Adapted with permission of Macmillan Publishing Co., Inc. from Everett M. Rogers and F. Floyd Shoemaker, Communication of Innovations, Second Edition. New York: Free Press, 1971. Copyright © 1971 by the Free Press, a Division of Macmillan Publishing Co., Inc.

require considerable consumer sophistication to put it to work. This takes some time to acquire.

A SOLAR INNOVATION DECISION-PROCESS MODEL

Reaching an innovation decision has been conceptualized as a dynamic process, with stages occurring over time.

The central theoretical questions of the National Study are:

- How well do the existing models of the innovation decision process fit the diffusion of solar energy? Is there anything unique about solar technologies that would set them apart from other technological innovations, requiring modification of existing theories?
- Given the diffusion process, what factors affect the ultimate positive or negative position an individual takes toward solar energy and his behavior to adopt or reject it?

The dynamic model used in this research is an adaptation of earlier models of the innovation decision process (Rogers and Shoemaker 1971; Berning and Jacoby 1974). Its stages are defined as follows:

Model of the Solar Decision Process

Initial Awareness
Knowledge-Evaluation
Decision-Intention
Action
Observation of Effects
Continuance/Discontinuance Decision

In this model, the initial awareness stage refers to having heard about solar technologies, but it is also intended to include an impetus to be interested in them. The impetus can take two forms: (1) perception that a personal or social problem exists requiring new solutions, or (2) interest stemming from a variety of other sources, such as professional interest, business interest, interest in do-it-yourself projects, social pressure, or interest based on social values. At the end of this stage an individual may be disinterested, eager to know more, or somewhere between.

The interest that comes into being in the initial awareness stage "propels" the person further into the process. In the knowledge-evaluation stage, the individual learns more about how the innovation works, how much it costs, where to get it, who is using it and with what results, who is for and who against it, and how it might "fit" his own situation. At the end of this stage, an individual has formed a favorable or unfavorable attitude—a position—toward the innovation, both in terms of its general use and its specific relevance to him. It is possible that he could be favorable to the new idea generally, yet unfavorable to his own involvement in it. If he is favorable to becoming involved, this "propels" him to the next stage of the process: decision-intention. In this stage, the individual decides to become involved with the innovation and makes his plans to adopt it within the foreseeable future. He has "behavioral intention," that is, he plans to act. He engages in prepurchase information-seeking behavior such as shopping around, getting bids and estimates, and planning schedules and financing. If no major obstacles or unanticipated events intervene, the likelihood is high that the individual will pass to the next stage.

In the action stage, the actual purchase is made or the project built. For solar energy, it is unlikely that a small-scale trial can be entered into; therefore, a more major commitment has to be made. This stage is not yet termed "adoption," since it is still possible that the individual's experience with the innovation may cause him to reject it. Once the action stage has been reached, the last two stages inevitably follow.

In the observation of effects stage, the individual lives with the consequences of the purchase action in all its manifest forms—its costs, benefits, what others say about it, its positive and negative psychic effects, and so on. After a time, the individual decides that he made a good choice and becomes satisfied, or he becomes regretful. If problems arise during this stage, he attempts to solve them.

In the continuance/discontinuance stage, the individual is either satisfied or dissatisfied with his choice, and he takes appropriate action. If he is satisfied, the adoption process is then complete. If he is strongly enough dissatisfied, he will take action to have the innovation removed or to sell the innovation to someone else.

This conceptualization of the solar decision process includes two ideas: (1) any individual may be unaware of an innovation and thus not involved at all in the process, and (2) any individual may stop or spend varying lengths of time at any of the first three stages of the process. There is no necessity for any individual to complete the process. Logically, innovators tend to complete the process before early adopters, early adopters before the early majority, and so on.

In the National Study homeowners survey, only the first three stages (awareness, evaluation, and intention) could be studied systematically, since the proportion of solar owners in the sample was anticipated to be too small for reliable results. The last three stages (action, observation, and adoption) are covered through the solar homeowners sample, from whom retrospective data on the earlier stages were also collected.

"NEXT TIERS"

We have described a dynamic process model of the adoption decision process. The National Study involves two concurrent national surveys, yet surveys are "cross-sectional." They measure constantly changing awareness, attitudes, and practices at a given moment of time. How, then, does the dynamic model relate to a static measurement of the population?

With respect to any given innovation, it is assumed that the relevant population falls somewhere along a continuum from ignorance about the innovation to adoption or discontinuance. The question is, how many individuals are in each stage of the process? A measurement of the numbers of the population at any one stage permits inferences to be drawn about the shape of the solar diffusion curve across time. This would permit, for example, more accurate estimates of the quads* solar energy will produce in the residential sector in the future and adjustment of planning and policies, accordingly. Further, if we can learn why people are at one stage and not another, we can more effectively anticipate how to inform, educate, devise policies for, and market to, the relevant segments in the population.

We anticipated that a cross section of the population would fall into the solar decision process approximately along the lines shown in Figure 1.3.

The shape of the pyramid suggests that most people would currently fall into the initial awareness and knowledge-evaluation stages. We anticipated that relatively fewer people would be totally unaware of solar technologies for residential use than would have been unaware in 1973, for example. A few people would be in the decision-intention stage, and fewer still in the action stage and beyond. In later years, when solar energy systems are more widespread (if that occurs), the shape of the pyramid would change as greater numbers of the population progressed toward an adoption decision. Finally, as solar energy systems approach "saturation," the figure would resemble an inverted triangle, with most of the population having taken action and just a few at the bottom tip of the triangle, still unaware or in the initial awareness stage.

---

*One quad is one quadrillion Btu of energy.

future energy needs. In this way, we gain a realistic assessment of how favorability to solar energy compares with favorability toward other energy possibilities. Each energy supply source has a set of concerns associated with it, for example, coal may cause air pollution, nuclear plants may be a source of radioactivity, and so on. Thus, preferences for different energy supply sources are probably the expression of a complex evaluation of the economic, environmental, social, and political desirability of these sources, along with an assessment of their reliability.

In addition, favorability was measured by asking consumers whether they favor or oppose the use of solar energy for homes and, in a separate question, its use in their own homes.

Favorable evaluation precedes the intention to adopt solar energy, yet favorability alone may not be enough to account for solar adoption. Perceived feasibility of retrofit for the particular home could be an important factor. A potential adopter could be knowledgeable and have all the "right" attitudes but, without a feasible solar retrofit site, might still not install a solar energy system. Similarly, if solar homes or systems are not locally available, there is little a potential adopter can do about the business infrastructure. Finally, a certain level of financial ability must be present to enable adoption.

Four types of barriers to adoption, once positive evaluation has been achieved, have been identified, following Leonard-Barton and Rogers' (1979) discovery that mechanical ability appeared to be an important predictor of solar adoption. These are:

- Availability; if solar homes and/or solar systems are not locally available, adoption may be desired but may not occur.
- Financial ability; if adoption is desired, but income is inadequate and incentives do not remedy this, adoption can be forestalled.
- Mechanical ability; "sweat equity" or do-it-yourself installation may be enjoyable, decrease cost, heighten the sense of self-reliance, and increase confidence in being able to handle systems. The ability to maintain systems, even if commercially installed, may enhance willingness to have a system.
- Structural feasibility; a homeowner may not adopt solar energy if he does not believe it can be retrofitted on his existing home. (Of course, perceptual data do not represent an assessment of actual structural feasibility of solar retrofit.)

Behavioral Intention

Behavioral intention refers to plans to invest in a solar energy system or solar home and the activity associated with these plans, such as shopping around. A 1979 national survey showed that 6 percent of the adult population were thinking of buying a solar energy system in the next two or three years (Roper 1979a). More recent data are presented in this study.

Favorable evaluation will likely lead to behavioral intention to adopt solar energy. Behavioral intention, in turn, will probably lead to solar adoption.

Adoption

Adoption of solar energy is the actual behavior or action that is the first outcome of the innovation decision process. Adoption distinguishes solar owners from nonowners. In this discussion, the term "adoption" is substituted for the term "action," defined in our model as the actual purchase of a system or the building of a solar home.

Included in this category are behaviors related to energy and resource conservation. We anticipated that solar adoption would be related to conservation activities. Conservation activities and behaviors seem to fall along a continuum from least costly, unfamiliar, risky, complex, and inconvenient to more costly, risky, complex, unfamiliar, and inconvenient. The effectiveness of these activities in actually conserving energy may be correlated with the continuum, with the least energy saved at the low end and most energy saved at the high end.*

Examples of behaviors at the low end of the continuum are simple conservation measures, such as turning off lights when not in use. This practice has already been adopted by most of the population (Farhar et al. 1979b). In the middle of the scale are such activities as weatherization, adding storm windows, and adding insulation. Considerably fewer than half of the population has reported taking these actions. At the high end of the scale are adoption of solar technologies and building solar homes. There may also be affiliated life-style adaptations, such as scheduling cooking, bathing, or laundry activities to off-peak hours.

We expected that these conservation activities would form a

---

*It should be noted that "energy saved" refers to energy from a conventional supply source. Further, many low-cost or no-cost energy conserving actions, such as lowering thermostat temperature settings, can be remarkably effective in decreasing energy use. The generalization is intended "loosely."

cognitive, affective, and behavioral "cluster," that is, they "go together." Consumers tend to enter the scale at the low end. Those who have advanced to the middle of the scale are more likely to become solar adopters.

Continuance/Discontinuance Decision

Just as there are thought to be complex motivations involved in the solar adoption-decision process, so satisfaction with that decision is likely to be a complex phenomenon. Satisfaction with solar adoption can be categorized as follows:

1. Economic: Present and anticipated future savings on utility bills and monthly costs of solar energy system
2. Physiological: Comfort, for example, more comfortable living zone in passive homes; more humid environment in dry climates from greenhouses
3. Psychological:
   - Increased sense of control over one's life; self-sufficiency and self-reliance
   - Self-esteem due to environmental protection; contribution to solving society's problems
   - Increased awareness and appreciation of the natural environment, for example, weather patterns
   - Increased awareness of energy consumption patterns
4. Social:
   - Increased prestige, status
   - Family routine unaffected or affected positively

Economic motivations for solar adoption have generally been considered paramount by policymakers, and are clearly important in the solar decision process. The degree to which expectations for economic benefits are satisfied by owning a solar energy system will be related to how satisfied owners are with it.

Here is what some solar homeowners reported in the field interviews.

- Utility bills are vastly lower than in our previous house.
- I know it's reduced my utility bills.
- The reason we put it in was to reduce the utility bill

and conserve energy because we were aware of the fact that the cost of energy was going up.

Physiological motivations involve having a more physically comfortable living environment. Many consumers complain in terms of health and comfort about drafty, "oily" heat that results from existing conventional furnaces.

Solar homeowners brought up the following physiological bases for satisfaction:

- It's healthier . . . you feel better when you're breathing cool winter air and you work better.
- Fresh, pure food from the greenhouse; a sense of independence; it adds humidity.
- We like the house better than conventional houses. We never run out of hot water.

Psychological motivations for using solar energy are more subtle but no less real or important than economic ones. If owners feel rewarded by experiencing psychological satisfaction, they will tend to evaluate their solar experience positively.

Solar owners expressed some of their psychological satisfaction in these words:

- It gives me a sense of security because energy bills are going up, independence, better heat, satisfaction from saving energy.
- It's morally right . . . it improves your life-style and relationship with earth.
- It's a status symbol.
- It helps keep the rent down; there is more reliable hot water than from the boiler; and it makes me happy to conserve.

Finally, social motivation to adopt involves what solar energy can contribute both to one's own family life and to the community or nation. Satisfaction can derive from these wishes being fulfilled. For some solar adopters, their action may be viewed as insurance against future life-style change. For example, those anticipating retirement on fixed incomes or high-energy consumers adopting out of guilt over high energy use may wish to maintain prior life-styles (Leonard-Barton 1978; Rogers et al.

1979, p. 41). Satisfaction for these adopters would thus be linked with a relatively unaffected life-style. For some solar adopters, life-style is not affected (Hamrin 1978). For owners of homes with passive features requiring routine interaction (for example, skylights, vents), the activity will likely become routine and, therefore, will not be perceived as negative or burdensome.

One solar owner reported social satisfaction in these words:

- It has given the neighborhood a real uplift. Communities from all over the world have come to see our model for saving energy while creating jobs.

Some of the comments from the pilot interviews reflected on living with a solar energy system:

- When people ask me, how do you like living in a solar house, we always say it's just like living in any other house. There really is no difference.
- It doesn't change the way we live one bit . . . it's just like having any old house.
- It's easier to keep a steady temperature in the house . . . spend more time reading and reflecting on the use of the sun. It happens to fit our life-style.
- We found all the things we'd been used to were just things we were used to living with; they weren't necessities, just conveniences.
- It's just like you get up in the morning and you open the blinds or the curtains, and you open the shutters. You shut them when it starts getting warm. It's just so obvious, you don't even have to think about it.

Consumer expectations for their experience with solar energy systems could be an important determinant of satisfaction after adoption. Positive experiences with solar energy systems in one study were not significantly correlated with satisfaction (Sawyer 1979). In this study, consumers with lower expectations (that is, those who anticipated problems with their systems) were more satisfied than those who had higher expectations. Those who anticipated some technical problems were more satisfied than those who did not. Expectations for system performance can be modified if consumers are aware of solar insolation in their climate, the length of experience of their contractor, and the relative complexity of the system installed. Realistic expectations have a better chance of conforming to actual experience.

Satisfaction with system ownership is the result of having varied expectations met, both economic and noneconomic. For many adopters, the noneconomic bases of satisfaction are probably the most meaningful; for others, saving money is the most important element of the entire experience. But even when this is the case, a solar owner with a low utility bill is likely to feel psychological gratification that he was smart to have made the decision he did.

The National Study sought a straightforward account of costs and benefits, problems encountered and how they were solved, from as broad a sample of solar homeowners as possible.* Theoretically, the decision to adopt a solar energy system is not considered complete until the owner has had an opportunity to evaluate the decision—to experience the system—and to become satisfied or dissatisfied with it. If the owner is satisfied over a period of time, then he has completed the adoption-decision process. If, however, he is dissatisfied, then he can reject the innovation by selling his solar home or having the system removed. Both of these outcomes are possible.

FACTORS AFFECTING SOLAR ADOPTION

In the initial awareness and knowledge-evaluation stages of the adoption-decision process, a number of factors that will affect the ultimate outcome come into play.† These factors speed up or slow down the rate at which the innovation is adopted. The study included items to measure the way that people vary in their perceptions about solar energy along dimensions known to hasten or slow down their progress toward adoption of innovations. In the National Study, these

---

*Other researchers have identified "cognitive dissonance" as potentially playing a key role in owner satisfaction. Once owners purchase a solar energy system they have, in effect, a "vested interest" in being satisfied with that decision. According to cognitive dissonance theory, owners would thus tend to seek out information favorable to the decision they have made and to avoid information unfavorable to that decision. Dissonance theory, however, does not account for the phenomenon of regret, which also occurs.

†The characteristics of an innovation that will affect its rate of adoption have been described in the literature as: compatibility with cultural values and norms, perceived complexity, perceived relative advantage, trialability, and observability (Rogers and Shoemaker 1971).

FIGURE 1.5. The National Study's Independent Variables

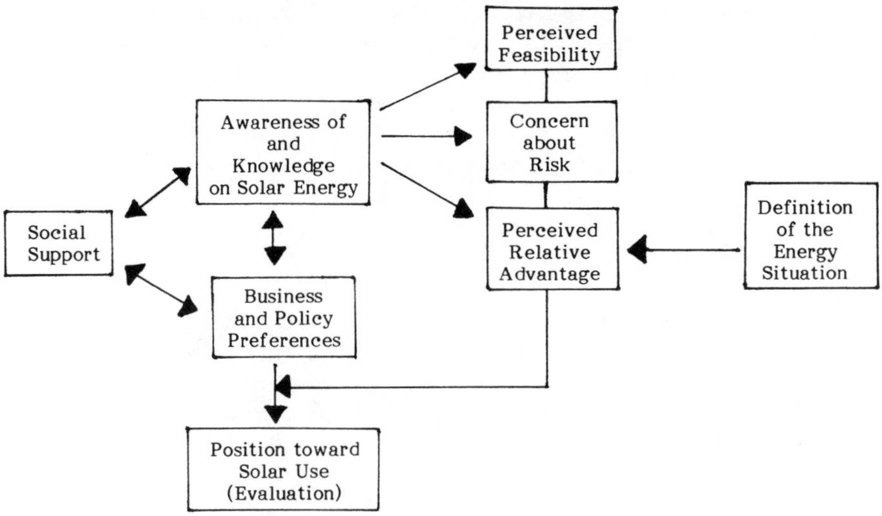

Source: Constructed by the authors.

characteristics are measured perceptually, that is, people were asked what they perceived the characteristics of solar energy to be.

Factors affecting solar adoption are the study's "independent variables." They are the reasons why individuals stand where they do in the solar innovation-decision process. These are:

- Definition of the energy situation
- Social support
- Knowledge and information sources
- Perceived feasibility of solar energy
- Concern about risk
- Motivation (perceived relative advantage of solar adoption)
- Business and policy knowledge and preferences

Each of these factors is expected to influence a consumer's position toward solar use and his ultimate decision on whether or not to become a solar user. The relationship between these factors and the dependent variables is shown in Figure 1.5. As each factor is described, points are illustrated by quotes from the pilot field interviews completed prior to preparing the survey instruments.

Definition of the Energy Situation

Belief in a national and/or personal energy problem may be an impetus to solar adoption. Perception of future energy supplies may be the best predictor of adoption behavior (Rogers and Leonard-Barton 1979). Existing empirical studies of solar owners reported mention of "fears of possible future energy shortages" as a factor affecting the solar adoption decision (Unseld and Crews 1979). Respondents expressed these concerns in the following words:

- In less than a decade the fossil fuel situation is going to be impossible; it is going to be so expensive and so scarce that we will have to go to alternatives.

A great deal of information exists on public perceptions of the reality and severity of the energy crisis (Farhar et al. 1979b). The public does not define energy as one of the most important national energy problems, but high proportions define the national energy situation as serious. In prior studies, "belief in an energy crisis" has not been consistently found to correlate to adoption of energy conservation behavior (Farhar and Mewes 1976).

The potential adopter may perceive a significant national energy problem and think that individual use of solar energy will help in solving that problem (altruistic view). In addition, the potential adopter may perceive a significant personal problem to which solar energy may be a solution (egoistic view). Rising household energy costs and/or other personal impacts of the energy supply situation may cause a homeowner to define the situation as a personal problem.

- I don't know if there is really a shortage, but whether it's real or not, we're excessive in our energy use. We're going to run out someday . . . that's very real.
- Sometimes people don't open their eyes and ears until things affect them directly.
- There is plenty [of energy] available, but society doesn't use what's available, like solar and coal. We use sources inappropriately, like wasting petroleum on electricity and gas.

We anticipated that:

- Solar adoption will be more likely among those who feel they have a significant personal energy problem.
- Solar adoption will be more likely among those who believe in a significant national energy problem or "crisis."

We judged that these relationships should be tested in the National Study.

Social Support

Is solar energy compatible with the "American way of life"? The "compatibility" of an innovation is "the degree to which an innovation is perceived as consistent with the existing values, past experiences and needs" of the potential adopters (Rogers and Shoemaker 1971). Compatibility may be a function of, and may be expressed by, solar's "image." Both social and personal normative beliefs are relevant to compatibility, that is, what individuals themselves think and what they think others think (Leonard-Barton and Rogers 1979). People are often persuaded to purchase products or adopt new ways of doing things as a result of what others say. If an individual's authority figures, reference group leaders, opinion leaders, close friends, and business associates favor solar energy, that person is also more likely to do so.

Social support for solar adoption can be provided by neighbors. The perception of a house as experimental appears to be related to the number of other similar houses constructed in the same neighborhood and to the aesthetic design of the house (Hamrin 1978). Greater appeal and acceptance of innovative housing technology and design occurs when such houses are part of a whole subdivision than when they are built as a single home among a group of conventional houses. Clustering of solar homes provides more support for the individual homeowner should he have technical problems, a broader base of information related to conservation behavior and technology, and more encouragement and enthusiasm for experimentation. With respect to solar homeowners in subdivisions, Hamrin reported that, "Their behavior really changed over just a couple of months to becoming fairly good conservers because they got training from their neighbors. They also got positive reinforcement for the good things they were doing."

Field interview respondents talked about the response of neighbors and friends to solar adopters. In their words:

- Living in this building—people sit up and take notice. It's a great topic of conversation. It's made me at least an interesting person . . . a celebrity.
- More and more, housing is maybe replacing the automobile as a symbol of personal identity.
- When people move into a solar community, there's a lot of awareness. One gets kind of wrapped up in energy conversation and therefore becomes a better energy conserver.
- Having the lowest utility bill is a status symbol around here.

We anticipated that probability of adoption will be increased by perceived compatibility of solar technology with existing community norms. Compatibility was measured in the study through items on solar's "image" and on whether "significant others" were favorable or opposed to the use of solar energy for homes. Significant others included neighbors and friends, co-workers, spouses, and all levels of government.

Knowledge and Information

When we reviewed prior research on consumer energy information needs we found an important research gap (Farhar et al. 1980). Little is known about what information sources on solar energy are used and perceived as credible, levels of exposure to information sources, levels of knowledge about solar energy, and information needs. Therefore, a major part of the National Study focused on information and knowledge variables.

Information Sources

Earlier knowers of an innovation have greater exposure to mass media and interpersonal channels of communication than do later adopters. Earlier adopters also seek information about innovations more than later adopters do (Rogers and Shoemaker 1971). We would thus expect information and knowledge levels to be related to stages of the innovation decision process, with those closer to adoption being the most well informed segments of the population about solar energy. How and where potential adopters get useful information about solar energy is crucial for policymakers and marketing experts to know for effective program design.

Pilot interview respondents put their own experiences with sources of information in these words:

- I have a friend who builds and he is into solar energy. He is the kind of person that I have a lot of confidence in, faith and trust in.
- If someone would just put a lot of money behind PR work and try to promote it . . . I think it would sell better.
- I would like to know how efficient solar homes are; I would want to buy one that someone lived in before and get feedback. I'd like to know how it operates.
- [Re a DOE publication] It's incomprehensible to the public.
- Actually installing stuff is the best way of communicating . . . they have the immediate reinforcement of lowered fuel costs.

The literature varies on whether word-of-mouth or mass media channels of communication are more important earlier or later in the innovation decision process. Some say advertising has no direct influence on ultimate product purchase (O'Brien 1971, cited in Berning and Jacoby 1974). Most would agree that mass media are most important in creating initial awareness of an innovation or product, while interpersonal channels are more important in attitude formation and behavioral intention stages (Rogers and Shoemaker 1971; Beal and Rogers 1957, cited in Berning and Jacoby 1974; Roessner et al. 1979). Interpersonal communication channels are thought to be more persuasive and more effective in overcoming feelings of excessive psychological risk among prospective adopters (Roessner et al. 1979).

The following have been mentioned as preferred solar information sources in previous studies (Farhar et al. 1979b; Unseld and Crews 1979):

- Federal energy agencies
- Local utilities
- Local heating/air conditioning contractors
- Local plumbing contractors
- Solar companies

- Community organizations
- Owners of solar energy systems
- Friends, neighbors, acquaintances
- Fairs and exhibits
- Workshops and lectures

We anticipated that likelihood of solar adoption would be increased by exposure to positive information from credible sources. Conversely, exposure to negative or mixed information from credible sources would tend to decrease likelihood of adoption. Several items in the homeowners survey examined information sources, including items on exposure to solar buildings in the neighborhood, information needs, knowledge of tax incentives, and likelihood of using (and thus the credibility) of a variety of organizational and substantive information sources.

Knowledgeability

Those who adopt innovations soonest have greater knowledge about them than do others in the population. Knowledge of an innovation spreads ahead of the actual adoption of that innovation. There appears to be a threshold of knowledge about an innovation in a social system, below which little adoption is likely to occur. This threshold tends to be about 20 percent of the total population of the social system (Roessner et al. 1979). The level of knowledge involved is greater than mere awareness of the innovation; it must be enough to permit evaluative processes to occur. Knowledge, however, does not guarantee favorability toward an innovation. For example, with respect to weather modification technology, more knowledgeable respondents are significantly more likely to be opposed to cloud seeding (Farhar and Mewes 1976; Farhar et al. 1978). The relationship between knowledge and evaluation may actually be curvilinear; that is, as knowledge about an innovation increases, evaluation is likely to become more favorable <u>or</u> more opposed, and less likely to remain neutral or undecided.

Knowledge is usually measured in two ways: (1) by subjective assessment in which the respondent indicates his extent of awareness ("very well informed" or "not at all informed"), and (2) by objective assessment, in which items are used to check accuracy of knowledge. The National Study primarily used subjective assessment to measure levels of knowledge about solar energy.

Prior research shows that consumers need more information about solar energy before they can make a decision to use it (Farhar et al. 1979b). It is important to know what types of information about solar energy are needed.

Awareness of solar technologies has been increasing. A 1974 national sample was asked if they had ever seen a "unit that transformed heat from the sun into energy," and 9 percent answered affirmatively. Two marketing studies reported that about two-thirds had read articles about solar energy. Awareness of solar heating was reported by 83 percent of a later national sample, and awareness of wind energy was reported by 75 percent of a marketing sample (Farhar et al. 1979b). More recent data are reported in this study.

Pilot field interviews that we conducted revealed a high degree of understanding among solar owners on how systems work, how to operate systems efficiently, and how to conserve energy in conjunction with the use of a solar energy system. Solar owners evidenced a degree of sophistication, which increased over time, on how to obtain the greatest performance with their systems.

We anticipated that awareness precedes understanding of solar energy systems and that likelihood of solar adoption is increased by several kinds of knowledge. Consumers have to know about kinds of systems and what they can do, where to buy systems, who is a reputable supplier, what is feasible for their homes, and what financial incentives exist.

Solar companies have frequently reported that educating consumers on solar energy in general is a time-consuming, and therefore costly, service they are forced to perform as part of their marketing efforts. Other methods of informing consumers about solar energy would thus be a direct aid to those marketing systems, who would benefit by having more informed prospects.

In the homeowners survey, included items measure subjective assessment of knowledgeability, awareness of solar technologies for residences and whether respondents have actually seen them, whether respondents feel informed enough to make a purchase decision, awareness of federal and state solar tax credits, perceived system costs, and awareness of where to buy solar energy systems and services locally.

Perceived Feasibility of Solar Energy

> Belief in effectiveness or perceived feasibility of innovations has been found to be a significant predictor of favorability toward them (Farhar and Mewes 1976; Krane 1977; Farhar et al. 1978). For solar energy, elements of feasibility include performance expectations for systems, reliability, and cost effectiveness. Perceived effectiveness thus includes the following components:

- Belief that a solar energy system can actually produce heat, hot water, or electricity
- Estimates of the percentage of the household's heat, hot water, or electricity that a solar energy system can provide or of the proportion of utility bills that can be saved; these would be expectations for solar energy system performance
- Beliefs about the need for a backup system for a solar energy system (is this perceived as part of a workable system or as a sign that systems are not presently feasible?)
- Perceived feasibility of solar energy systems for use at present, or perception of solar energy as a future rather than as a presently feasible technology
- Perception of system durability over time

Respondents in the pilot field interviews commented on the effectiveness of solar energy systems in these words:

- I think there are probably more efficient collectors than they're using . . . hell, they're using beer cans. It's an experiment because they are cheap . . .
- To take on the added cost of unproven systems leaves it less than desirable.
- I don't think you should expect everything to be 100 percent perfect the first time . . .
- I've found they haven't worked the kinks out of solar . . . people won't invest 'til they do.
- Not sure how practical they are . . . I am not so darn sure solar energy is the answer . . . there may be a combination of answers.

We expected that favorability and solar adoption will be more likely if solar energy systems are perceived to be effective in the ways described above.

Concern about Risk

All energy supply sources have components of perceived risk (Farhar et al. 1979b). Air pollution and other environmental damage are risks associated with coal; core meltdown, waste storage, and release of radioactivity are perceived risks of nuclear energy; foreign relations and the balance of trade are perceived risks of oil use. Energy conservation and solar energy may well have perceived risks associated with their adoption.

"Acceptable risk for a new technology is defined as that level of safety associated with ongoing activities having similar benefit to society" (Slovic 1977, p. 10). The concept of risk has traditionally included only physical safety and health (for example, incidence of injury, disease, and death). This concept resulted in three propositions termed "laws of acceptable risk." Based on the evidence, acceptability of risk seems related to:

- The third power of the benefits
- The voluntariness of the activity (for example, risks 1,000 times greater from voluntary activities such as skiing will be accepted more than those from involuntary activities such as food preservatives)
- The number of persons exposed to the risk, inversely (Slovic 1977)

In pilot field interviews, respondents spoke about risk of solar adoption in these words:

- I don't want to be a guinea pig—to spend $12,000 and find out I should have done it differently.
- One fear I had before we moved here [solar development] was that there was going to be some nut out here who was an ecofreak radical type who is going to be my neighbor, and when I'm out planting my vegetation in front of my house, he's going to stroll by and look down disdainfully at some of the ornamental plants that I particularly like . . .
- When you start building 50 houses that are solar, they are no longer perceived in the same light. There is not as much risk associated, because if something goes wrong, they've got neighbors that can act as a backup.
- I took a gamble, but so far I'm glad about the decision. I want to see how effective it's going to be . . . how reliable.
- My wife thinks collectors are ugly. She's going to have bridge club and these people are coming there and they look at that and say, my God, Dorothy, what happened to your roof?

Slovic noted that the following variables are thought to influence acceptability risk:

- Voluntariness of risk taking
- Familiarity of the risk
- Perceived controllability
- Potential for catastrophe (multiple fatalities)
- Immediacy of consequences
- The degree to which risks are known to the public and to scientists
- The extent to which risks are common (as opposed to dreaded)
- The likelihood that a mishap would prove fatal (severity of consequences)

Empirical evidence suggests that respondents will tolerate higher risk levels for more beneficial activities and that they are willing to accept greater risks from voluntary activities than from involuntary activities that provide the same level of benefit. Compared with low-risk perceivers, high-risk perceivers are less likely to purchase a new product (Arndt 1967). Earlier adopters have a more favorable attitude toward risk than do later adopters (Rogers and Shoemaker 1971).

The implication from prior research is that programs mandating solar use when solar adoption is widely perceived as risky could have a negative impact on diffusion of solar technologies. Where risking solar adoption is voluntary, higher levels of uncertainty will be tolerated.

Six general types of risk have been identified (Lutz and Reilly 1973; Kaplan, Szybillo, and Jacoby 1974; Roselius 1971). These are listed below with examples of how each type could relate to solar adoption:

1. Economic or financial (Will solar energy end up costing too much?)
2. Physical (Is a solar energy system on my house safe?)
3. Psychological (Will I feel like an idiot or fool if I adopt solar energy?)
4. Social (Will my friends laugh at me if I adopt solar energy?)
5. Performance (Will a particular solar energy system actually deliver what I expect it to?)
6. Time loss (Will the system fail, requiring time, inconvenience, and effort on my part to get it adjusted, repaired, or replaced?)

"Socioeconomic risk" has been defined as "the possibility of significant social or economic injury associated with making an incorrect purchase decision in a particular situation of consumer uncertainty" (Perry and Hamm 1969). The higher the socioeconomic risk involved in a particular purchase decision, the greater the importance of personal influence in the decision as compared with other sources of influence. Social risk was found to be more germane than economic risk in forming an overall index of risk. The findings suggest that promotional strategies in a high-risk purchase situation should try to reach the consumers through personal channels (opinion leaders, word-of-mouth) rather than through general media. These strategies should emphasize the social benefits more than the economic ones (Perry and Hamm 1969; Lutz and Reilly 1973).

Concern about the individual risks involved in purchasing a system are expressed in questions people have concerning system reliability and maintenance. Concern that solar energy use might be environmentally damaging is also a possibility. Each of the concerns included in the National Study was mentioned in prior research or in pilot field interviews.

We expected that favorability and solar adoption would be less likely, the greater the concern about risk; conversely, the lower the concern, the higher the favorability and likelihood of adoption.

Motivation: Perceived Relative Advantage

> Motives to adopt solar energy can be classified using a four-cell table, with two dimensions: the first dimension is whether the benefits are economic or noneconomic in nature; the second dimension is whether they accrue to the individual (egoistic) or society (altruistic).
>
> | | Nature of Benefits | |
> |---|---|---|
> | To Whom Benefits Accrue | Economic | Noneconomic |
> | Egoistic | 1 | 2 |
> | Altruistic | 3 | 4 |
>
> The first cell relates to financial advantages of solar energy to the individual adopting it, such as lower energy costs over time. The second cell pertains to the psychological or other rewards of solar adoption, such as feeling happy about it, having fun, and getting attention from other people. The third cell relates to community or national economic well-being, including such advantages as improving the balance of trade by reducing reliance on imported oil. The fourth cell relates to advantages that benefit the community such as conserving natural resources and protecting the environment.
>
> Perceived relative advantage includes the idea of the net effect of positive and negative consequences of adoption, relative to not adopting or taking some other course

of action. Most diffusion studies support the idea that the relative advantage of a new idea, as perceived by members of a social system, is positively related to its rate of adoption.

The following economic advantages of solar energy ownership were described in pilot field interviews of solar owners, in their own words:

- . . . in the long run, it decreases your costs.
- It would give us a good resale value.
- I conserve because I'm poor, not because I'm scared . . . couldn't care less about saving gas.

The following noneconomic advantages were also described:

- . . . my main concern is not cost, it's the fact that oil is just going to be harder and harder to get.
- It probably will create a lot more jobs; technicians and people like that.
- Could save resources for other uses.
- Lessen dependence on utility company name; gain autonomy from all controlled energy sources.
- Conserving energy is important to our economy . . . I don't like to see dollars leaving this country.
- Fuel is running out . . . why not take advantage of something that's easy to get?
- It's silly to use electricity for space heating when the sun's available and more efficient.
- There is no such thing as cheap energy. Everything is a trade-off. The government has seemed to be much too heavily pro-nuclear and pro-coal, hastening the deterioration of the environment.
- The whole idea of us being the first ones appealed to us . . . to have pioneered this thing—solar energy in a tenement building!
- Motivated chiefly by desire to be energy self-sufficient in old pioneer spirit. Willing to pay a certain amount to indulge myself, but it still has to work economically. I get a kick out of it every day.

The limited data collected in marketing studies on perceived advantages and disadvantages of solar energy systems are reviewed in detail in Farhar (1979b, pp. 142-54). In summary, while the perceived cost of solar energy was mentioned most frequently as a disadvantage of solar energy systems (by up to 82 percent of samples), savings in fuel and costs were mentioned most frequently as an advantage (by up to 76 percent of samples). Other perceived advantages of solar energy identified in these studies include: (1) environmentally desirable, (2) aesthetically pleasing, (3) plentiful, (4) decreased dependence on utilities, (5) saving resources, (6) safe, and (7) higher resale value of homes. Other perceived disadvantages were: (1) maintenance cost and reliability/performance, (2) climate problems, (3) storage, (4) aesthetically displeasing, (5) lower resale value of homes, and (6) dangerous.

The amount spent for heat and electricity may be a motivator for solar adoption. The higher the utility bill, the higher may be the perceived relative advantage of solar adoption from an economic standpoint.

We anticipated that favorability and solar adoption would vary positively with perceived relative advantage.

Business and Policy Knowledge and Preferences

During the knowledge-evaluation stage of the innovation decision process, the individual seeks or comes across information about who is favorable and unfavorable to the innovation, where and how to get it, who is involved in it, what it costs, and so on. A number of items were included in the survey to measure both knowledge and preferences regarding business and policy questions.

In planning for the study, the following research question was defined: What roles does the public prefer for institutions in the promotion of solar energy (for example, federal government, solar industry, utilities)? What role do citizens prefer for themselves?

Energy self-sufficiency at a national level and individual autonomy at the household level may be preferred situations, which the adoption of solar energy is seen to enhance. In addition, considerable debate exists concerning appropriate roles for industry, the federal government, communities, and individual citizens in solving energy problems.

Locus of control of technological decision making has been found to be important in the social acceptability of a technology (Farhar et al 1978; Farhar and Mewes 1976). Solar technologies are amenable to individual, organizational, and community-level adoption. If citizens perceive that they actually have control over the solar adoption decision and, thus, increased control over their own energy supply, they may evaluate the solar alternative more favorably than other energy alternatives (for example, developing synfuels).

In other studies, conservationism and environmentalism were found to be associated with faith in citizen action "rather than in experts or the federal government" (Brewer and Brunner 1978; Brunner 1978). "Efforts to achieve self-reliance" was a factor affecting the solar adoption decision found in existing studies of solar owners (Unseld and Crews 1979). Another study found that attendees at public hearings on solar policy preferred individual ownership of solar energy systems to utility leasing schemes or centralized use of solar power (Farhar et al. 1979a). About two-thirds of the sample indicated they would be willing to pay more than their present utility bill for a privately owned solar energy system; about a third said they would pay more for each utility-owned solar energy system option. About a quarter responded they would pay nothing more for privately owned solar energy, over a third would pay nothing more for the utility-leasing option, and 42 percent for the central utility option. Virtually no one said they did not want privately owned solar energy systems. These data support the notion that autonomy and personal action may have widespread appeal.

In pilot field interviews, respondents commented on this issue. Some examples of what they said are:

- Individuals have to do what they can, or at least make some effort. Individual homes or small businesses are the ones that have to use solar energy, turn down thermostats. It just makes sense on a small-scale individual level.
- The problems are all so big that we need some kind of centralized force to order a control of them . . .
- We are going to have to have government-controlled energy.
- . . . Waste and fuel are problems that are best dealt with at the place of use.
- We're being manipulated by the corporate structure and the government for the profit motive.
- I think it's great for each person to take responsibility and their their own little collectors and their own little systems.

We expected that those preferring individual and community decision making on energy would tend to prefer solar energy.

Knowledge of existing tax credits, both federal and state, was measured in this survey, as were preferred financial incentives. Interest in home energy audits was assessed, as was willingness to pay for audits. The perceived installed costs for solar domestic hot water and solar space heating systems were assessed. Preferences among a variety of financing mechanisms were indicated. Whether people perceived solar energy products and services to be available to them was assessed. Whether respondents felt that it would be easy or difficult to retrofit their own particular homes with solar energy systems was considered. Finally, respondents were asked to indicate what various types of organizations (such as private industry, utility companies, and local government) should be doing to promote the use of solar energy. Preferred organizational roles in informing consumers, warranting products, and several other activities were assessed.

If the organizations and policies actually promoting solar use are seen by the public as appropriately doing so, the diffusion of solar energy will be advanced, and, if these activities are not regarded as appropriate, it will be retarded.

SUMMARY

To recapitulate, adoption of solar energy for residences is a process occurring over time. Some consumers complete this process sooner than others; every homeowner falls along a continuum from ignorance about residential solar technologies to having a solar application on his home. At the present time, we anticipated that most homeowners would be aware of solar technologies but would not know enough about them to evaluate the potential of solar applications for their own particular situations.

We expected that favorability to solar use would precede behavioral intention, which, in turn, would precede purchase and ultimate adoption of solar energy. Favorability to solar energy, we anticipated, would be affected by a homeowner's definition of the energy situation, social support, knowledge and information sources, perceived feasibility of solar energy, concern about risk, motivation (perceived relative advantage), and business and policy knowledge and preferences (such as knowledge of existing tax credits). In the National Study, data were collected from both homeowners and solar homeowners on each of these dependent and independent variables to make possible the "next tiers" analysis. Findings from the homeowners survey in Chapter 3, and from the solar homeowners survey in Chapter 6, are organized using the variables we have described as a framework.

2

METHOD

Up until now, virtually nothing was known about public opinion on solar energy at the national level. Compared with the wealth of survey data on nuclear energy and energy conservation, the paucity of information on solar energy is striking. Yet if solar energy is to make a significant contribution to the nation's energy supply, millions of individuals will have to make decisions to adopt it, and hundreds of firms will have to be involved in the design, manufacture, distribution, and installation of the many kinds of solar energy systems and homes possible. It seems obvious that, if solar energy is to be a viable option, industry, government, and consumers all need to know more about consumer knowledge, attitudes, and practices with respect to solar energy. When it comes to residential solar energy, consumers are the energy decision-makers.

In this chapter, we describe how the National Study came into being and the sampling techniques and data collection methods used in it. In the first part of the chapter, we briefly describe the preliminary work behind the National Study, including literature reviews, pilot interviewing, meetings with study users, and a small longitudinal study. In the second part, we describe the national sampling of homeowners and solar homeowners and summarize sample characteristics. Data collection techniques are also described. The National Study presented no new and unusual problems in survey research.

PREPARATION FOR THE NATIONAL STUDY

A good deal of research was completed before we began designing the National Study. The major preparatory activities included literature reviews, pilot field interviews, meetings with study users, and a six-month study of nine solar households in Arizona.

We reviewed the existing survey data on solar energy (Farhar et al. 1979b; Unseld and Crews 1979).* Solar energy has received little attention in surveys of the American public; one or two items about it have been included in about 20 national surveys. Most of the information available about public opinion on solar energy is derived from marketing studies of special, localized samples and is not generalizable to the public at large.

Most of the public has heard about using solar energy for heating and generating electricity. Motivations for using solar energy can be both economic and noneconomic. Solar ownership in the United States is quite limited. Advantages perceived are environmental and economic. Initial cost has been defined as an important barrier.

In identifying research gaps on solar energy, there are insufficient national data on perceived effectiveness, relative advantage, knowledge and information sources, behavioral intention, and action. Almost no data exist on policy preferences (Farhar et al. 1980). This lack of data occurs in every theoretically meaningful category.

The few existing studies of solar owners were limited to relatively small samples in localized areas, the largest of which was a sample of 800 Florida solar owners. Unfortunately, these studies were conducted as surveys rather than as exploratory in-depth field studies, which would have been more appropriate, given how little is known about the phenomenon of solar adoption.

We also discovered that most of the attitude and behavior research that had been done on solar energy was conducted without regard to theory. Theory is supposed to describe and explain the "real world." Using theory aids in knowing the important questions to ask. The lack of theory in existing studies prevented them from being highly useful for policy and marketing purposes, because data on theoretically relevant variables were frequently not collected or analyzed.†

We sought to fill these research gaps with national-level data on solar energy that would employ and test theory and would, thus, produce results useful for industry, government, other researchers, and the attentive public—for everyone involved.

Sound field work is an indispensable "reality check" for a study in its planning stages. We conducted focused interviews with solar owners, nonsolar owners, near-adopters, solar leaders, and commu-

---

*Findings from these reviews are summarized in Appendix A.

†There are a few exceptions. Some significant research on solar adoption has been conducted in California by the Institute for Communication Research, Stanford University (Leonard-Barton and Rogers 1979).

nity leaders in four communities across the United States, selected for their geographic diversity, known solar activity, varying income levels, and different community sizes. Davis, California, was selected as a small western community where a good deal of solar action has occurred, including a solar subdivision, innovative building designs, city ordinances requiring recycling and energy conservation, and continued community planning in these areas. The San Luis Valley of Colorado is a largely low-income agricultural area, not noted for its avant-garde style, while evidencing widespread solar adoption. Carbondale, Illinois, was the small town scene of considerable midwestern solar activity, including on-farm and in-town applications, and community organizations to promote solar energy. Finally, New York City was a good example of solar use in the urban setting, with several early installations associated with "sweat equity" housing rehabilitation efforts.

By tape recording interviews that were open-ended conversations, guided by a list of generic topics, and content-analyzing the results, we were able to identify the language and thoughts of the people with the most direct and relevant experience: those who had thought about solar energy and those who had done something about it. The interviews were a rich data source, both on factors affecting individual decisions to adopt solar energy and on community dynamics. Quotes from the interviews illustrating the relationship between theory and real world experience were included in Chapter 1 to illustrate the study's guiding ideas.

We met with study users to learn what they defined as the most important questions they would like to see answered through the study. May 22, 1979, was spent in a day-long session with DOE officials,* and May 23 was spent with representatives of large and small firms manufacturing solar equipment, builders of solar homes, and public interest groups.

The degree of overlap in interest between the industry and government groups was remarkable. Their many points can be classified into five key questions:

1. What factors affect consumer decisions to invest in solar energy?
2. What do consumers know about solar energy, and what are their misperceptions?

---

*Representing Conservation and Solar Applications, the Office of Consumer Affairs, the Office of Commercialization, and the Office of Policy and Evaluation.

3. What kinds of people are likely to adopt solar energy next, how do they differ from others, and how can they be encouraged?
4. According to consumers, what roles in solar promotion should various organizations play? What specific incentives, financial arrangements, and so on should these organizations provide to encourage solar adoption?
5. What experiences have consumers had with solar energy—reasons for adopting, life-style effects, purchase problems?

Each of these questions is addressed in the National Study.

Nine households in Tucson, Arizona, participated in a six-month study in which respondents learned to be personal investigators of their own energy behavior, using techniques of self-observation. Data were also collected through videotaped interviews, questionnaires, and repeated observations. The study was designed to document the relationship between values and life-styles and solar adoption (Zimmerman 1980). The study found that solar use does not interfere with daily routines or require alteration of habits. Nevertheless, solar use itself appears to affect homeowners. Families in the study exhibited such changes as greater awareness of the natural environment, participation in community energy fairs and conferences, and exchanging information with others about their experiences. The in-depth Arizona study deepened our understanding of the potential importance of values and life-style preferences in the solar adoption decision.

When the results of these preliminary research efforts were completed, we designed the homeowner interview schedule and solar homeowner questionnaire, replicating between them the items on factors affecting the solar adoption decision.* These study instruments and the research plan were reviewed and approved by the Energy Information Administration and the Office of Management and Budget.

SAMPLING AND DATA COLLECTION

Homeowners Survey

The universe from which respondents were selected consisted of all owner-occupied year-round housing units in the 48 contiguous

---

*The two study instruments are not identical since some questions appropriate for homeowners would not be appropriate for solar homeowners and vice versa. Study instruments may be found in Appendixes B and C.

states, including mobile homes, but excluding town houses, condominiums, and military bases.*

The most recent year for which estimates of the size of this population are available from the Bureau of the Census, Annual Housing Survey, is 1978. The 1978 estimate for year-round owner-occupied housing units in the entire United States is 50,283,000. To arrive at an estimate of the number of housing units, excluding Alaska and Hawaii, in November 1980 (the time of the survey), Gallup used population counts from the 1980 census and analysis of the relationships between population and the number of households and between population and the number of owner-occupied housing units (and, indirectly, between number of households and number of owner-occupied housing units). These analyses resulted in an estimate of 52.2 million year-round owner-occupied housing units in the 48 continental states for the fall of 1980.†

The Gallup Organization, Inc., maintains a national probability sample frame of 362 interviewing areas. This is the Gallup Master Sample and is used for all Gallup Omnibus surveys as well as for special surveys of the nation's general population. All interviewing areas were included for the National Study's homeowner survey. The design of the sample is that of a replicated, probability sample down to the block level in the case of urban areas and to segments of townships in the case of rural areas.

The sample size of approximately 2,000 (the actual number of completed interviews was 2,023) was dictated by analytic needs for subgroups of sufficient size and by the overall level of accuracy required. The sampling techniques used resulted in data determined to be self-weighted by Gallup; that is, the sample was assessed to be representative of U.S. homeowners.‡

---

*These dwelling units were excluded due to the fact that their owners do not have decision control over the external appearance of the building. Owners of attached houses having decision control over external appearance (ascertained through a screening question) were included in the sample.

†The information used in deriving the estimate was obtained from Census Bureau Publications Current Population Reports, Series P-25, Nos. 708, 729, 805, 807, and 888; Annual Housing Survey H-150-78-A and prior issues; and figures released by the Census Bureau to the media on the 1980 census of population, New York Times, December 17, 1980, and subsequent issues.

‡The specific standard errors for estimates derived from the homeowners survey would vary, depending on the estimate in question. Standard errors were computed by Gallup for a variety of estimates,

Interviewers were trained by Gallup through the use of written instructional materials, a tape-recorded summary of key information, and telephone briefings with each interviewer. A "hotline" to headquarters was available to all interviewers to deal with any on-the-spot problems, as they took to the field.

Interviewers entered the field on October 16, 1980. The national election was held on November 4, 1980. Because the weekend preceding the presidential election is key to polling the outcome, Gallup concentrated its efforts on measuring voter opinion. Field work for the survey was, therefore, suspended during the weekend prior to the election and on Monday and Tuesday (Election Day). Gallup reported that there was no reason to believe that the national election had any adverse impact on the process of data collection for the National Study. Data collection was completed on November 24, 1980.

The average length of the personal interviews was approximately 56 minutes. However, despite its length, Gallup reported that the level of respondent interest and cooperation was high. A number of respondents requested additional information and, in soliciting comments from respondents selected for validation, Gallup received favorable reactions to the survey. During pretesting, Gallup had found that the survey generated a great deal of interest and a number of questions. For this reason, interviewers were given pamphlets on solar energy and how to get more information about it that were left with respondents at the end of the interview. The use of the pamphlet material was beneficial for two reasons. First, if questions were raised during the course of the interview, it allowed the interviewer to continue without delay by informing the person that material would be given them after completion of the questioning. Second, it created a positive feeling at the close of an interview since the respondent received some information in return for his or her time.

Validation of completed interviews was conducted through telephone interviews when a phone number was available or by mail, if unavoidable. Approximately 30 percent of all completed interviews, selected randomly, were validated.

---

using several different computational methods, each employing the replication feature of the sample design. As expected, the estimated standard errors ranged from .009 to .018, or 0.9 percentage points to 1.8 percentage points. The 95 percent confidence interval for estimates with standard errors in the middle of this range would be about 2.6 or 2.7 percentage points, plus or minus. At the high and low ends of this range, the errors would be smaller, under 2 percentage points. These figures, of course, are based on estimates for the full sample.

The level of refusal encountered in the survey was approximately the same as that that Gallup has encountered in similar work. Reasons for refusal ranged from "I don't believe in surveys" to "I'm not interested," "I don't have time," to "If you don't get out of here, I'll call the cops." Gallup reported that the problems encountered in fielding the homeowners survey were in no way unusual or different from difficulties normally encountered in such work. The difficulties that did occur were judged not to have a significant negative impact on the validity or reliability of procedures used in the study. Gallup concluded that the respondents appeared to understand and respond properly to the survey questions. The one item for which response is not believed to be valid is related to ownership of certain solar energy systems; this problem is discussed in Chapter 3.

Gallup edited and encoded the data and provided a clean data tape to SERI, along with frequency distributions on all items. After the data were transmitted to SERI, all respondent identifying information was destroyed by Gallup, thus ensuring confidentiality and anonymity to respondents. Data analyses were conducted at the SERI and the University of Colorado computing centers.

Solar Homeowners Survey

The mail survey of solar homeowners resulted in completed questionnaires by 3,809 owners of residential solar energy systems. Because the universe of solar homeowners is unknown, a probability sample of this population was impossible to achieve. Instead, a purposive sample was compiled. An effort was made to identify as large a sample of solar homeowners as possible to permit planned analyses to be feasible. The resulting 3,809 valid responses represent by far the largest and most comprehensive survey of this population.

"Solar homeowners" were defined as owners of homes employing any type of residential renewable energy system. These included the variety of active and passive solar applications for space and water heating and/or cooling and wind energy systems. Homeowners with wood-burning stoves or fireplaces as their only renewable energy system were specifically excluded from the sample.

A list of 6,911 names of probable solar homeowners was compiled from various sources. A few existing directories and lists were available. Other sources for the list were researchers, regional solar energy centers, state energy offices, local solar energy societies, consumer protection offices, and solar firms. Press releases requesting volunteer participants were issued to over 1,500 publications nationwide. Of the total list, approximately 40 percent were self-selected to participate in the survey; about a quarter had participated in one of three previous surveys.

However, no claim can be made regarding the sample's representativeness of the universe of solar homeowners. Similarly, it is impossible to judge how well the distribution of respondents by state represents the distribution of actual installations by state. To provide a better understanding of the sample's geographic distribution, the following state-by-state breakdown is offered:

State	Percentage of Sample
Florida	23
Hawaii	19
Connecticut	17
California	7
Colorado	5
New Mexico	2
Iowa	2
Pennsylvania	2
Massachusetts	2
Nevada and Louisiana	0
All other states	1

The sample and its distribution by state are artifacts of the methods used to gather names of potential respondents rather than representative of the distribution of systems in the nation. The relatively large proportion of Connecticut respondents, particularly when compared with California, resulted from the availability of a list of participants in the HUD Domestic Hot Water Demonstration Program and illustrates the manner in which the sample was influenced by available information. Nevertheless, the sample size compensates to some degree for its nonrepresentative and nonprobabilistic nature. Until a more comprehensive list of solar homeowners is identified and properly sampled, we are confident that this sample represents the best available body of information on that population. Nevertheless, one cannot generalize from these findings to the population of solar homeowners in the nation.

Of the 6,911 original names on the mailing list, 767 were identified through the mail survey as nonowners of solar homes, as deceased, or as bad addresses. This left a total of 6,144 potential respondents. From them, 3,809 valid responses were received, representing a 62 percent response rate to the survey. This was probably due to two factors: (1) the follow-up procedures employed, and (2) high interest on the part of potential respondents.

Solar homeowners were initially mailed a questionnaire and cover letter with a postage-paid return envelope. Potential respondents were assured of the strict confidentiality of their responses. This first mailing occurred during the first week of September 1980. After a two-week interval, nonrespondents were sent a follow-up postcard requesting their completion of the survey. A second postcard was sent to nonrespondents after another two weeks. After a final two-week interval, a final mailing included a second questionnaire and cover letter. All mailings were completed prior to the end of October 1980. Completed questionnaires were accepted through the end of December 1981; after this, they were considered late and response data were not entered.

The 62 percent response rate is high for a survey of this nature, particularly since pretests indicated an estimated one hour needed to complete the 135-item questionnaire. Thirty-five percent (2,157) of potential respondents did not respond to the survey. Ninety-six indicated their refusal to participate by replying with a message to that effect. Of these refusals, 44 offered no explanation, 21 indicated displeasure with the survey, 9 offered problems with their solar energy systems as a reason for refusal, 6 were unhappy with the federal government, 5 considered the survey a waste of time or tax money, 4 thought the questionnaire was "too personal," 5 were "too busy," and 2 were physically disabled. Smaller numbers of invalid responses or nonresponses were categorized as incomplete (23), late (18), unidentifiable due to return address label being removed (3), and blank questionnaire returned (33).

SERI subcontracted with a local marketing firm to edit and code all data from completed questionnaires. In the process of entering data, return address labels on envelopes, which had been used as a control procedure for sending follow-up mailings to nonrespondents, were destroyed. All respondents were thus assured anonymity and confidentiality of their responses.

Completed questionnaires were edited and coded by Marketing Management Concepts, Inc., and a data tape, code book, and frequency distributions were provided to SERI. Used primarily for data analyses were various SPSS factor analysis, cross-tabulation, and scale programs. Analyses were performed at the SERI and University of Colorado computing centers.

3

FINDINGS FROM THE NATIONAL HOMEOWNERS SURVEY

INTRODUCTION

Who are the homeowners included in the national homeowners survey?* The sample represents homeowners in the United States. Most of the sample (90 percent) live in single-family detached residences, while 6 percent live in attached residences, and 3 percent live in mobile homes. These homeowners have the right to change the exterior appearance of their homes without going to a board or other authority. Homeowners represent about 65 percent of all occupied housing units in the United States. Gallup estimates that the size of the population represented by this sample is 52.2 million year-round owner-occupied housing units.

The sample is 55 percent male and 45 percent female heads of household. Almost 6 out of 10 are between 35 and 64 years of age. Almost 1 in 5 are 65 or older, with the same proportion under 34.

Forty percent of the sample have at least some college training, while 1 in 5 has less than a high school education. Over half of the households have incomes between $15,000 and $45,000 annually, while about a third live on under $15,000 a year. Reflecting the conservative victor in the November 1980 election—when the survey was conducted—78 percent of the sample characterize themselves as middle-of-the-road to strongly conservative in political orientation.

The sample is mostly white, with over half living in families with children. Almost 1 in 3 are married couples without children. About half live in cities of over 50,000, while almost one-third are rural in residence and one-fourth live in small towns. The majority heat their homes with natural gas.

---

*Statistics on the demographics, values, and life-styles of the homeowners sample are presented in Appendix D.

The findings from the national homeowners survey* are organized along the lines of the study's guiding ideas, discussed in Chapter 1. There we presented the three dependent variables we are attempting to explain—evaluation (favorable or unfavorable attitudes toward solar energy), behavioral intention (plans to adopt or not to adopt solar energy), and action (actually purchasing or building a solar energy system or home). Responses on the first of these variables, favorability toward solar use, are described first in this chapter. Responses on behavioral intention and on action taken, and how these responses relate to other study responses, are reported toward the end of the chapter. Included with them are other data on energy conservation and other behaviors. In between are findings on the independent variables—those factors that we expected would affect a consumer's attitudes and behavior toward solar energy. As a reminder, these major factors are social support, definition of the energy situation, perceived feasibility of solar energy, knowledge and information, concern about risk, and motivation. We discovered during the data analysis that policy preferences can also affect likelihood of solar adoption, so responses on business and policy knowledge and preferences are presented along with those for other independent variables.

As we proceed through the chapter, we begin to present results of analyses that show how these independent variables relate to each other and how they might affect homeowners' decisions on the use of solar energy in their own homes. At the end of the chapter are findings from a special analysis on regional differences in attitudes toward solar energy. We complete the chapter with a summary of findings and conclusions.

FAVORABILITY TO SOLAR ENERGY

Consumers were asked how they feel about solar energy, both in comparison with other energy supply sources and by itself for residential use.

Comparative Preferences for Energy Supply Sources

Prior research has shown extensive favorability to the use of solar energy in the United States. This study gauged solar favorability

---

*The homeowners interview schedule that was used is reproduced in Appendix B.

TABLE 3.1. Comparative Preferences for Energy Supply Sources

Here is a list of several energy sources available to us. Please rank the top three energy sources you would prefer to see developed to meet our future energy needs. (1 = most preferred, 2 = next most preferred, 3 = third most preferred.) Now, please indicate the source you least prefer.

Energy Source	Most Preferred %	Top 3 %	CEQ Data[a]	Least Preferred %	Top 3 Minus Least	CEQ Data
Solar energy (includes solar, wind, wood)	31	66	61	3	+63	+55
Energy conservation (reducing energy needs by adding more home insulation, driving a fuel-efficient car, etc.)	14	45	35	4	+41	+32
Synfuels (for example, industrial plant converts oil shale or coal to a liquid or gas)	10	38	26	5	+33	+16
Water power from dams or falls (hydropower)	6	34	31	8	+26	+21
Coal	12	36	36	13	+23	+27
Oil and natural gas	14	34	28	12	+22	+19
Nuclear energy	8	27	23	45	-18	-10
Don't know	6			9		
No answer			2	1		
N	(2022)					
Missing data	(1)					
Total N	(2023)					

[a]Data from the Roper and Cantril national survey of adults 18 and over conducted for the Council on Environmental Quality, January-March, 1980.

in a more stringent fashion than have most studies by asking people to compare it to other potential sources of energy.

Solar energy is by far the most preferred energy source, as data in Table 3.1 show. When asked which of seven energy sources they preferred to see developed to meet our future energy needs—coal, nuclear energy, energy conservation, hydropower, solar energy, oil and natural gas, or synfuels—31 percent ranked solar energy as the most preferred choice and 66 percent ranked it in the top three. Second choice was energy conservation—almost half selected it in the top three. Preferred next in the ranking were synfuels, selected by 38 percent, coal, by 36 percent, oil and natural gas, 34 percent, hydropower, 34 percent, and nuclear energy, 27 percent.

The least preferred source was nuclear energy (45 percent), followed by coal (13 percent), oil and natural gas (12 percent), hydropower (8 percent), synfuels (5 percent), energy conservation (4 percent), and solar energy (3 percent).

When the "least preferred" proportion is subtracted from the proportion rating a power source in the "top three," the overall ranking remains almost the same. Solar energy is first, followed by energy conservation. Synfuels remain in third place and hydropower replaces coal in fourth place. Coal, oil and natural gas, and nuclear energy are in fifth, sixth, and seventh places, respectively. Nuclear energy is the only power source with a negative proportion at -18 percent.

These findings are borne out by another national survey. The President's Council on Environmental Quality (CEQ) sponsored a national survey of adults 18 years of age and older on environmental opinion (CEQ 1980). Data were collected by Roper and Cantril between January and March, 1980. Included in the CEQ survey was this same item on preferred power sources, except that respondents were asked to indicate "which two or three we should concentrate on most." Solar energy was ranked first in the CEQ survey, selected by 61 percent, followed by coal at 36 percent, and energy conservation at 35 percent. Hydropower was chosen by 31 percent, oil and natural gas by 28 percent, synfuels by 26 percent, and nuclear energy by 23 percent. In the overall rankings (where the "least preferred" proportion was subtracted from the most preferred), CEQ found solar energy first, energy conservation second, coal third, hydropower fourth, oil and natural gas fifth, synfuels sixth, and nuclear energy last. This pattern is similar to the SERI data, despite the differences in populations studied (homeowners, adults) and time of data collection (early, late 1980). Coal ranks higher in the CEQ survey and synfuels rank higher in the SERI survey, but solar energy and energy conservation are the most preferred energy supply sources in both surveys.

Position on Residential Solar Use

Survey results that show majority favorability toward most energy supply sources are achieved by asking simply, "Do you favor or oppose the use of nuclear energy or solar energy or coal?" Prior polls had shown massive proportions indicating solar popularity, so this survey attempted to make it more difficult to express favorability, both by using the comparative approach and by bringing solar energy from an abstract idea to a more concrete reality. This was done to obtain as realistic and conservative an assessment as possible of favorability to solar energy among U.S. homeowners. Two items were used to assess a favorable overall attitude or position toward residential solar use. The first asked about solar energy in the abstract, the second in the concrete. Data are presented in Table 3.2.

TABLE 3.2. Position on Solar Use

Based on your understanding of solar energy for homes, how do you feel about it—do you strongly favor, favor, oppose, or strongly oppose its use?

Response	%		N
Strongly favor	32	77	(644)
Favor	45		(905)
Unsure/neutral	18		(370)
Oppose	4	5	(81)
Strongly oppose	1		(14)
	100		(2014)
		Missing data	(9)
		Total N	(2023)

Given what you know about solar energy right now, do you strongly favor, favor, oppose, or strongly oppose the idea of using it on your home?

Response	%		N
Strongly favor	21	58	(419)
Favor	37		(754)
Unsure/neutral	22		(448)
Oppose	16	20	(318)
Strongly oppose	4		(78)
	100		(2017)
		Missing data	(6)
		Total N	(2023)

When asked how they felt about the use of solar energy in homes, 77 percent of homeowners are favorable or strongly favorable, with 5 percent opposed. However, when asked how they feel about using solar energy on their own homes, 58 percent are favorable and 20 percent are opposed. This constitutes a reduction in favorability of 19 percentage points, an increase in uncertainty of 4 points, and an increase in opposition of 15 points. These are not minor shifts. Clearly there is a marked difference in attitude when the object is distant or abstract compared with when it is more concrete and immediate—when it is closer to home. While there is widespread favorability to the idea of solar use, attitudes become more conservative when contemplating actual personal use of solar energy. Nevertheless, homeowners in general are quite favorable to the use of solar energy in homes, and a majority favor its use on their own homes.

SOCIAL SUPPORT

How consumers view those already using solar energy and how they think others feel about solar use can affect their own opinions and actions. If homeowners are in a favorable social environment, their own attitudes and actions are likely to be affected in the direction of solar use.

Imagery is a subtle yet real phenomenon. Solar advocacy has frequently been equated with the "radical fringe" or countercultural elements of our society, particularly by nonsolar enthusiasts. People will often reveal how they personally feel about something themselves when they describe others who are doing it or who own it.

Consumers were asked to select from a list of phrases those, if any, that accurately describe people who have solar energy systems. Data are presented in Table 3.3. The modal response to this item (50 percent) is "economy-minded people"; also chosen frequently are "scientific types" (46 percent), "upper income people" (46 percent), "environmentalists" (44 percent), and "do-it-yourselfers" (41 percent). Response codes that might be equated with fringe groups receive relatively few choices. "Back-to-the-land types" is chosen by 23 percent as is "young people," while "political radicals" is at the bottom of the list, with 3 percent. These findings debunk the myth that solar energy is widely perceived as a favorite of iconoclasts and fringe groups.

Being economy minded has recently become more fashionable, as the growing market for fuel-efficient cars attests. A frequently heard idea is that the reason rich people are rich is that they are careful with their money. The notion of upper-income people being interested in solar energy probably relates both to the economy-minded idea and to their ability to afford solar features. Environmentalists are probably chosen because solar energy is viewed as an environmentally benign energy source. Do-it-yourselfers have obvious interests in remodeling, tinkering, and innovating. In general, solar imagery would seem to be reasonably acceptable to significant portions of the population. One cautionary note in the results is that 1 in 10 identifies solar users as "not very different from my friends," which suggests that solar owners are not yet widely known and that they are still perceived as somehow "different." However, if more opinion leaders become solar owners, this image will change dramatically.

TABLE 3.3. Solar's Image

Which of the following statements, if any, would you say accurately describes people who have solar energy systems? (CHECK ALL THAT APPLY)

Response	%	N
Economy-minded people	50	(1010)
Scientific types	46	(931)
Upper-income people	46	(922)
Environmentalists	44	(883)
Do-it-yourselfers	41	(832)
Young people	23	(464)
Back-to-the-land types	23	(459)
Socially responsible people	19	(390)
People who enjoy being different	17	(333)
Not very different from my friends	10	(205)
Retired people	6	(116)
Political radicals	3	(55)
None of these	1	(15)
Other	--	(9)
Don't know	6	(117)
N	(2014)	
Missing data	(9)	
Total N	(2023)	

Consumers will be encouraged to adopt solar energy if those around them are favorable to the idea. They will also be encouraged if government is supportive of solar adoption. An item was included in the survey on perceptions of favorability toward solar energy on

TABLE 3.4. Perceived Government Favorability toward and Social Support for Solar Energy

Now, as I read you the following, please tell me whether each is favorable to, or opposed to, using solar energy for homes.

	Favorable	Mixed/ Neutral	Opposed	Don't Know	Total	N Answering
The federal government	69%	8	8	15	100	(2015)
Your state government	59	11	8	21	99	(2012)
Your local government	49	18	8	25	100	(2006)
Your neighbors and friends	42	21	7	30	100	(2001)
People you work with	43	17	5	34	99	(1913)
Your spouse/housemate	58	15	8	19	100	(1826)

the part of these "significant others." Data are presented in Table 3.4.

Federal and state governments are perceived as favorable by a majority (69 percent) and local governments by almost half (49 percent). The proportion unsure about governmental positions was lowest for the federal government (15 percent) and highest for local government (25 percent). These findings suggest that local governments can play an increasingly important leadership role in encouraging solar use in the future.

A majority of spouses and housemates are perceived as favorable (58 percent), with almost a fifth unsure about spouse's opinion. About a third are unsure what neighbors, friends, and co-workers think about solar energy, while 42 and 43 percent, respectively, assess neighbors/friends and co-workers as favorable. Perceived levels of opposing sentiment are low for all these groups, varying from 5 to 8 percent.

The findings make clear that use of solar energy is perceived to be "compatible" with existing norms and values by most homeowners. Greater favorability and increased incidence of solar use among opinion leaders would increase the proportion of co-workers, neighbors, and friends perceived as favorable. Favorable word-of-mouth communication by these groups would heighten a supportive social milieu in which to adopt solar energy.

DEFINITION OF THE ENERGY SITUATION

The data presented in Table 3.5 make clear that most homeowners are optimistic about the nation's future energy situation. The step number response to the question on where the United States is now was subtracted from the step number response to where the United States will be five years from now. The resulting score was recoded. For example, if a respondent indicated that we are presently at step 5 and anticipated that in five years we will be at step 8, his resulting score would be 3. If we said that we are at step 5 and will be at step 2, his resulting score would be -3.

Scores were recoded as follows:

-10 to -6 = Much worse
-5 to -1 = Worse
0 = Same
1 to 5 = Better
6 to 10 = Much better

When the rankings are recoded into the five-step scale from

"much better" to "much worse," 27 percent of the sample anticipate harder energy times, while 57 percent think the situation will improve. These findings do not suggest a widespread impetus for adoption of residential solar energy based on worry about future energy supplies. However, among those most likely to adopt solar energy, the anticipation of future problems might be more frequent than among homeowners at large.

TABLE 3.5. Perceived Present/Future National Energy Situation

Here is a picture of a ladder. Let's suppose the top of the ladder represents the best possible energy situation for out country, and the bottom, the worst possible energy situation for our country. Please show me on which step of the ladder you think the United States is at the present time.

Just as your best guess, if things go pretty much as you now expect, where do you think the U.S. will be on the ladder, let us say, about five years from now?

Step	% Present	% Future	
10	2	5	Best
9	1	6	
8	7	16	
7	10	14	
6	15	13	
5	25	13	
4	18	7	
3	11	8	
2	4	6	
1	2	3	
0	2	3	Worst
Don't know	4	7	
Total	101	101	
	(N = 2014)	(N = 2016)	

Perceived Future Energy Situation

Recode. Response	%	N
Much better	1	(11)
Better	56	(1044)
Same	16	(305)
Worse	27	(495)
Much worse	--	(7)
	100	(1862)
Missing data		(161)
Total N		(2023)

TABLE 3.6. Impact of Energy Situation on Households

In different parts of the nation, people have reported a variety of effects of the energy situation on their daily lives. How has the energy situation affected your family (household)? Would you say . . .?

Response	%	N
It really has had no effect on us.	14	(279)
We have had to make a few adjustments, but our lifestyle has not been affected.	47	(942)
Our life has been less comfortable and convenient, but it is not serious.	30	(610)
We have had to make serious changes in our daily habits.	9	(179)
Don't know.	--	(10)
	100	(2020)
Missing data		(3)
Total N		(2023)

Similarly, while most homeowners say they have been affected by the nation's energy problems, they do not define their inconvenience as serious. Responses to the item on impacts of the energy situation are shown in Table 3.6. A minority of 9 percent indicate a serious impact on their households; 90 percent have not been seriously impacted. This finding also suggests little impetus for adoption of solar energy based on a significant personal energy problem.

Earlier research has shown that Americans do not define energy as among the most significant national problems, a designation so far reserved for such issues as inflation and crime (Farhar et al. 1980). The fact that most homeowners do not perceive a serious impact of the energy situation on their households reveals clearly that they do not link the energy situation to inflation, which is widely believed to have a serious impact. Although economists debate the extent to which inflation is driven by energy prices, it is clear from these findings that homeowners do not perceive inflation as a function of increased energy costs.

PERCEIVED FEASIBILITY OF SOLAR ENERGY

The perceived feasibility of solar energy will likely influence adoption behavior. Considerable controversy has occurred over the

TABLE 3.7. Perceived Feasibility

Item	Proportion responding . . .						
	Strongly Agree	Agree	Total Agree	Unsure	Strongly Disagree	Disagree	Total Disagree
In general, solar energy systems are technically and economically practical today for homes.	10%	35	45	35	18	2	20
If they are practical at all, solar energy systems are only practical for homes in warm, sunny climates.	5	29	34	22	39	5	44
Solar energy systems in general will perform satisfactorily over a long period of time.	9	43	52	43	4	—	4

technical readiness of systems now on the market, the economics of solar energy, and the cost effectiveness of systems in all parts of the nation. Three items were included in this survey to collect information on the highly complex matter of perceived feasibility. The results of these items, presented in Table 3.7, suggest that a majority may still view solar energy as a technology of the future rather than as a technology ready for use now.

The first item stated, "In general, solar energy systems are technically and economically practical today for homes." While a plurality of 45 percent agree, over half of the sample is either unsure (35 percent) or they disagree (20 percent). The second item explored climate concerns, stating that "If they are practical at all, solar energy systems are only practical for homes in warm, sunny climates." While a plurality of 44 percent disagree, the majority view was either to agree (34 percent) or not to know (22 percent). The third item asked about system durability, resulting in a majority (52 percent) who agree that "solar energy systems will perform satisfactorily over a long period of time." Most of the rest are unsure (43 percent).

This hesitance over feasibility constitutes a major barrier to solar use. It can only be overcome by increased experience with effective systems through demonstrations, model homes, and persons with systems that perform well sharing their experiences with others.

KNOWLEDGE AND INFORMATION

A series of items in the survey explored the level of awareness and knowledgeability of respondents on residential solar energy. Information sources used and solar information needs are also defined.

Awareness of Solar Technologies

The innovation-decision process cannot even begin if people are unaware of an innovation's existence. Therefore, an item was included to assess consumer awareness of a variety of solar technologies that can be used for residential applications. The technologies included were:

- Solar panels or collectors
- Solar panels or collectors to heat water for homes
- Solar panels or collectors to heat homes
- Wood stoves to heat homes
- Solar swimming pool, hot tub, or spa heaters

- Attached solar greenhouses to heat homes
- Solar cells (photovoltaics) to produce electricity
- Small wind machines (or windmills) to produce electricity or pump water
- Bioconversion (such as gasohol and rising farm wastes to produce gas for fuel)
- Passive solar design (such as using south-facing windows to get heat from the sun, with few windows on the north; using water containers, walls, floors, or ceilings to collect and store the sun's heat, and so forth)

Homeowners were asked whether they had heard of each technology, whether they had physically seen it,* and whether they owned it.† Table 3.8 shows a high level of awareness of solar technologies for residential applications among homeowners. The CEQ survey further supports this finding with a similar pattern of results.‡

A majority is aware of virtually all the solar technologies listed. Wood stoves (79 percent**) and small wind machines (77 percent) are the technologies the sample is most aware of, a finding identical to that of CEQ, where wood stoves (90 percent) and small wind machines (80 percent) were most frequently mentioned. This result is not surprising, since neither wood stoves nor windmills are unfamiliar American products. In the government lexicon, they are included as solar technologies, since "solar" is used generically to denote renewable energy technologies.

Bioconversion is the next most known technology in the SERI survey (78 percent), while 68 percent were aware of it in the CEQ survey.

Following closely in both surveys are solar domestic hot water and space heating systems. In the SERI survey, 74 percent are aware of solar panels to heat homes and 72 percent are aware of solar domestic hot water systems. In the CEQ survey, the percentages were 78 and 79, respectively.

Also well recognized is passive solar design. Two-thirds of

---

*Interviewers were instructed as to the meaning of "having seen" a technology. Respondents had to have actually seen it, not seen it on TV or in a magazine.

†Results are reported on page 107.

‡In the CEQ survey, respondents were asked, "Please tell me which you have heard or read about and which you haven't."

**This includes the 67 percent listed in Table 3.8 plus 12 percent indicating ownership of wood stoves.

TABLE 3.8. Awareness of Solar Technologies

Solar energy is a very broad term that includes many ways of using the sun to produce energy. Here is a list of solar features that can be used by individual homeowners to produce energy. Please tell me for each solar feature whether you have heard of it, have seen it, or own it.

Solar Technology	% Have Seen	% Have Heard Of	% Total	CEQ Data[a] % Have Heard About
Bioconversion (such as gasohol and using farm wastes to produce gas for fuel)	12	66	78	68
Small wind machines (or windmills) to produce electricity or pump water	37	40	77	80
Solar panels to heat homes	24	50	74	78
Solar panels to heat water for homes	23	49	72	79
Wood stoves to heat homes	50	17	67	90
Passive solar design (such as using many south-facing windows to get heat from the sun with few windows on the north; using water containers, floors, or ceilings to collect and store the sun's heat, etc.)	17	50	67	74
Solar swimming pool, hot tub or spa heaters	13	38	51	58
Solar cells (photovoltaics) to produce electricity	8	36	44	8
N	(2001)			
Missing data	(22)			
Total N	(2023)			

[a]Data from the Roper and Cantril national survey of adults 18 and over conducted for the Council on Environmental Quality, January-March, 1980.

the SERI sample are aware of passive design and 74 percent of the CEQ sample reported awareness. Solar swimming pool, hot tub, or spa applications are known of by 51 percent of the SERI sample and 58 percent of the CEQ sample.

More homeowners have seen wood stoves and small wind machines than the other solar technologies. About one in four have actually seen solar panels, but fewer have seen passive solar features and fewer still photovoltaic systems. As will be discussed, consumers consider seeing solar technologies an important way to learn about them.

TABLE 3.9. Exposure to Solar Buildings

About how many solar homes and buildings have you noticed in your own neighborhood?

Actual number	%		N
None	78		(1568)
1	10	23	(201)
2	5		(98)
3	3		(57)
4	1		(18)
5	1		(18)
6	1		(17)
7 or more	2		(31)
Don't know	1		(12)
	102		(2020)
	Missing data		(3)
		Total N	(2023)

A marked difference in awareness of photovoltaic technology occurred between the two samples. While 8 percent of the CEQ sample indicated awareness, 44 percent of the SERI sample does. It is unclear why this difference occurred. The CEQ item read "solar cells to produce electricity," while the SERI item read "solar cells (photovoltaics) to produce electricity." Different item wording might account for some of the difference in response, although it seems unlikely that more would be aware of "photovoltaics" than of "solar cells." Perhaps the double definition offered in the SERI wording triggered more recognition. The two surveys were taken about nine months apart, so it is possible that awareness of photovoltaics increased in the population during that time. However, an increment of 36 points is quite marked for a less than one year increase in technological awareness. Finally, it is possible that homeowners (SERI sample) are markedly more aware of photovoltaics than are adults (CEQ sample). No basis exists for choosing one or more of these interpretations, yet it is possible that awareness of photovoltaics increased dramatically during 1980. A more definitive answer awaits further evidence from other studies.

Consumers are not well exposed to solar energy systems on buildings in their own neighborhoods, as shown in Table 3.9. Almost a quarter have noticed one or more solar homes or buildings nearby, while 78 percent have not noticed any buildings. This finding suggests that various kinds of solar installations are not widespread or commonly seen. For some, such installations might be a source of information and a contributor to a sense of familiarity with solar features. For most homeowners, however, solar buildings are not part of their everyday lives, and are, therefore, probably still viewed as unique and innovative.

Knowledgeability

Homeowners are not well informed about solar energy. When asked, "How well informed would you say you are about solar energy?" most respondents indicate a low level of knowledge, with over a fifth responding that they are not at all informed; 7 percent say that they are "very well informed" (Table 3.10).

When asked whether they presently know enough about solar energy to make a decision about using it in their homes, 73 percent indicate that they do not or are unsure. About a quarter of the sample say they know enough to decide (Table 3.11).

Responses to these two items were cross-tabulated, with significant results. As would be expected, consumers indicating that they are more well informed about solar energy are significantly more likely to express confidence in making a decision to use it. In fact, 85 percent of those indicating they are well informed about solar energy indicate that they probably or definitely know enough to use solar energy in their homes. Conversely, 92 percent of those indicating they are not at all informed say they probably or definitely do not know enough to make such a decision.

TABLE 3.10. Knowledgeability (Subjective Assessment)

How well informed would you say you are about solar energy? Would you say . . .

Response	%	N
Very well informed	7	(147)
Somewhat informed	33	(666)
Slightly informed	39	(785)
Not at all informed	21	(415)
Don't know	--	(8)
	100	(2021)
	Missing data	(2)
	Total N	(2023)

TABLE 3.11. Knowledgeability (Informed Enough to Make Solar Adoption Decision?)

Do you feel you presently know enough about solar energy to make a decision about using it in your home (or buying a solar home)?

Response	%	N
Yes, definitely	11	(212)
Yes, I think so	17	(335)
Unsure	9	(179)
No, probably not	33	(672)
No, definitely not	31	(624)
	101	(2022)
	Missing data	(1)
	Total N	(2023)

Knowledge Related to Perceived Feasibility

We examined what knowledgeable consumers perceive about solar feasibility. Those indicating that they were well informed about solar energy are much more likely than others to believe systems are technically and economically practical today. Of those knowledgeable, 59 percent agree with the item on practicality, while 52 percent of nonknowledgeable homeowners report that they are unsure. Knowledgeable consumers are also more likely than others to disagree: 30 percent do not perceive systems as presently practical, compared with 16 percent of uninformed respondents.

Those somewhat and very knowledgeable are most likely not to see solar applications as limited to warm, sunny climates. Over half of them disagree with the statement on climate, while 41 percent of uninformed respondents agree with the statement that solar practicality is climate-dependent. However, about a third of knowledgeable consumers take the position that solar energy is practical only in warm, sunny climates.

Agreement that systems are durable increases with knowledgeability. The majority of informed consumers agree that systems will perform satisfactorily over a long period of time (70 percent); uninformed consumers are unsure (61 percent).

In sum, knowledgeability tends to be associated with perception of solar energy as technically feasible and cost effective. A minority of knowledgeable consumers may not perceive solar energy to be technically effective at present. Uninformed consumers are simply not sure and hesitate to venture an opinion one way or the other on solar energy's feasibility.

Consumers who feel they know enough to adopt solar energy are significantly more likely than others to define it as feasible. Those who do not know enough are more likely to be unsure about feasibility. Similarly, the more knowledgeable homeowners are significantly more likely to disagree on solar climate limitations than are others. Perception of system durability increases with having enough knowledge to decide on solar adoption. Being unsure about feasibility decreases significantly with having enough knowledge to decide. Of those with enough knowledge to decide, a smaller proportion tend to be cautious on solar feasibility (about 25 percent) than among those indicating they are very well informed about solar energy (about 30 percent).

Information Needs

To define information needs on residential solar applications, two survey items were used. The first asked whether more information was desired. The second asked of those giving an affirmative response to the first item what information they would like to have. A majority, 58 percent, indicated they would like to have more information (see Table 3.12). A sizable minority of 42 percent said they would not. The negative response can probably be accounted for by four factors:

- Some respondents were afraid they would be contacted by salesmen after the interview if they responded affirmatively; marginal notes to this effect were made by interviewers.
- Some respondents may feel they already have enough information to make a decision, as discussed earlier.
- Some respondents are not favorable to solar energy or have already decided not to use it for their homes and are, therefore, not interested in becoming more informed about it.
- Some respondents, while favorable to solar energy, are not yet ready to purchase and do not feel they presently need information.

Information needs were defined by respondents in an open-ended fashion. Interviewers probed for accurate and complete responses, which were then used by Gallup and SERI to develop detailed response categories. A high level of detail in these categories is useful to those designing information and marketing products and services.

TABLE 3.12. Information Needs

Would you like to get more information regarding solar energy systems?

Response	%	N
Yes	58	(1166)
No	42	(854)
Don't know	--	(3)
	Total N	(2023)

What information would you like to have? (Open-ended; response categories developed from verbatim responses.) (N = 1166)

General Information	%	N
Initial cost—system, installation	39	(454)
Operating cost—maintenance, repairs	29	(333)
General basic information on solar energy for homes (don't know much)	26	(299)
Feasibility of retrofit for my home (NOT climate); renovation needed, type of home, what's involved in installation	18	(208)
How it works, operates	18	(207)
How-to-do-it technical information on materials, design, and installation	17	(195)
Payback; lifecycle costs; costs compared to benefits	10	(119)
Performance, efficiency, effectiveness (how much of my heat, hot water, electricity will it provide?)	10	(118)
What kinds of systems are available?	9	(109)
Savings in money or conventional energy	9	(99)
Solar energy in my climate—does it work here?	6	(71)
Who supplies/installs materials/systems in my area? Availability in my area?	4	(43)
Advantages and disadvantages (e.g., who benefits? How does it benefit me or the nation?)	3	(39)
State-of-the-art developments	3	(38)
Financial incentives and arrangements, how to pay for it	3	(33)
Durability, life expectancy, will it last a long time?	3	(31)
Reliability, dependability, will it be reliable?	3	(30)
Is maintenance service available, reliable, too costly?	2	(28)

TABLE 3.12 (continued)

General Information	%	(N)
What are weather effects on solar (e.g., snow on panels, storms, etc.)?	2	(20)
Safety	2	(19)
How difficult is it for me to maintain, live with, etc.?	2	(17)
What government is doing about it	2	(17)
Aesthetics	1	(16)
Reputability of solar firms, architects, etc.	1	(12)
Technologies	**%**	**N**
Solar panels, collectors	9	(108)
Windmills	5	(58)
Passive solar energy	4	(41)
Greenhouses	3	(36)
Photovoltaics	3	(30)
Storage (how long; thermal mass)	2	(25)
Back-up systems	2	(19)
Woodstoves	1	(12)
Applications	**%**	**N**
Space heating	15	(179)
Water heating	13	(154)
Electricity	3	(33)
Cooling	2	(22)

Data presented in Table 3.12 can be grouped into four major categories: (1) technical information, (2) cost and savings information, (3) information on living with a solar energy system, and (4) political and business information. Under technical information are: how systems work, how-to-do-it information, performance, state-of-the-art developments, feasibility of retrofit for the particular home, technologies available, and residential applications. Under cost and savings information are: initial cost for the system and installation, operating costs, pay-back or life-cycle costs, savings in money or energy, and financial incentives and arrangements. Under information on living with a system are: durability and life expectancy, reliability and dependability, availability of maintenance service, effects of weather on solar energy systems, safety, difficulty of maintaining and living with systems, and aesthetics. Under political and business information are: who is benefitted by solar energy, what the government is doing about it, who supplies and installs systems, and what is the reputability of solar firms.

Given these categories, 46 percent of the responses are for technical information, 32 percent are for cost and savings information, 5 percent are for information on living with a system, and 3 percent are for political and business information.

Estimating to the population as a whole from these findings, in all, between 28.5 and 31.6 million American households would like to receive more information on solar energy.

TABLE 3.13. Information Sources

I am going to read you a list of possible sources of information about solar energy. I would like you to indicate how likely it is that you would use each source to get information about solar energy, assuming these were available to you. (Response categories: Very likely, Likely, Unsure, Unlikely, Very unlikely.)

Response	% Likely	% Very Likely	% Total
People who have solar energy systems	43	37	80
Demonstration buildings or model solar homes	45	30	75
Books, journals, and reports about solar energy	47	26	73
Magazines and newspapers	51	19	70
Energy fairs, exhibits, home shows	48	18	66
Television and radio programs	48	17	65
Friends, relatives, neighbors and acquaintances	37	16	53
Local contractors (heating, plumbing, etc.)	42	11	53
Homebuilders, architects	40	11	51
Home energy audits	36	10	46
Seminars/workshops	30	13	43
Adult education classes	27	9	36

TABLE 3.14. Organizational Information Sources

Now, as I read another list of organizations that have information about solar energy, please indicate for each organization how likely it is that you would contact each source for information, assuming these were available to you. (Response categories: Very likely, Likely, Unsure, Unlikely, Very unlikely.)

Response	% Likely	% Very Likely	% Total
Solar companies	50	19	69
State energy or solar office	47	15	62
Public library	42	14	56
Federal information organizations (e.g., Government Printing Office, National Solar Heating and Cooling Information Center, etc.)	42	14	56
Federal energy agencies or laboratories (e.g., Department of Energy, Solar Energy Research Institute, Regional Solar Energy Centers, etc.)	38	16	54
Utility companies	43	11	54
State or county agricultural extension service agents	37	13	50
Consumer groups (e.g., Consumers Union, etc.	37	12	49
Local government	35	7	42
Trade and professional organizations (e.g., Solar Energy Industries Ass'n.)	32	7	39
Solar or alternative technology clubs/ associations	28	9	37
Environmental organizations (e.g., Sierra Club, Wilderness Society, etc.)	23	6	29
Lobbying organizations (e.g., Solar Lobby, etc.)	13	3	16

Information Sources

Two items on information sources were included in the survey. Consumers were asked to indicate how likely it was that they would use each source to get information about solar energy. The first item focused on types of sources such as media, friends, solar homeowners, and home energy audits (see Table 3.13). The second item focused on organizational sources, such as utility companies, solar companies, and governments (see Table 3.14).

All types of sources are likely to be used by over 36 percent of the population. Most consumers would like to see solar energy technologies: solar owners (80 percent) and demonstrations (75 percent) are the most frequently chosen likely and very likely sources. Also selected frequently are energy fairs, exhibits, and home shows (66 percent). Reading about solar energy is also important: print media are chosen by 73 percent of the sample. Television and radio are selected as likely sources by 65 percent.

Friends, relatives, neighbors, and acquaintances are selected as likely sources as frequently as are local heating, ventilating, and air conditioning (HVAC) and plumbing contractors (53 percent). Home builders and architects are seen as likely sources by 51 percent. Home energy audits are mentioned by 46 percent; seminars and workshops by 43 percent; and adult education classes by 36 percent.

The most popular organizations for solar information are solar companies (69 percent) and state energy offices (62 percent). Following closely are the public library (56 percent) and federal information organizations (56 percent), such as the Government Printing Office. Utility companies and federal energy agencies are each chosen by 54 percent of the sample, and agricultural extension services are selected by half the sample.

Consumer groups are selected by 49 percent, local government by 42 percent, trade organizations by 39 percent, solar energy associations by 37 percent, environmental organizations by 29 percent, and lobbying organizations by 16 percent. These results suggest a need for relatively unbiased information, even though solar companies head the list.

Information Sources Factor Analysis

The 25 items on information sources were factor analyzed* and five dimensions emerged. Results are briefly summarized in Table 3.15.† The need for relatively unbiased consumer information on

---

*Oblique factor analysis with Kaiser normalization.

†For those interested in doing further research on solar information sources, the items defining these factors were also scaled, with excellent results.

TABLE 3.15. Information Sources Factors

Dimensions	Factor Loadings
Factor 1: Federal organizations	
Federal energy agencies or laboratories (e.g., DOE, SERI, RSECs)	.86495
Federal information organizations (e.g., GPO, NSHCIC, etc.)	.79756
Factor 2: Media	
Magazines and newspapers	.83349
Television and radio	.60339
Books, journals, and reports about solar energy	.58779
Factor 3: Local expertise	
Local contractors (heating, plumbing, etc.)	-.62760
Homebuilders, architects	-.59186
People who have solar energy systems	-.50314
Factor 4: Educational Settings	
Seminars/workshops	.76030
Adult education classes	.67554
Factor 5: Advocacy organizations	
Lobbying organizations (e.g., Solar Lobby, etc.)	.64737
Environmental organizations (e.g., Sierra Club, Wilderness Society, etc.)	.63347

solar energy is underscored by the emergence of federal laboratories and information organizations as the most important type of information source. Media, and especially print media, are also very important sources of information. Three other major possibilities are local expertise, educational settings, and advocacy organizations. Local expertise, as represented by heating or plumbing contractors, architects, and solar homeowners, provides the interactive information exchange that is highly important in being persuaded about solar energy (as compared to simply being informed). Educational settings, such as classes and workshops, offer the kind of information particularly useful to the do-it-yourselfer. Advocacy organizations can provide information on the advantages of solar energy as well as relevant political and sometimes business information. It is interesting to note that solar companies do not define any of these types but do fall within the local expertise type more than any other.

Audits as Information Sources

Utility companies, through the Residential Conservation Service (RCS) Program, make home energy audits for both energy conservation and renewable resource measures available to most utility customers in the nation. A $15 ceiling has been placed on the cost of the audits by Congress. Two survey items were included to measure levels of interest in the audit and willingness to pay for an audit.

A home energy audit for solar measures can be an important information source for potential solar owners, one that provides information homeowners identify as needed. The list below provides examples of information needs consumers identified (Table 3.12), compared with information an RCS audit is required to provide:

Examples of information needs:

1. What systems are available?
2. Is solar energy appropriate for my home?
3. Who installs systems in my area?
4. What are the costs and savings?
5. What financial incentives are available?

RCS audit provides:

1. Information on active solar domestic hot water, active solar space heating, combined active systems; passive systems, including direct gain and indirect gain systems for heating and thermosyphon water heaters; and small wind energy systems
2. Information on solar retrofit potential for the specific building audited
3. List of contractors and suppliers
4. Estimates of contractor- and resident-installed costs and of first-year savings
5. Information on Low Income Weatherization Assistance Program; information on federal and state tax credits that are applicable to each measure; and example calculations on savings from tax credits

The first survey item on audits stated:

> Planning is underway to make home energy audits available to all homeowners desiring them starting early in 1981. The audits will cover both energy conservation measures (such as weather stripping) and solar energy measures (such as solar panels). How interested would you say you are in haveing such an audit done in your home?

Data are presented in Table 3.16. Over half of the sample indicate interest, with 20 percent saying they are "very interested"

TABLE 3.16. Audits

Planning is underway to make home energy audits available to all homeowners desiring them starting early in 1981. The audits will cover both energy conservation measures (such as weatherstripping) and solar energy measures (such as solar panels). How interested would you say you are in having such an audit done in your home?

Response	Percent		(N)
Have already done it	2	51	(38)
Very interested	20		(403)
Interested	29		(591)
Unsure	11		(217)
Not very interested	18	38	(356)
Not at all interested	20		(415)
	100		(2020)
		Missing data	(3)
		Total N	(2023)

What is the most you would be willing to pay for such an audit? (Asked of those who responded very interested, interested, or unsure.)

Response	Percent		(N)
Nothing at all	33	33	(402)
Under $25	30	62	(367)
$25-49	17		(202)
$50-74	7		(83)
$75-99	4		(45)
$100-149	2		(28)
$150-199	1		(16)
$200 or more	1		(11)
Don't know	5		(51)
	100		(1205)
		Missing data	(6)
		Total N	(1211)

and 29 percent saying they are "interested." Two percent indicate they have already received an audit. Almost half are unsure (11 percent) or not interested (38 percent).

Respondents who expressed interest or were unsure were asked what they would be willing to pay for such an audit. A third would be unwilling to pay anything, while 62 percent indicate willingness to pay from under $25 to $200 or more for an audit; 5 percent do not know. Thus, approximately 37 percent of homeowners are likely to be interested in the audit and willing to pay the $15 audit fee. This figure is relatively close to the RCS audit participation rate estimated by DOE for the RCS Program. Thus, estimating from these findings, between 17.8 and 20.8 million American homes might have home energy audits done by their utility companies over the five-year length of the program. Not enough information is yet available from utility companies on actual audit participation rates to validate estimates; however, anecdotal information is that response rates are as high as or higher than expected where advertising is used to make consumers aware of the audits.

Summary

To summarize findings on knowledge and information, consumers have a high level of awareness of solar technologies for residential applications, yet they do not know in detail about solar energy. Three out of four have not seen any solar buildings in their own neighborhoods. Most do not know enough to make a decision about using solar energy for their own homes. Knowledgeability tends to be associated with perception of solar energy as technically feasible and cost effective at present.

Consumers indicate they want technical information, cost and savings information, information on living with solar energy, and political and business information. Between 28.5 and 31.6 million U.S. households would like to receive more information on solar energy. Most consumers would like to <u>see</u> systems; reading about solar energy is also important.

The five most important types of solar information sources are, in order: federal organizations, media (especially print media), local expertise (contractors, people who own systems), educational settings, and advocacy organizations.

About 37 percent of homeowners are interested in receiving a home energy audit for energy conservation and solar alternatives. Competently performed audits may be a significant solar information source since they supply the kinds of information consumers say they need.

CONCERN ABOUT RISK

A number of concerns about the use of solar energy have been expressed in prior studies and field interviews (Farhar et al. 1980). Moreover, the relatively small number of systems currently installed suggests that significant barriers to solar adoption remain. In 18 survey items, concern about the risk of solar adoption was measured. For each item, respondents were asked to indicate the degree of importance to them, if they were thinking about using solar energy in their homes. Data are presented in Table 3.17.

The concerns mentioned most frequently as "important" or "very important" form a set of items regarding acquiring a solar

TABLE 3.17. Perceived Barriers to Solar Use (Concern About Risk)

I am going to read you a list of some things you might see as problems with solar energy. For each item on the list, please tell me how important a concern it would be for you, if you were thinking about using solar energy in your home. (Response codes: Not at all important, Slightly important, Somewhat important, Important, Very important)

Responses	% Important	% Very Important	Total %
Warranty coverage for solar energy systems	28	58	86
Initial cost of a solar home or solar energy systems	25	60	85
Operating reliability of solar energy systems (need for frequent repair and maintenance)	29	56	85
Dependability and reputability of solar firms	25	58	83
Possible damage to solar energy systems by dirt, vandalism, storms, corrosion, etc.	33	46	79
Safety of solar energy systems	25	53	78
Cost and difficulty of getting solar energy systems covered by homeowner's insurance	32	44	76
Expense of maintaining solar energy systems	32	41	73
Possibility that solar energy systems now on the market will soon be obsolete and better systems will be available later	32	34	66
Problems of obtaining a loan to finance the solar energy system	30	34	64
Climate too cold or cloudy or not enough wind	26	34	60
Codes or covenants might prohibit it	30	28	58
Possibility of lowering resale value of house	23	29	52
Worry over whether getting a solar energy system would be time-consuming and difficult	29	14	43
Effects of solar energy systems on attractiveness of house	28	14	42
Possible problems with the utility company	20	14	34
Might interfere with lifestyle	13	7	20
What friends and neighbors would say	5	2	7

energy system. These are warranty coverage, defined as an important concern by 86 percent; initial cost of a solar home or solar energy systems (85 percent); operating reliability of solar energy systems (85 percent); and dependability and reputability of solar firms (83 percent). The proportions expressing these concerns are so close—within 3 percentage points of each other—that they are virtually interchangeable as the most important concern.

The next four items form a second set of concerns relative to living with a solar energy system. These are the possibility of damage to the system, defined as an important concern by 79 percent; safety of solar energy systems (78 percent); cost and difficulty of getting solar energy systems covered by homeowner's insurance (76 percent); and expense of maintaining a system (73 percent).

Five other concerns are defined by a majority as important: the possibility that present systems will soon be obsolete (66 percent); problems of obtaining financing (64 percent); climate (60 percent); local codes and covenants (58 percent); and the possibility of lowering resale value of house (52 percent).

A minority of the sample define the remaining concerns as important. The "hassle factor" involved in purchasing a system is noted by 43 percent. A barrier frequently mentioned in government reports as important is aesthetics; it is so defined here by 42 percent. A third of the sample is worried about possible problems with their utility company if they adopted solar energy. Another barrier frequently mentioned by the government, life-style effects, is defined by a fifth of the sample as important, while seven percent express concern over their personal reputations if they were to become solar owners.

These results provide a relative ranking of the importance of these frequently mentioned barriers to solar adoption. Clearly, the initial cost* and reliability of systems are of primary importance, along with having a reputable firm backing a purchased system with a warranty. The provision of low-interest loans would obviously relieve the initial cost barrier, while the availability of good quality systems marketed by reputable firms and backed by warranty would obviate the most central concerns.

Concern about Risk Factor Analysis

The 18 "concern-about-risk" items were subjected to factor analysis.† The items form three major dimensions. The first dimen-

---

*The importance of initial cost as a barrier to solar energy is emphasized by perceptions of system costs, discussed in the section on Business and Policy Knowledge and Preferences.

†Oblique rotation with Kaiser normalization.

sion is concern about safety and reliability. Items defining this dimension and their factor loadings are:

Factor 1: Concern about safety	Factor Loadings
Cost and difficulty of getting solar energy system covered by homeowner's insurance	.66860
Safety of solar energy systems	.64689
Operating reliability of solar energy systems (need for frequent repair and maintenance)	.63509
Codes or covenants might prohibit it	.62800
Possible damage to solar energy system by dirt, vandalism, storms, corrosion, etc.	.60595

The second dimension is concern about life-style and aesthetic effects. Items defining this dimension and their factor loadings are:

Factor 2: Concern about life-style and aesthetic effects	Factor Loadings
Might interfere with life-style	.61863
Effects of solar energy on attractiveness of house	.48744

The third dimension is concern about system purchase. Items defining this dimension and their factor loadings are:

Factor 3: Concern about system purchase	Factor Loadings
Initial cost of solar home or of solar energy system	.63210
Dependability and reputability of solar firms	.49540
Expense of maintaining solar energy systems	.49441
Warranty coverage for solar energy system	.43288

Thus, we can conclude that most concern about the risk of solar adoption can be subsumed under three major categories: (1) safety of solar energy systems, (2) concern about life-style and aesthetic effects, and (3) concern about system purchase.

In addition, the 18 concern-about-risk items were subjected to a scale analysis, with highly favorable results. The items scale with high reliability (0.943) and sound scale statistics. The Concern about Risk Scale was, therefore, used in the study's multivariate analyses, including the "next tiers" analysis.

Perceived System Costs

Initial cost of solar energy systems is an important concern to most homeowners (see Table 3.18). In evaluating this concern, it is important to know how much homeowners think systems actually do cost. System costs vary markedly by quality, performance, size,

TABLE 3.18. Perceived Cost of Solar Energy Systems

About how much do you think it would cost to buy a solar water heater for your house including the installation cost?*

Response	%	(N)
Less than $500	8	(151)
$500-999	13	(264)
$1000-1499	10	(194)
$1500-2499	13	(252)
$2500-3499	7	(140)
$3500-4999	3	(64)
$5000 or over	5	(92)
Don't know	42	(855)
	101	(2012)
	Missing data	(11)
	Total N	(2023)

About how much do you think it would cost to buy a solar energy system to heat your home, including installation cost?*

Response	%	(N)
Less than $2500	6	(125)
$2500-3999	9	(172)
$4000-5499	13	(262)
$5500-6999	7	(131)
$7000-8499	6	(112)
$8500-9999	3	(59)
$10,000 or over	14	(279)
Don't know	43	(868)
	101	(2008)
	Missing data	(15)
	Total N	(2023)

*Documentation for system costs is very difficult to locate. According to Hot Water From the Sun, HUD/DOE, Franklin Research Center, Philadelphia, 1980, the usual cost for solar domestic hot water is $2000-3000. According to the Residential Conservation Service Program Audit Procedure Data Base, the usual cost for solar space heating is $6,000-10,000. Actual costs may now be somewhat higher.

and locale. A national cost estimate for a "usual" active solar domestic hot water system, contractor-installed, is between $2,000 and $3,000 (Franklin Research Center 1980). A national cost estimate for an active solar space heating system, based on DOE research, is between $6,000 and $10,000.

Given these figures, data presented in Table 3.18 show that a plurality of 42 percent do not know the cost of an active solar domestic hot water system and 43 percent do not know the cost of solar space heating. Responses on perceived cost also are dispersed, suggesting that respondents were guessing rather than informed. By dividing in half the responses in the $1,500 to $2,499 category, we can estimate that 6.5 percent underestimate cost and 6.5 percent are correct in cost estimates. By dividing in half the responses in the $2,500 to $3,499 category, we can estimate that 3.5 percent are correct in cost estimates, while 3.5 percent overestimate costs. Given these estimates, about 37 percent underestimate the actual cost of solar domestic hot water, approximately 11 percent overestimate it, and 10 percent estimate costs in the correct range. For active solar space heating, 28 percent underestimate the cost and about 30 percent are in the correct range.

Most homeowners are not knowledgeable about the actual costs of installing solar energy systems, and, if anything, they tend to underestimate costs. This fact, coupled with their already high degree of concern about solar installation costs, constitutes a significant barrier to solar adoption.

MOTIVATION

Motivation is a complex and frequently murky phenomenon in human affairs. In a way, the entire National Study is about motivation to adopt solar energy. Conceptually, however, motivation is here narrowed to include the perceived advantages of solar adoption. Both economic and noneconomic motives are explored. Findings on perceived relative advantage, willingness to pay more for solar than for conventional energy, and perceived financial advantage of solar ownership are discussed.

Perceived Relative Advantage

The perceived relative advantage of any innovation is known to affect the rate at which it will be adopted. In our discussion of motivation to adopt solar energy in Chapter 1, we mentioned that both egoistic and altruistic motives are very likely involved. Further, both economic and noneconomic motives exist. These different kinds of motives can be classified in a four-cell table as follows:

Types of Motivations

	Economic	Noneconomic
Egoistic		
Altruistic		

A 15-part item was included in the survey to measure the importance of various types of solar advantages to homeowners. The list of advantages was generated from prior research and pilot field interviews. Results are presented in Table 3.19. Most of the advanages on the list (13 items) are termed "important" or "very important" by almost two-thirds of the sample or more. At the top of the list in importance are saving money over the long term (84 percent), reducing utility bills now (82 percent), and protecting against rising costs (82 percent). Two other economic advantages are also consid-

TABLE 3.19. Perceived Benefits of Solar Use (Relative Advantage)

I am going to read a list of possible advantages of using solar energy that could enter into your decision about using solar energy in your home. For each item on the list, please tell me how important it would be for you in making such a decision. (Response categories: Not at all important, Slightly important, Somewhat important, Important, Very important)

Item	% Important	% Very Important	% Total
Saving money over the long term	31	53	84
Reducing utility bills now	28	54	82
Protecting against rising costs	36	46	82
Having a more reliable supply of energy	36	43	79
Conserving natural resources; protecting the environment	35	42	77
Increasing comfort of home	41	30	71
Easing energy shortage	37	33	70
Increasing overall self-reliance	36	32	68
Increasing independence from utility company	28	37	65
Availability of income tax credits	36	29	65
Reducing need for more large power plants	35	29	64
Increasing independence from federal government policies	28	35	63
Increasing resale value of home	33	28	61
Getting to innovate or experiment	20	7	27
Increasing status, prestige, self-esteem	9	5	14

ered highly important. These are the availability of income tax credits as a motivation (65 percent) and increasing the resale value of the home (63 percent). These are the economic motivators for adoption that add credence to the significance of cost factors and financial incentives in impeding or hastening solar adoption among homeowners.

However, three percentage points behind is "having a more reliable supply of energy" (79 percent). This motivation for solar adoption may have increased with concern about blackouts, rolling brownouts, and threats of these interruptions in utility-supplied energy. Along with the impacts of the oil embargo and other events, the reliability of more conventional energy supply sources is perhaps more in question than it used to be. The fear that we may run out of certain fuels could also play a part here.

Within 2 percentage points of reliability, mentioned by 77 percent of respondents, is "conserving natural resources and protecting the environment." Several other "noneconomic" solar advantages are also held to be widely important. Increasing comfort of home is mentioned by 71 percent. Easing the energy shortage is defined as an important solar advantage by 70 percent. Increasing overall self-reliance is chosen by 68 percent and increasing independence from utility companies by 65 percent. Other important advantages are reducing the need for more large power plants (64 percent) and increasing independence from federal government policies (63 percent).

The final two items are not defined as highly important by large portions of the population. These are getting to innovate or experiment (27 percent) and increasing status, prestige, or self-esteem (14 percent).*

Relative Advantage Factor Analysis

The 15 relative-advantage items were factor analyzed.† The items form three major dimensions. The first dimension reflects economic motivations. Items defining this dimension and their factor loadings are:

Factor 1: Responsible self-interest	Factor Loadings
Saving money over the long term	.75605
Reducing utility bills now	.70107

*This latter reason may in actuality be more important than many customers are willing to admit.

†Oblique rotation with Kaiser normalization.

The second dimension reflects noneconomic motivations. Items defining this dimension and their factor loadings are:

Factor 2: Advantages to society	Factor Loadings
Conserving natural resources; protecting the environment	-.80446
Easing the energy shortage	-.70519
Reducing need for more large power plants	-.65503
Increasing overall self-reliance	-.62275

The third dimension also reflects noneconomic motivations. Items defining this dimension and their factor loadings are:

Factor 3: Personal satisfaction	Factor Loadings
Getting to innovate or experiment	.61646
Increasing status, prestige, self-esteem	.58877

Combining these factors with our four-cell table on motivation, we have:

Types of Motivations

	Economic	Noneconomic
Economic	Factor 1 Responsible self-interest	Factor 3 Personal satisfaction
Altruistic	—	Factor 2 Advantages to society

The altruistic-economic cell is empty owing to the fact that no items representing this type of motive were included in the survey. Such an item might be "improving the balance of trade." Empirical verification of this motivational typology for adoption of solar energy must await further research. The present findings show that economic self-interest is the most important perceived advantage of solar energy, followed by advantages to society (an altruistic-noneconomic motive), and finally by the fact that using solar energy is perceived to have psychological gratification—in a word, it is fun.

The 15 relative advantage items were also subjected to scale analysis, with highly favorable results. The items scale with high reliability (0.915) and sound scale statistics. The Relative Advantage Scale was used in the study's multivariate analyses, including the "next tiers" analysis.

Willingness to Pay More for Solar Energy

The assumption is often made that cost factors are the overriding concern in solar adoption. In solar market penetration modeling efforts, price is used as the "figure of merit"—the criterion that drives the entire model and accounts for its projections. While this approach has been criticized, little has been done to expand modeling capabilities to include noneconomic variables that can affect the rate of diffusion.

To assess in preliminary fashion how widespread noneconomic motivation might be, we included in the survey an item asking: "How much more (if anything) than your present total monthly bill for utilities (heat, hot water, and electricity) would you be willing to pay per month for your own solar energy system at your home?" Response categories included: (a) nothing more per month, (b) $8, (c) $25, (d) $40, (3) $85, (f) even more, (g) don't want solar energy system at my home, and (h) don't know. Data are presented in Table 3.20.

Of the total sample, 86 percent imply they might be interested in owning a solar energy system, while 14 percent indicate they are not interested. About half report they would pay nothing more per month, while 29 percent would be willing to pay more than present

TABLE 3.20. Perceived Noneconomic Value of Solar Ownership

How much more (if anything) than your present total monthly bill for utilities (heat, hot water, and electricity) would you be willing to pay per month for your own solar energy system at your home?

Response	%		(N)
Nothing more per month	49	49	(993)
$8 per month	8		(159)
$25 per month	12		(241)
$40 per month	6	29	(131)
$85 per month	2		(31)
Even more	1		(17)
Don't want solar energy system at my home	14		(279)
Don't know	8		(160)
	100		(2011)
	Missing data		(12)
	Total N		(2023)

utility costs for system ownership. The modal amount chosen was $25 more per month, while the average amount more is $28.70, among those willing to pay more.

These findings suggest that, although immediate economic considerations are undeniably important for large segments of the population, an important minority is willing to pay somewhat more for solar over conventional energy for the other values they perceive it to offer.

Willingness to pay more for solar energy was cross-tabulated by perceived system costs. For solar domestic hot water systems, 39 percent of those willing to pay more underestimate the system cost (more than any other group), while 28 percent judge cost correctly. No difference is found in estimate of system cost among those not willing to pay more. Those who do not know or do not want a system tend not to know what systems cost. The same pattern is found for solar space heating, with 46 percent of those willing to pay more underestimating system cost and 11 percent assessing it accurately.

One of the financing alternatives for solar purchase presented in the survey involved increasing monthly payments to the utility company to pay for the system. About 5 percent of all respondents choose this option. Responses were divided between those choosing and not choosing this option and cross-tabulated by willingness to pay more for solar energy. This gives a measure of whether those willing to pay more for solar energy are willing to pay it to the utility company. Of those willing to pay more for solar energy, 92 percent do not choose the utility-

FIGURE 3.1. Willingness to Pay More for Solar Energy by Income

Willingness to pay more for solar energy than present monthly utility costs is greater among high and medium income respondents than among low income respondents.

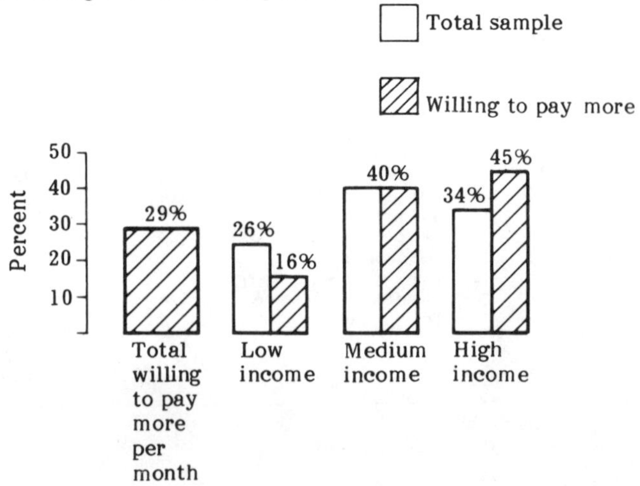

Source: Constructed by the authors.

payment option. Of those choosing the utility-payment option, 49 percent are willing to pay more for solar energy, while 38 percent are unwilling to pay more than present utility bills. Thus, consumers willing to pay more to have solar energy systems are not willing to pay it to a utility company.

Willingness to pay more for solar energy was also cross-tabulated by income (Figure 3.1). Among those willing to pay more, 45 percent have high incomes, 40 percent have medium incomes, and 16 percent have low incomes. Of high-income consumers, about half are not willing to pay more and about 40 percent are willing to do so. Among low-income consumers, over half do not want solar energy systems, while almost half are unwilling to pay more for solar energy.

Primary heating fuel was also cross-tabulated by willingness to pay more for solar energy. Natural gas users are not likely to be willing to pay more; 51 percent say they are not willing, while 29 percent say they are. Of fuel oil users, 18 percent are not interested in getting a solar energy system, and 44 percent indicate willingness to pay nothing more for one. Among electricity users, 31 percent are willing to pay more for solar energy, and 55 percent are unwilling to do so. Propane users are least likely to be willing to pay more (17 percent) and most likely willing to pay nothing more (61 percent).

Perceived Financial Advantage of Solar Ownership

Consumers were asked whether they thought owning a solar energy system would result in sufficient utility savings to make it a good investment. This is one measure of the perceived relative advantage of solar ownership. About a third indicate that, at present costs, owning a solar energy system would be a good investment. Another third are unsure, and the final third indicate that it would not be. Data are presented in Table 3.21.

No items were included in the survey on perceived dollar savings on utility bills if systems were owned. However, it is possible that, since system costs tend to be underestimated, savings might tend to be overestimated, perhaps owing to wishful thinking. These findings make obvious the need consumers have for sound information on system costs and savings.

Perception of solar energy as a good investment was cross-tabulated with willingness to pay more (Figure 3.2). Of consumers who perceive solar energy as a good investment, 45 percent are willing to pay nothing more per month and 43 percent are willing to pay something more (between \$8 and \$85+). Of those who do <u>not</u> think solar energy is a good investment, 21 percent indicate willingness to pay something more for it and 54 percent are willing to pay nothing

TABLE 3.21. Perceived Financial Advantage of Solar Ownership

At present costs, do you think that owning a solar energy system will result in sufficient utility savings to make it a good investment?

Response	%		(N)
Yes, definitely	11	33	(230)
Yes, I think so	22		(445)
Maybe, not sure	33	33	(670)
No, probably not	22	33	(443)
No, definitely not	11		(223)
Don't know	1		(12)
Total N	100		(2023)

more. Apparently for some homeowners, whether or not solar energy is a good investment is not the deciding factor.

In sum, both economic and noneconomic reasons to adopt solar energy exist. At the present time, "responsible self-interest" is seen as the most important advantage of solar use, followed by altruistic-noneconomic advantages, or "advantages to society." Finally, the personal satisfaction of solar ownership is the third most important perceived advantage.

FIGURE 3.2. Perception of Solar Investment Potential by Willingness to Pay More

Almost half of those who think solar energy is a good investment are willing to pay more per month for it; the other half are not.

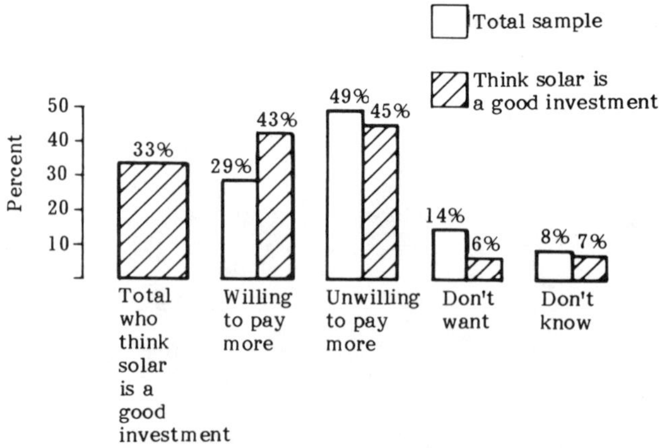

Source: Constructed by the authors.

What do these findings imply about solar energy's future prospects? At the present time, consumers are conserving both residential and transportation energy in an unprecedented manner. The usual progression of residential conservation appears to begin with low-cost or no-cost actions, such as caulking, weather stripping, lowering the temperature on the thermostat, and insulating the hot water heater, and proceed to somewhat more expensive measures such as adding insulation or double-pane windows. Once a household has moved through the sequence of conservation actions appropriate for its building structure and family life-style, the only way for it to further reduce its utility-purchased energy use is to add solar features. Some households—the ones that started conservation action early—have probably already reached the point of having completed all feasible conservation retrofits. The next step for them is probably to add a solar domestic hot water system.

Utility bills are increasing rapidly; for example, the Public Service Company of Colorado predicted in the spring of 1981 that utility bills during the 1981-82 heating season would probably be 65 percent higher than those of the previous year. If the forecast is correct, a family paying a $100 utility bill would be facing a bill of $165 in less than a year's time. These rises in utility prices may serve to increase the importance of the perceived egoistic-economic advantages of solar adoption. Thus, two complementary pressures toward solar adoption could develop in the years just ahead. The first is the possibility that a "conservation ethic" will become more and more widespread, moving energy conservation from the merely fashionable to the desirable, and finally to a more mandatory moral position. A similar process has occurred with respect to smoking. The advantages of solar energy to society that are perceived now could form the basis of future social norms governing residential energy use, which encourage (or even mandate) solar adoption and sanction its nonuse.

The second trend that could develop is that, as residential energy prices rise, solar options may become increasingly attractive financially. This is much more likely to occur than is reduction of per-unit cost, due to improved production efficiencies.*

The combination of these two trends—the development of social norms that strongly encourage solar use and the increasing cost of

---

*The reasons for this are complex; solar energy systems are not like hand-held calculators. They must be designed for specific functions, sized for specified loads, and properly installed on specific buildings.

utility-supplied energy—could well provide very strong motivation to adopt solar energy in the coming years.

BUSINESS AND POLICY KNOWLEDGE AND PREFERENCES

A number of items were included in the survey to assess knowledge of and preferences for appropriate roles in promoting the use of solar energy. The preferred locus of energy-decision control was explored, as were preferences for organizational roles, such as what actions are appropriate for utility companies or the federal government to take.

Consumers were asked their preferences for financial incentives to adopt solar energy, and they were questioned on their knowledge of existing federal and state solar tax credits. They were asked how they would pay for systems and which financing mechanism they preferred. Perceived availability of solar products and services was explored, as was the perceived difficulty of adding solar energy systems to their homes. These items are discussed in this section. Finally, findings from analyses of perceived solar energy system costs with other key items are presented.

Preferred Locus of Energy-Decision Control

One of the guiding ideas of the study was to explore attitudes on the preferred locus of energy-decision control. Consumers have been used to centralized conventional energy sources, and behaviorally their involvement with residential energy has amounted to setting thermostats, throwing switches, and paying utility bills at the end of the month. In a word, consumers are not accustomed to thinking of themselves as responsible for their own energy supply. If consumers do not perceive that individual action can affect energy supply, they will be less likely to engage in behavior to conserve energy or to adopt solar energy. A related notion is where responsibilities are seen to lie. Individuals can avoid taking responsibility for energy supply by fixing that burden on industry and government.

A set of four items was used to assess to what extent homeowners feel that solutions to energy problems are primarily the responsibility of individuals, private industry, or government. Data on these items are presented in Table 3.22. The first of these asked whether "people like me have a responsibility to help resolve our country's energy problems by cutting back on consumption, even if this means making some sacrifices in the way I live." A large majority, 86 per-

TABLE 3.22. Preferred Locus of Decision Control

Item	Proportion responding . . .						
	Strongly Agree	Agree	Total Agree	Unsure	Strongly Disagree	Disagree	Total Disagree
People like me have a responsibility to help resolve our country's energy problems by cutting back on consumption, even if this means making some sacrifices in the way I live.	29%	57	86	5	7	1	8
People like me should trust the federal government to find a solution to the energy crisis.	9	28	37	20	32	11	43
Our energy problems will be resolved when energy companies have a free hand to find the energy we need	8	34	42	25	26	7	33
We should pay more attention to the particular energy needs of each city or town--and to meeting those needs through local resources wherever possible.	19	66	85	9	5	1	6

cent, agree or strongly agree with this statement, while 8 percent disagree, and 5 percent are unsure. This supports the notion that individual responsibilities are perceived to exist.

The second item stated: "People like me should trust the federal government to find a solution to the energy crisis." Attitudes are divided on this question, with 37 percent in agreement and 43 percent in disagreement; 20 percent are unsure. Although there is division on the question, the results suggest that the federal government is not viewed as totally responsible for solving the nation's energy problems. However, some degree of federal government responsibility is implied.

The third item stated: "Our energy problems will be resolved when energy companies have a free hand to find the energy we need." Results again indicated a division of opinion, with 42 percent agreeing and 33 percent disagreeing. A quarter of the sample is unsure. This suggests that energy companies are seen as having a role in the solution of problems but are not held solely or even primarily responsible for solutions.

Finally, the fourth item stated: "We should pay more attention to the particular energy needs of each city or town—and to meeting those needs through local resources wherever possible." A large majority of 85 percent agree or strongly agree with the statement; 6 percent disagree and 9 percent are unsure. The results suggest that a potentially strong role for local governments in solving energy problems would be socially desirable. Bringing energy decision making closer to home also brings it closer to individual control.

The overall pattern of response suggested by these findings is that while roles are seen for the federal government and the energy industry, a strong positive response is found to the idea of more decentralized energy decision making at the local and individual levels. The data support the notion that consumers are beginning to view energy decisions as something in which they can appropriately, and perhaps effectively, be involved. The preferred roles for various types of organizations are discussed below.

Preferences for Organizational Roles

Policy preferences on solar energy were a notable gap in prior research. Two aspects of policy are covered in the present study: (1) whether a given activity or policy should be pursued, and (2) if so, by what type of organization.

A set of items was introduced by the following statement:

> I will read you a list of things that might be done to promote the use of solar energy in homes. For each item I

read to you, please tell me which of the groups on the card, if any, should be doing these things. (CHECK ALL THAT APPLY.)

Respondents were handed a card with the following list of response codes:

Federal government
State government
Local government
Utility companies
Private industry
Banks
It doesn't matter
Should not be done
Don't know

If a respondent answered any of the last three responses, then no organizational preference was specified for him. Results are presented in Table 3.23. To interpret the data matrix, it is useful to read one column at a time to see preferences regarding a particular policy, then one row at a time to understand preferred roles for each organizational type.

The following policies were listed, followed by the proportion implying endorsement of them:

- Educate and inform consumers (91 percent)
- Provide financial incentives, such as grants, low-cost loans, or other kinds of financial assistance (87 percent)
- Inspect systems to see how they are working (85 percent)
- Identify and make a list available of qualified solar contractors (83 percent)
- Conduct home energy audits, in which an inspector would look over a house and make recommendations on how energy can be saved and what kinds of solar energy could be used (83 percent)
- Cover guarantees or warranties for solar energy systems if a company goes out of business (67 percent)
- Require by law or ordinance passive solar design and/or solar domestic hot water in new home construction (54 percent)

For educating and informing consumers, relatively unbiased information sources are preferred: the federal government is chosen by 49 percent. State and local governments are chosen by 39 and 38 percent, respectively, and utility companies by 38 percent. No one thinks it should not be done.

TABLE 3.23. Preferences for Organizational Roles in Solar Diffusion

I will read you a list of things that might be done to promote the use of solar energy in homes. For each item I read to you, please tell me which of the groups on the card, if any, should be doing these things. (CHECK ALL THAT APPLY)

	Inform Consumers[a]	Provide Financial Incentives[b]	Inspect Systems[c]	List Solar Contractors[d]	Conduct Home Energy Audits[e]	Cover Solar Warranties[f]	Mandate Passive Design or Solar DHW in New Homes[g]
Federal government	49%	60	11	19	15	36	18
State government	39	40	16	27	17	19	19
Local government	38	26	35	37	27	11	24
Utility companies	38	22	32	24	41	8	8
Private industry	26	17	22	20	22	13	6
Banks	5	29	2	5	3	7	2
It doesn't matter	2	2	2	2	3	2	2
Should not be done	—	3	3	2	4	10	27
Don't know	7	8	10	13	10	21	17
N	(1992)	(1994)	(2000)	(2007)	(2013)	(2002)	(2008)

[a]Educate and inform consumers.

[b]Provide financial incentives, such as grants, low cost loans, or other kinds of financial assistance.

[c]Inspect systems to see how they are working.

[d]Identify and make a list available of qualified solar contractors.

[e]Conduct home energy audits, in which an inspector would look over a house and make recommendations on how energy can be saved and what kinds of solar energy could be used.

[f]Cover guarantees or warranties for solar energy systems if company goes out of business.

[g]Require by law or ordinance passive solar design and/or solar domestic hot water in new home construction.

The federal government is chosen by 60 percent to provide financial incentives to promote the use of solar energy, followed by state government (40 percent).

Responsibility for inspecting systems should primarily be borne by local government (35 percent) or utility companies (32 percent), according to these findings. Listing solar contractors is also seen as a governmental function: 37 percent view it as a local and 27 percent as a state responsibility. Utilities are mentioned by 24 percent. Utilities are viewed as the most appropriate organizations to conduct home energy audits by 41 percent, followed by local government at 27 percent.

Interestingly, covering solar warranties if a solar company goes out of business is selected as a federal government role by 36 percent and as a state government function by 19 percent. The highest proportion for all policy items, 21 percent, are unsure who should cover these warranties, while 10 percent think they should not be covered.

For mandating the use of passive solar design and/or solar domestic hot water in new homes, a majority of 54 percent choose one or more organizations, implying that this policy should be adopted. Local government is the modal organizational choice for mandating solar applications. Together, 46 percent indicate that it does not matter who does it (2 percent), they are unsure (17 percent), or that it should not be done (27 percent). A higher proportion is opposed to mandating solar use than to any other policy listed.

The most frequently mentioned roles for each organization are as follows: the federal government is seen to have primary responsibility for informing consumers, providing financial incentives, and backing solar warranties if companies go out of business. State government is chosen for primary responsibility in listing solar contractors. States are also frequently mentioned for informing consumers and providing financial incentives. The primary responsibilities preferred for local governments are in mandating the use of solar energy in new home construction, inspecting installed systems, and listing solar contractors. Another preferred local government role is to educate and inform consumers.

The primary responsibility chosen for utility companies is to conduct home energy audits, with other selected roles in informing consumers, inspecting systems, and listing solar contractors.

Since the items addressed are policy-related, it is not surprising that private industry and banks are not selected as frequently as governments. However, banks are frequently mentioned as appropriate organizations to provide financial incentives, probably in reference to providing low-cost loans, which were mentioned in the item.

Private industry roles mentioned frequently are conducting audits, inspecting systems, informing consumers, and listing contractors.

Policy Preference Factor Analysis

The 63 "policy preference" items were factor analyzed for heuristic purposes.* Preference for different organizations (in contrast to preference for specific policies) formed factors or dimensions. For example, one factor represented preference for federal government involvement while another represented preference for utility company involvement. This serendipitous discovery led to the construction of scales measuring preference for government and industry involvement in solar promotion. These scales were used in the study's multivariate analyses.

The results of the factor analysis on policy preferences show that consumers vary in preference less on what particular programs are implemented to increase solar use and vary more in preference for the particular kinds of organizations that are responsible for these programs.

Preferred Financial Incentives

Federal and state governments have been exploring policy alternatives for solar financial incentives that would be effective in increasing solar penetration in the residential sector. Government incentives for solar energy were favored in several previous studies, indicating strong citizen support for tax incentives to promote solar use (Farhar et al. 1980).

Another policy initiated by the Energy Security Act of 1980 was the Energy Conservation and Solar Bank. The Bank would provide low-interest loans to lower- and middle-income homeowners to aid them in dealing with the initial cost of solar installation, something tax credits do not do. Tax credits also are of no assistance to those paying little or no taxes.

In the survey, an item to gauge public preferences for various forms of financial incentives was included. The item stressed that each type of incentive (low-interest loan, tax credit, direct grant, rebate) would provide the same financial benefit. Therefore, consumers were not guessing which type of incentive would provide the

---

*Oblique factor analysis with Kaiser normalization.

TABLE 3.24. Preferred Financial Incentives

A number of alternatives for federal or state programs have been proposed for aiding homeowners in financing solar energy systems. I would like you to look at each of these options and tell me which one you would prefer, assuming that each provided the same financial benefit.

Response	%	N
Low interest loans (apply to a bank for a loan guaranteed by the government)	30	(588)
Income tax credits (the amount of credit is subtracted from what you owe the government)	24	(468)
Direct grant for a portion of the system cost (apply to the government before the system is purchased)	13	(256)
Direct rebate for a portion of the system cost (apply to the government after the system is purchased)	11	(219)
Other (please specify)	--	(1)
Prefer no financial assistance from government sources	8	(166)
Makes no difference	7	(133)
Don't know	7	(150)
	100	(1981)
	Missing data	(42)
	Total N	(2023)

most assistance, but instead were selecting the form of incentive they preferred.

As data presented in Table 3.24 show, low-interest loans are the most preferred type of incentive, selected by 30 percent, followed by income tax credits, selected by 24 percent; 13 percent choose a direct grant and 11 percent a rebate. No financial assistance from government sources is preferred by 8 percent, while 7 percent said the form of incentive makes no difference. Based on these findings, low-interest loan programs such as the Solar Bank would be well received by homeowners as an incentive to adopt solar energy.

TABLE 3.25. Knowledge of Tax Incentives

Knowledge of Federal Tax Incentives

Did you happen to know that there is a federal credit on your income tax for purchase of solar energy systems for the home?

Response	Percent		(N)
Yes, aware of federal tax credit	46		(934)
No, unaware of federal tax credit	17	54	(341)
Don't know/unsure	37		(745)
	100		(2020)
		Missing data	(3)
		Total N	(2023)

Knowledge of State Tax Incentives

As far as you know, does your state have an income tax credit for the purchase of solar energy systems for the home? (Response codes: Yes, No, State does not have income tax, Don't know)

Response	Percent		(N)
Yes	14		(276)
No	20		(401)
State does not have income tax	10		(196)
Don't know/unsure	56		(1148)
	100		(2021)
		Missing data	(2)
		Total N	(2023)

Accuracy of State Incentive Awareness

Response*	Percent		(N)
Yes (accurate)	8		(155)
Yes (inaccurate)	6		(121)
State has no income tax	10		(196)
Don't know (state has credit)	19		(380)
Don't know (state has no credit)	38		(765)
No (inaccurate)	5		(97)
No (accurate)	15		(304)
	100		(2018)
		Missing data	(5)
		Total N	(2023)

*Responses were recoded for accuracy based on state of respondent residence and current information on states with income tax credits (see Appendix E).

Knowledge of Solar Tax Credits

The federal government provides a 40 percent tax credit on the purchase of most solar energy systems, and many states provide credits on state income taxes as well. For example, in Colorado the combined federal and state tax credits cover 70 percent of system costs, up to $10,000. Tax credits can affect solar adoption decisions in two ways. First, they increase the relative advantage of solar adoption by making it less costly. Second, their existence implies a federal and state government milieu favorable to the use of solar energy systems for homes.

While tax incentives may be a preferred policy to promote solar use, they cannot truly serve as an impetus to solar adoption if potential users are unaware of their existence. Two items were included in the survey to measure respondent awareness of existing incentives; data are shown in Table 3.25.

The majority of homeowners (54 percent) are unaware of the federal tax credits for solar energy; 46 percent are aware of them. Furthermore, an even smaller proportion are aware of state tax credits. About 31 percent of the sample live in states that have tax credits; one in four of them are aware of it. A majority of 56 percent do not know whether there are credits in their state. Of those who don't know, a third live in states with solar tax credits. Of the respondents, 8 percent know accurately that there is a solar tax credit in their state, while a quarter know correctly that there is not; 6 percent incorrectly believe that their state has a solar tax credit, while 5 percent wrongly think that their state does not. A perceived state tax credit that does not exist could perhaps be an incentive to adopt solar energy up to the point of purchase. An actual tax credit that is not perceived can be little incentive until a purchaser is actually buying a system and learns of it.

While more are aware of the federal tax credit for solar energy than of state credits, consumers are not sufficiently informed about the credits for them to speed the solar adoption process.

Preferred System Financing

Two items were included in the survey to explore preferred payment and financing arrangements for solar energy systems. Most consumers would need to finance the cost of installing a system: 64 percent report they would finance all or part of the system cost, while 11 percent would pay cash for a system.

Those interested enough to consider the question were asked

TABLE 3.26. **Financing a Solar Energy System**

If you were thinking of investing in a solar system, how would you pay for it?

Response	%		N
Pay cash	11		(217)
Finance through a loan	39	64	(781)
Both (part cash, part financed)	25		(504)
Not interested in solar energy system	16		(317)
Don't know	9		(189)
	100		(2008)
	Missing data		(15)
	Total N		(2023)

(If pay cash, finance, both, or don't know): which financing arrangement would you prefer?

Response	%	N
Take out regular home improvement loan.	39	(596)
Don't know.	10	(155)
Have your utility company install and maintain the system and lease it to you for a monthly rate.	10	(152)
Makes no difference.	8	(122)
Rent the system from a solar contractor at a monthly rate with an option to buy.	8	(120)
Receive an extension on your mortgage with the same monthly payments.	8	(119)
Increase monthly payments to your utility company until the system is paid for.	6	(99)
Pay increased monthly mortgage payments, but with no extension.	5	(74)
Have your utility pay for the system; you would repay the total cost when you sell your house.	5	(73)
Have an automatic deduction from your paycheck.	1	(19)
	100	(1529)
Missing data		(162)
Total N		(1691)

which financing alternative they would prefer from a list of eight options. These included:

- Take out a regular home improvement loan
- Receive an extension on your mortgage with the same monthly payments
- Pay increased monthly mortgage payments, but with no extension
- Have an automatic deduction from your paycheck
- Increase monthly payments to your utility company until the system is paid for
- Have your utility company install and maintain the system and lease it to you for a monthly rate

- Rent the system from a solar contractor at a monthly rate with an option to buy
- Have your utility pay for the system; you would repay the total cost when you sell your house

Respondents could also indicate that they had no preference among the various options or that they did not know which option they preferred.

Results, reported in Table 3.26, indicate a diversity of response. A plurality of 39 percent choose a regular home improvement as their preferred financing arrangement. Second most preferred is the utility company leasing arrangement, chosen by 10 percent. One in five choose one of the utility financing options. Mentioned least often is an automatic payroll deduction. Ten percent are unsure which financing arrangement would be best.

The popularity of the loan option, coupled with the preferred financial incentive of low-interest loans, suggests that these types of financing programs could do much to hasten the residential use of solar energy.

Perceived Availability of Solar Products and Services

Potential users of solar energy cannot adopt it if it is not available in their area. Availability and perceived availability are not necessarily one and the same—whether consumers know something exists will be a function of advertising, word-of-mouth communication, and how hard they look for it. Some potential solar adopters have not yet become solar homeowners, however, because when they were in the market for a home, no solar homes were available.

To check how widespread the perceived availability of solar products and services is, we included a survey item asking which (of a list) "do you know for certain are available in this area?" The use of the phrase "for certain" was included to reduce guessing about possible availability. The list included: new solar homes, solar energy systems to purchase, reliable solar distributors and installers, small wind machines (windmills), architects who can do solar design, contractors who can build or convert homes to solar energy, materials and products for solar energy systems, and wood stoves.

As data presented in Table 3.27 show, 79 percent report local availability of wood stoves. From there, the perceived availability of solar products and services drops off markedly. About one in four thinks contractors, materials and products, systems, solar architects, and new solar homes are locally available. One in five believes

TABLE 3.27. Perceived Local Solar Infrastructure

Which of the following do you know for certain are available in this area? (CHECK ALL THAT APPLY)

Response	%	(N)
Wood stoves	79	(1607)
Contractors who can build or convert homes to solar energy	30	(596)
Materials and products for solar energy systems	28	(567)
Architects who can do solar design	26	(528)
Solar energy systems to purchase	26	(520)
New solar homes	23	(470)
Small wind machines (windmills)	21	(412)
Reliable solar distributors and installers	13	(257)
None of these	5	(96)
Don't know	9	(173)
N (2013)		
Missing data (10)		
Total N (2023)		

small wind machines are available; 13 percent say reliable solar distributors and installers are locally available to them. (The use of the qualifier "reliable" undoubtedly decreased the proportion reporting local availability of solar installers, since reputability of solar firms is a highly salient concern.) Of the sample, 5 percent report that none of the listed solar products or services are locally available to them, while 9 percent do not know about their availability.

The relatively sparse consumer recognition of solar products and services as locally available suggests an important barrier to solar adoption. In part, the data may reflect that, in many locales, most or all of these products and services actually are unavailable. Further, the data suggest that even where solar products and homes are locally available, significant portions of the buying public are unaware that they exist. Finally, findings reflect the fact that the marketing of solar energy is not yet fully institutionalized; that is, the norms governing who is responsible for marketing what products and services have not yet been sufficiently defined and are, thus, not widely understood.

Perceived Difficulty of Retrofit

A key barrier to the widespread use of solar energy in homes is the difficulty of retrofit. Since most of the housing stock that will be in existence in the year 2000 has already been built, solar retrofit is essential if solar energy's contribution to overall residential energy supply is to be significant.

Homes must have adequate space available to mount collectors, they must have proper orientation and adequate solar access, and roofs must be structurally adequate to sustain the system's weight. Adequate space for storage is also essential for some system types. It is not presently known what proportion of housing meets these fundamental prerequisites for solar retrofit. Moreover, most homeowners are probably unaware of what they are. As a source of information for homeowners, home energy audits for solar measures include an assessment of each home's suitability for retrofit.

The survey item on perceived feasibility of solar retrofit gives a measure of how difficult homeowners think it would be to retrofit solar energy systems, whether or not they are correct in their assessment. If retrofit is thought to be difficult and, therefore, probably costly, this belief serves as a barrier to solar adoption.

Consumers were asked, "Based on what you know about using solar energy (domestic hot water, solar greenhouses, home heating systems), would you say your house could easily use solar energy, could use it with some difficulty, could use it with major renovation, or could not use it?"

As shown in Table 3.28, 15 percent perceive that solar retrofit would be easy and 17 percent think they could retrofit with some difficulty. The plurality of 39 percent, however, think they could use solar energy only with major renovation, and 10 percent feel it cannot be used at all on their homes. These findings on perceived feasibility of retrofit demonstrate a major barrier to solar adoption.

TABLE 3.28. Perceived Feasibility of Retrofit

Based on what you know about using solar energy (domestic hot water, solar greenhouses, home heating systems), would you say your house could easily use solar energy, could use it with some difficulty, could only use it with major renovation, or could not use it?

Response	%	(N)
Could use easily	15	(306)
Could use with some difficulty	17	(349)
Could use but only with major renovation	39	(783)
Could not use	10	(204)
Don't know	19	(374)
	100	(2016)
	Missing data	(7)
	Total N	(2023)

Information programs, and particularly home energy audits, appear to be the best way to deal with this particular problem.

Perceived System Cost Analyses

Consumers were asked to estimate costs for installed solar domestic hot water (SDHW) and solar space heating (SSH) systems. As reported, 42 percent do not know the cost of a solar domestic hot water system and 43 percent do not know the cost of solar space heating. When usual system costs are estimated at $2,000 to $3,000, about 37 percent underestimate the cost of solar

FIGURE 3.3. Feeling Informed by Perceived System Costs

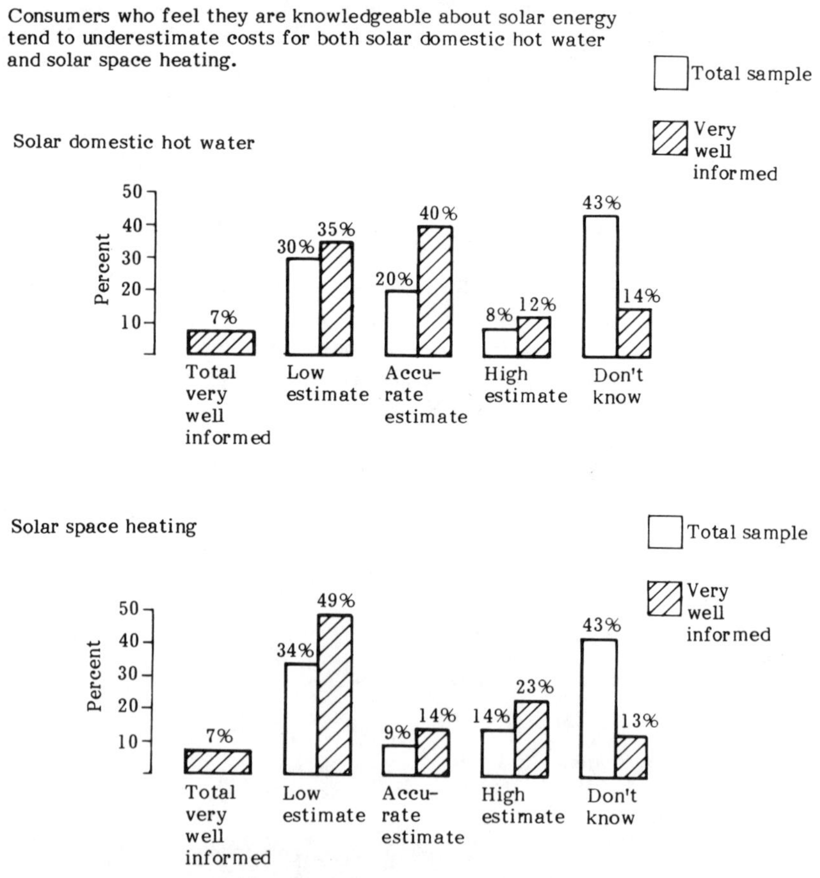

Source: Constructed by the authors.

domestic hot water, approximately 11 percent overestimate it, and 10 percent estimate costs in the correct range. When usual system costs are estimated at $6,000 to $10,000 for active solar space heating, 28 percent underestimate the cost and about 30 percent are in the correct range.

The responses on cost estimates were recoded as follows:

SDHW	Recode	SSH
<$1,500	Low estimate	<$7,000
$1,500-2,500	About right	$7,000-10,000
>$2,500	High estimate	>$10,000

These dollar groupings were the closest approximations to accuracy the study's response codes would permit. The "low," "about right," and "high" categories are used in the analyses employing the solar cost estimate variables.* Perceived system cost was analyzed by knowledge, feeling informed enough to make a solar adoption decision, concern about initial system cost, perceived investment potential, and behavioral intention.

Subjective assessment of solar knowledgeability was cross-tabulated with perceived system costs (Figure 3.3). Among consumers who say they are very well informed, 40 percent—the highest proportion of all groups—indicate a SDHW system cost in the correct range; 12 percent—again, the highest proportion—overestimate system cost; and 35 percent underestimate cost. Those not at all informed about solar energy are far more likely than others to indicate that they do not know what SDHW systems cost.

Informed consumers even more seriously underestimate the cost of solar space heating systems. Among those indicating that they are very well informed, 14 percent (the highest proportion) give estimates in the correct range. Almost half underestimate system cost and 23 percent overestimate it, while 13 percent are unsure. Those uninformed are far more likely (69 percent) to indicate that they do not know what solar space heating systems cost.

Feeling informed enough to make a solar purchase decision was

---

*A system cost of $7,000 to $10,000 pertains to a system designed to provide most of a home's heating. Some homeowners may know correctly that less expensive systems to supplement conventional heating can be obtained and may be a good investment. Additionally, any respondent might have adjusted cost estimates downward to reflect known tax credits.

cross-tabulated by perceived system costs. Of consumers who indicate they probably or definitely know enough, 31 percent correctly estimate SDHW system cost, 10 percent overestimate cost, and 36 percent underestimate. About a quarter are unsure. Of those who do not know enough to decide, 51 percent are unsure of system cost, 15 percent correctly estimate it, 28 percent are low, and 6 percent are high in their estimates.

Of those who indicate they probably or definitely know enough, 10 percent estimate solar space heating costs in the correct range. Underestimating the cost are 46 percent and overestimating are 20 percent. One in four say they are unsure of system costs. In comparison, of consumers who do not know enough to decide, 52 percent are unsure of solar space heating costs, 30 percent underestimate cost, 11 percent overestimate, and 8 percent are about right.

Thus, consumers who may feel they are knowledgeable about solar energy, in fact, tend to underestimate system costs for both domestic hot water and space heating, the latter more seriously. Of those who think they know enough to use solar energy, over a third are likely to be unpleasantly surprised when they learn what solar domestic hot water systems actually do cost, and nearly half will likely be shocked to learn the actual cost of solar space heating systems. It is also interesting to note that of consumers indicating they are well informed about solar energy, 79 percent are males and 21 percent are females.

These findings, coupled with the virtually ubiquitous concern about high initial system cost expressed by the sample as a whole, are a cautionary note. The concern about high initial cost comes from a sample uncertain about and underestimating of actual installed cost of solar energy systems. These facts contribute to understanding why there are not more installed systems in the context of such widespread public favorability toward the use of solar energy. When it is noted that a majority of homeowners are unaware of the 40 percent federal tax credits for solar energy, the situation is even easier to understand.

Listed with possible problems related to solar adoption, and defined as important by 85 percent of the sample, was "initial cost of a solar home or solar energy systems." For the analysis, respondents were divided into two groups: those defining initial cost as an important concern and those not doing so. These responses were cross-tabulated by perceived system costs, with no significant associations found, and with whether or not a respondent wanted a solar energy system, again with no significant results.

When perceived cost is cross-tabulated with perception of solar energy as an investment, significant results obtain. Of consumers who think the initial cost of solar energy is "somewhat important,"

FIGURE 3.4. Perception of Solar Investment Potential by Perceived System Costs

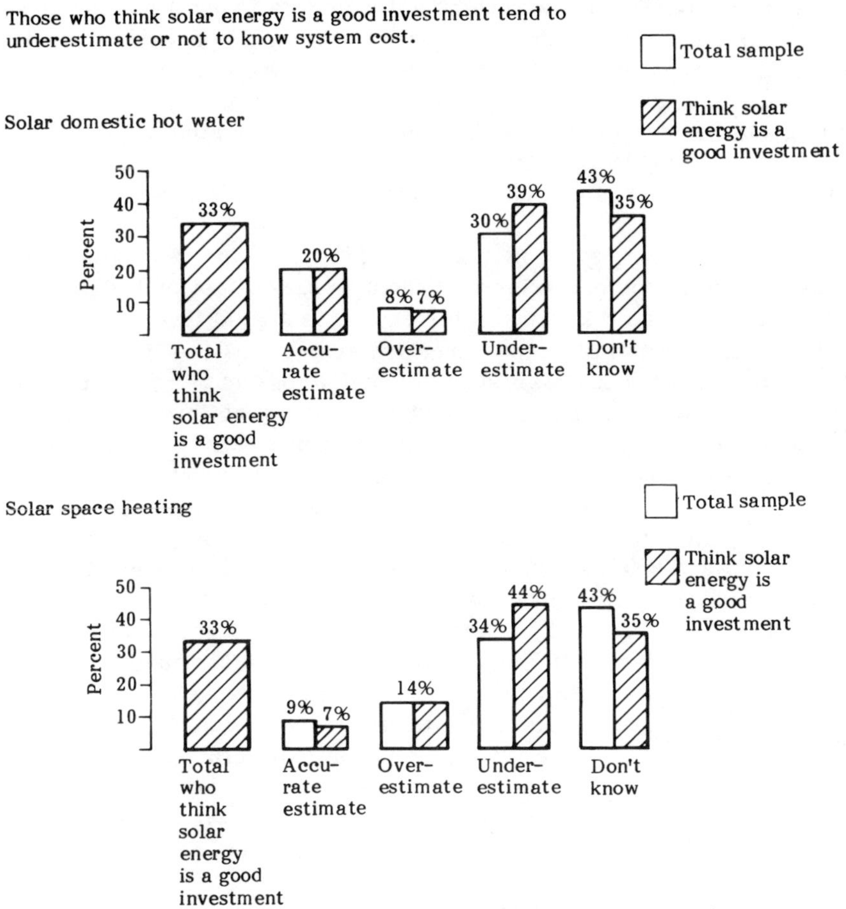

Source: Constructed by the authors.

nearly two-thirds think owning a solar energy system will result in sufficient utility savings to make it a good investment.

Another significant result occurred when concern about initial cost was cross-tabulated by how a consumer would pay for a system. Of those who think the initial system cost is an important concern, the majority would finance the cost through a loan, as compared with paying cash or both financing and paying cash.

The perceived investment potential of solar energy was compared to perceived system costs (Figure 3.4). Of consumers who think solar energy systems are a good investment, 20 percent accurately estimate the cost of a solar domestic hot water system, 7 percent overestimate, 39 percent underestimate, and 35 percent do not know the cost. Of those who think solar energy is not a good investment, 25 percent accurately estimate the cost, 12 percent overestimate, 28 percent underestimate, and 35 percent do not know the cost. Of those accurate in their cost assessment of solar domestic hot water, 43 percent think it is not a good investment and 35 percent think it is, while 13 percent are unsure. Of those estimating low, 42 percent think it is a good investment, while 30 percent think it is not.

Precisely the same pattern is found with respect to solar space heating system costs, except that the accuracy of estimates is less. Of consumers who think that solar energy is a good investment, 7 percent accurately estimate space heating system cost, 14 percent overestimate, 44 percent underestimate, and 35 percent do not know its cost. Of those who think solar energy is not a good investment, 12 percent accurately estimate system cost, 20 percent overestimate, 33 percent underestimate, and 35 percent do not know. Of those accurate in their cost assessment of solar space heating, 46 percent think it is not a good investment and 29 percent think it is, while 25 percent are unsure. Of those estimating low, 43 percent think solar energy is a good investment while 32 percent think it is not.

These findings suggest that, for some consumers, initial cost is an important factor in their consideration of solar energy as an investment. Thus, if they knew that systems cost more than they realize, they might not see solar energy as such a good investment. For others, however, the findings suggest that defining solar energy systems as a good investment may not be solely related to initial system cost. For example, some consumers are willing to define solar energy as a good investment even though they are unsure of system cost.

Behavioral intention was also cross-tabulated by perceived solar costs (Figure 3.5). Of those with definite plans to invest, 41 percent correctly assess the cost of solar domestic hot water systems, 38 percent underestimate, 9 percent overestimate, and 13 percent are unsure. In contrast, 29 percent of those with no plans to invest underestimate SDHW cost and 47 percent are unsure. Of those with definite plans to invest, 54 percent underestimate the cost of solar space heating, while 52 percent of those who may invest do likewise. In contrast, 31 percent of those with no plans to invest underestimate solar space heating cost.

Perceived solar cost was cross-tabulated by behavioral intention (Figure 3.5). Of those accurately assessing SDHW cost, 70 percent have no plans to invest, 22 percent may invest, and 7 percent plan to

FIGURE 3.5. Behavioral Intention by Perceived System Costs

Of those planning to invest in solar energy, about half know what solar domestic hot water systems cost, and a majority underestimate solar space heating costs.

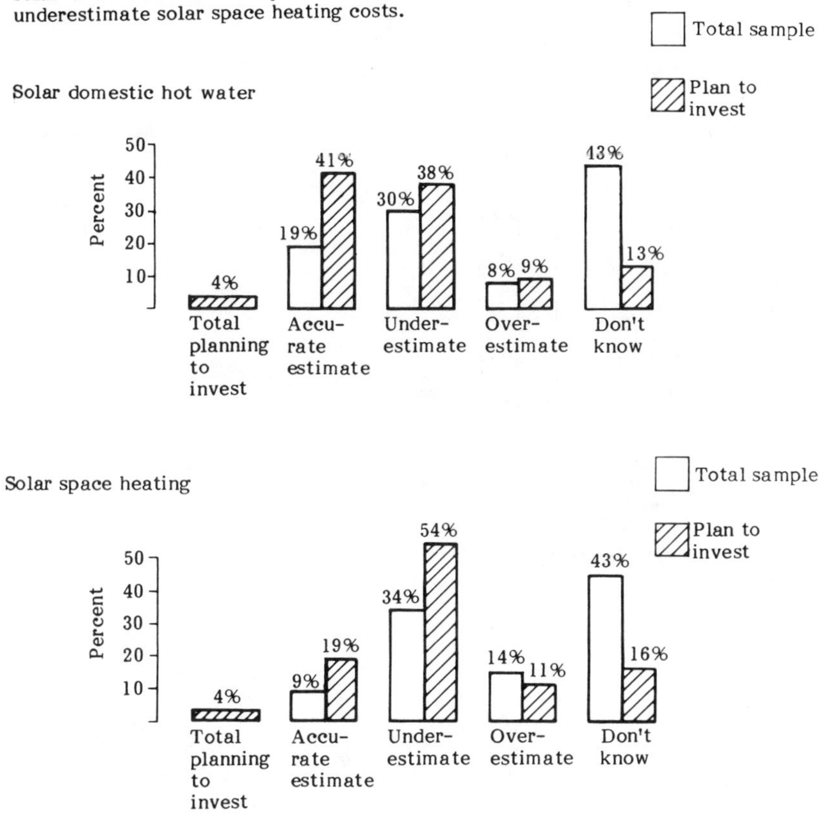

Source: Constructed by the authors.

invest. Of those accurately assessing solar space heating cost, 76 percent have no plans to invest, while 16 percent may invest and 8 percent plan to invest.

Thus, of consumers with plans to invest in solar energy, about half have a fairly good idea of what SDHW systems cost; however, a majority underestimate solar space heating cost. Of consumers with a good idea of what systems actually cost, 70 percent or more have no plans to invest in solar energy. These findings suggest that some of those planning to invest ultimately may not do so, once they learn more about actual costs. The findings are further evidence of initial system cost as a barrier to solar adoption.

FIGURE 3.6. Perceived System Costs by Behavioral Intention

Those who know what systems actually cost tend not to have plans to invest in them.

Solar domestic hot water

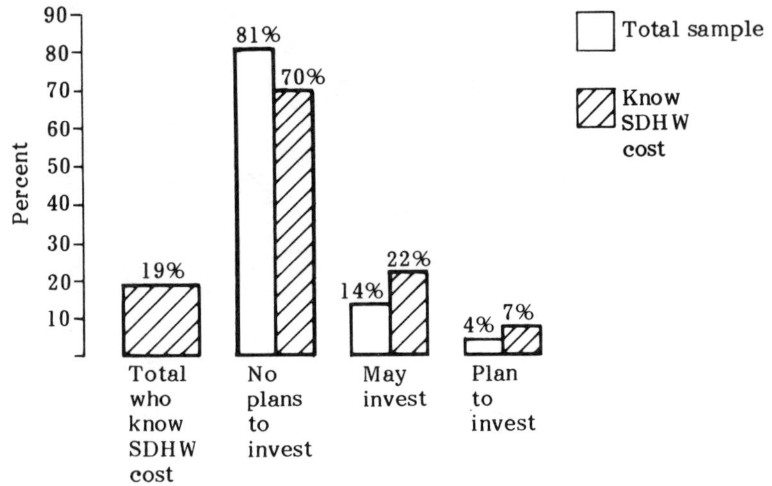

Solar space heating

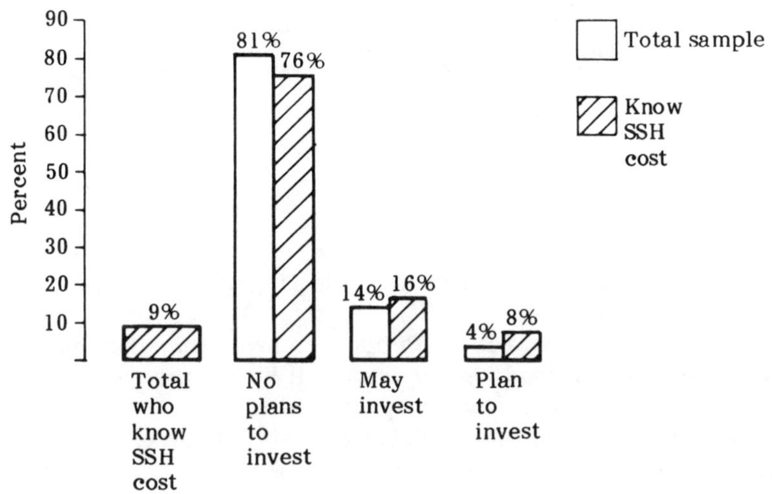

Source: Constructed by the authors.

BEHAVIOR

In this section, findings are presented on behavioral intention to purchase solar energy and on the kinds of systems in which potential purchasers are most interested. Several study variables are also related to behavioral intention. Data on reported solar ownership among homeowners are presented. Energy conservation actions taken are described, and differences in conservation by region and by primary heating fuel are also described. Behaviors affiliated with the "voluntary simplicity" life-style are also reported.

Behavioral Intention

Respondents were asked, "To what extent, if any, have you considered investing in a solar energy system of any kind for your house in the next 2-3 years?" This item was replicated by CEQ in its national survey of environmental opinion, so comparable results are available. Data are reported in Table 3.29. The first response category—have considered, definitely plan to invest (that is, have obtained cost estimates and/or equipment)—refers to action taken on the part of the potential buyer. Of the SERI sample, 4 percent indicate plans to invest in solar energy in the next 2 to 3 years, as did 3 percent of the CEQ sample. The second response category indicates that the person has considered it and may invest in solar energy. This is a less stringent category than the first one. Of the SERI sample, 14 percent indicate that they may invest, while 16 percent of the CEQ sample indicated the same.

A third response possibility is that the person has considered solar energy and has decided not to invest in it. Of the SERI sample, 13 percent have decided against solar investment, as did 8 percent of the CEQ sample. A final possibility is that the consumer has not considered investing in solar energy. About two-thirds of each sample indicate this response (68 percent of the SERI sample and 69 percent of the CEQ sample). Finally, a small proportion in each sample indicate that they do not know their plans with respect to solar investment.

The potential near-term market for solar energy systems suggested by these findings is around 18 or 19 percent of the nation's households.*

Those in the SERI sample indicating that they definitely plan to

---

*It should be remembered that the SERI sample is homeowners and the CEQ sample is adults 18 years of age and older.

TABLE 3.29. Behavioral Intention: Interest in Solar Applications

To what extent, if any, have you considered investing in a solar energy system of any kind for your house in the next 2-3 years?

Response	%	CEQ Survey[a] %
Have considered, definitely plan to invest (i.e., have obtained cost estimates and/ or equipment)	4	3
Have considered, may invest	14	16
Have considered, and will not invest	13	8
Have not considered	68	69
Don't know	1	4
	100	100

[a]Data from the Roper and Cantril national survey of adults 18 and over conducted for the Council on Environmental Quality, January-March, 1980.

Interest in Solar Applications

What kind of solar energy system have you thought of investing in? (CHECK ALL THAT APPLY). (Asked of those who definitely plan to or may invest.)

Response	%	(N)
Solar panels (or collectors) for hot water	56	(200)
Solar panels (or collectors) for home heating	50	(179)
Small wind energy conversion system	18	(63)
Passive solar homes or features	18	(65)
Solar greenhouses	15	(54)
Swimming pool, hot tub or spa heater	6	(23)
Solar cell (photovoltaic) system	5	(18)
Other (please specify)	1	(3)
Don't know	7	(26)
N	(354)	
Missing data	(13)	
Total N	(367)	

invest or may invest were asked what kinds of systems they have considered. Findings are reported in Table 3.29. A majority of 56 percent have considered solar domestic hot water systems, while 50 percent mentioned active space heating. These are the two most frequently considered applications. Other solar applications with their frequency of mention are: small wind machines (18 percent), passive solar homes or features (18 percent), solar greenhouses

(15 percent), swimming pool heaters (6 percent), and photovoltaic systems (5 percent).

Preferred System Function

Of those planning to invest in solar energy, the majority (55 percent) have considered investing in domestic hot water. The same holds for those who may invest (57 percent). About 44 percent of those planning to invest have considered space heating, while 52 percent of those who may invest have done so.

More of the consumers planning to invest have considered small wind energy systems (34 percent) than have those who may invest (14 percent). Similarly, 32 percent of those planning to invest have considered passive solar homes and features, compared with 15 percent of those who may invest.

Almost a third of those with more definite plans have considered a solar greenhouse, compared with 12 percent of those who may invest. Of those who plan to or may invest, most have not considered pool, spa, or hot tub applications. Of those with plans, 14 percent have considered these systems, as have 5 percent of those who may invest.

Much smaller proportions of both groups have considered photovoltaic technology; 9 percent of those with plans and 4 percent who may invest have considered photovoltaic systems.

In summary, the most likely applications to be selected by the next solar buyers are, in order: solar domestic hot water, solar space heating, small wind energy conversion systems, and passive solar homes and features.

Perception of Solar Investment Related to Behavioral Intention

Respondents were asked whether they thought a solar energy system would result in sufficient utility savings to make it a good investment. As reported, about a third of respondents indicated that, at present costs, owning a system would be a good investment, another third are unsure, and the final third think it would not be a good investment. Behavioral intention was cross-tabulated by responses to this item (Figure 3.7).

Of those who think solar energy is a good investment, 7 percent plan to invest, 24 percent may invest, and 69 percent have no plans to invest. Even more of those who think it is not a good investment have no plans to invest—88 percent. Of those planning to invest,

FIGURE 3.7. Behavioral Intention by Perception of Solar Investment Potential

Those planning to invest in solar energy tend to assess systems as a good investment.

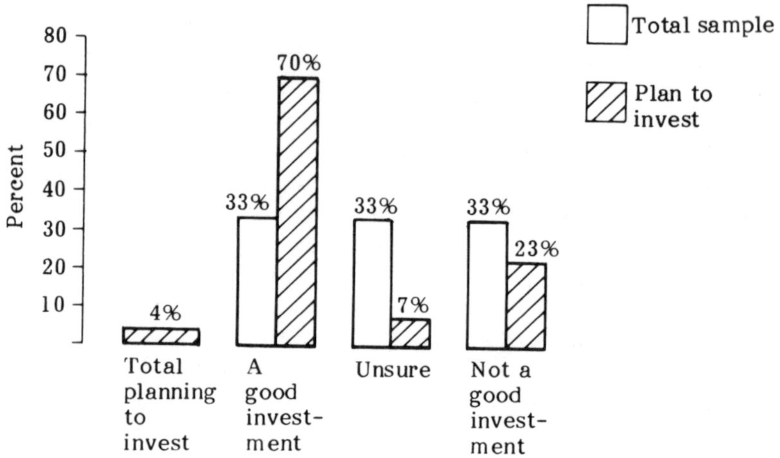

Most of those who think solar is not a good investment have no plans to invest in it.

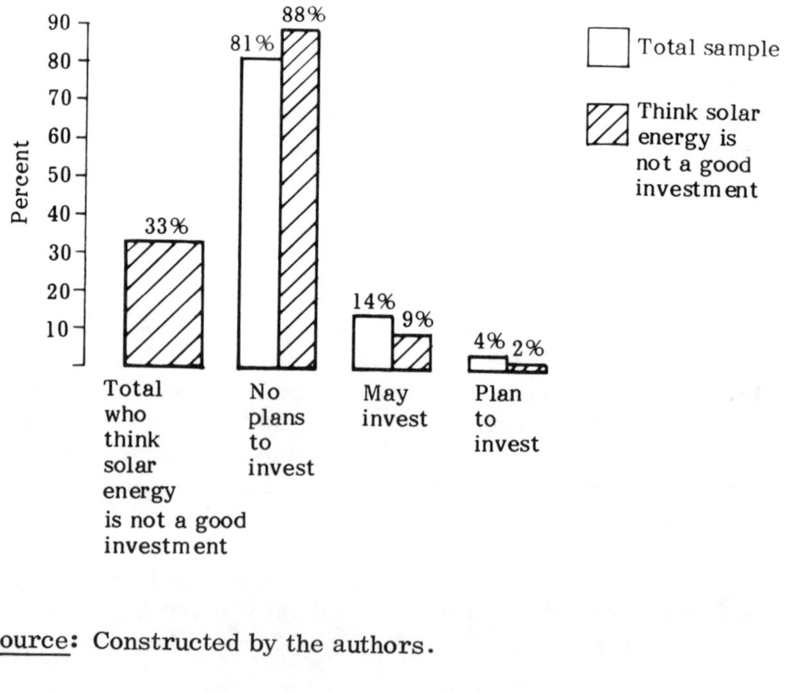

Source: Constructed by the authors.

70 percent assess solar energy as a good investment, while 58 percent of those who may invest and 28 percent of those with no plans to invest think the same.

Present Utility Costs Related to Behavioral Intention

Respondents were asked to look up or estimate their average monthly utility costs for December, January, and February (winter costs) and for June, July, and August (summer costs). These costs were recoded into three categories: (1) low, for costs less than one standard deviation below the mean cost for the sample, (2) average, for costs falling between ±1 standard deviation from the mean, and (3) high, for costs greater than one standard deviation above the mean. In winter, 1 percent have low costs, 88 percent have average costs, and 11 percent have high costs. In summer, 2 percent have low costs, 84 percent have average costs, and 14 percent have high costs. This result shows that utility costs do not form a normal distribution.

These utility cost categories were cross-tabulated by behavioral intention. No significant differences are found in these cross-tabulations. High monthly utility costs might be conducive to solar adoption if consumers thought their use of solar energy would sooner or later reduce utility bills. However, these results suggest that present utility costs are not high enough to provide a significant motivation for solar adoption.

Knowledge Related to Behavioral Intention

Feeling knowledgeable enough to decide about solar energy was cross-tabulated with behavioral intention (Figure 3.8). High levels of knowledge about solar energy are associated with plans to invest in it. Of consumers considering themselves knowledgeable enough, 10 percent definitely plan to invest and 30 percent may invest, while 60 percent have no plans to do so. Of those not knowing enough, 90 percent have no plans to invest. Feeling knowledgeable is clearly associated with purchase intentions.

Reported Solar Ownership

We expected only a few solar homeowners to turn up in the national sample of homeowners since solar ownership in the United States is still quite limited. Nevertheless, it would be an important

FIGURE 3.8. Feeling Knowledgeable Enough to Decide on Solar Use by Behavioral Intention

Knowing enough to decide on solar use is associated with plans to invest.

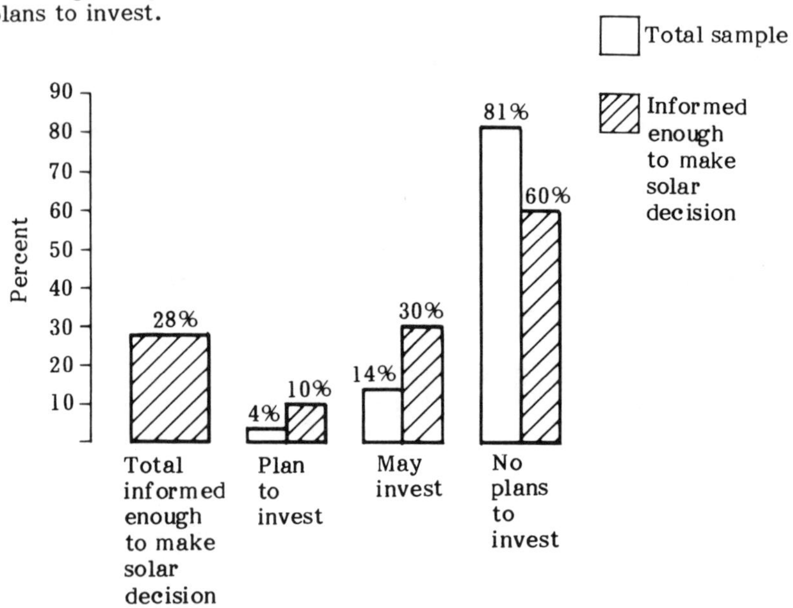

Source: Constructed by the authors.

baseline measure of solar penetration to know how many people presently report solar ownership. In future national surveys, data on solar ownership can be collected and compared with already existing data to measure the rate of solar diffusion, that is, how rapidly it is occurring.

What appears to have been measured with the item on solar ownership is probably perceived rather than actual solar ownership. Although no problems arose during item pretesting, the results make clear that, owing to misperceptions on the part of both interviewers and respondents, more homeowners claim ownership of certain solar technologies than probably exist in the nation. The findings appear to reflect a desire to be considered a solar owner, that is, perceived ownership may be considered another measure of favorability, a further confirmation of positive attitudes toward solar energy.*

---

*Another possibility is that a few respondents may have told interviewers what they thought interviewers wanted to hear.

TABLE 3.30. Reported Ownership of Solar Energy Systems

Solar Technology	%	N	N Answering
Solar panels (or collectors)	0.6	(10)	(1731)
Solar panels (or collectors) to heat water for homes	0.5	(8)	(1774)
Solar panels (or collectors) to heat homes	0.1	(2)	(1771)
Wood stoves to heat homes	14.9	(247)	(1662)
Solar swimming pool, hot tub, or spa heaters	0.3	(5)	(1734)
Attached solar greenhouse to heat homes	0.3	(6)	(1737)
Solar cells (photovoltaics) to produce electricity	0.3	(5)	(1753)
Small wind machines (or windmills) to produce electricity or pump water	0.8	(13)	(1731)
Bioconversion (such as gasohol and using farm wastes to produce gas for fuel)[a]	1.2	(22)	(1842)
Passive solar design (such as using many south-facing windows to get heat from the sun, with few windows on the north; using water containers, walls, floors, or ceilings to collect and store the sun's heat, etc.)	3.2	(56)	(1770)

[a]Data are probably invalid; see discussion in text.

Since the number of homeowners reporting ownership of solar technologies is so small, no reasonable estimates of the number of systems actually in existence can be made on the basis of these data, except in the case of wood stove ownership, which will be discussed later.

Respondents were asked whether they owned each of ten types of solar applications.* Data on solar ownership are reported in Table 3.30.

Ownership of "solar panels" was asked because the item was also measuring awareness of solar technologies. A total of 10 respondents, or one-half of 1 percent, report ownership of solar panels; 8 of them, solar domestic hot water systems; and 2, solar space heating systems.

Wood stove ownership is reported by 247 respondents, or 14.9 percent of those responding to the item (12.2 percent of the total sample). This is a sufficient number to use as a basis for estimating wood stove ownership in the 48 states. Translated to the approximate number of owner-occupied housing units, this results in an estimate of between 6.73 million and 8.82 million households with wood stoves, at the 95 percent confidence level.

Those who report owning solar swimming pools, hot tubs, or spa installations number 5, or about one-third of 1 percent. Owner-

---

*For actual item phrasing, see Table 3.8 or Item 13 in the homeowner interview schedule, Appendix B.

ship of solar cells (photovoltaics) to produce electricity is reported by 5 respondents, or about one-third of 1 percent. In all probability, there would not be 5 actual photovoltaic system owners in the sample, given the fact that there are only about 40 experimental photovoltaic systems on unoccupied houses in the United States. These respondents may have been referencing calculators, wristwatches, or toys powered by solar cells as "photovoltaic systems."

Small wind machines, or windmills, to produce electricity or pump water are reportedly owned by 13 respondents, or eight-tenths of one percent of the sample. Some of the old, nonfunctioning agricultural windmills frequently seen in agricultural areas were probably included in this count.

The data on bioconversion system ownership are almost certainly invalid. We had anticipated that, particularly in rural areas, some farmers might be producing gasohol, methane, or other fuels for their own use. Even though the bioconversion item had been pretested by SERI and Gallup and no problems had arisen, some interviewers marked bioconversion system "ownership" if the respondent had "used gasohol." The interview schedules were screened for written comments to this effect, and the data were corrected not to reflect bioconversion ownership in these cases. Nevertheless, the remaining 22 respondents reporting ownership of bioconversion systems probably include several who said they "owned" a bioconversion system by virtue of having used gasohol, and this fact was not noted by their particular interviewers. Therefore, far too many "bioconversion owners" are reported to represent accurately the number of such systems in the nation.

Ownership of passive solar design features is reported by 56 respondents, or 3.2 percent of the sample. When it is noted that the item included in its examples "using many south-facing windows to get heat from the sun, with few windows on the north . . .," some respondents may have reported "solar ownership" of houses that would not technically qualify as passive solar homes. The data probably reflect perceived ownership of passive features, regardless of the technical accuracy of the perceptions.

The tallies of how many systems are owned to not tell us how many solar owners there are, since one household could accommodate several kinds of systems. When asking how many solar owners there are, it is important to distinguish exactly what kinds of solar applications are included. Table 3.31 presents data on the number of reported solar owners in the homeowners sample. Those reporting bioconversion system ownership are excluded from the data.

If wood stove ownership is considered solar ownership, a total of 349 respondents, or 17.4 percent of the sample, are solar owners. If, however, wood stove ownership is excluded from the count, then

TABLE 3.31. Reported Solar Owners in the National Homeowners Sample[a]

	%	N
Total solar owners, excluding bioconversion but including wood stoves	17.4	(349)
Total solar owners, excluding both bioconversion and wood stoves	5.1	(102)
Total solar owners, excluding bioconversion, wood, and passive solar design	2.3	(46)
(N = 2001)		

[a]Not all respondents answered each item on the list; see Table 3.30 for the specific N for each item. Twenty-two respondents did not answer the question for any of the solar technologies, leaving a balance of 2001 who answered for one or more of the technologies. It is unlikely that those reporting ownership of systems would neglect to answer for that technology. Therefore, 2001 is used as the percentage base for data in this table.

102 respondents report solar ownership, or 5.1 percent of the sample. If the figures on passive solar home ownership are also omitted from the count, then 46 respondents report solar ownership, or 2.3 percent of the sample. Clearly, solar ownership, even "perceived" solar ownership, is still quite limited in the continental United States.

About 2 percent of the sample, or 47 respondents, report owning more than one solar energy system. Data in Table 3.32 show that the most frequently occurring combination is a passive solar feature or home with a wood stove. Owners of small wind energy conversion systems also tend to have wood stoves.

To summarize perceived solar ownership, wood stoves are the most widely owned "solar" technology, followed by passive solar design features. Small wind machines are the third most widely owned solar technology, and active solar domestic hot water and space heating systems are in fourth place. Very small numbers report ownership of solar greenhouses, photovoltaic systems, and solar swimming pool heaters. About half of households using solar energy are using

TABLE 3.32. Ownership of More than One Solar Energy System

SYSTEM TYPE	Wood stove	Passive design	SDHW[a]	SSH[b]	Collectors	Photovoltaics
Wood stove		22	1	and passive[c] 1	1	
Passive design	and wind 1		and wood stove 1	and solar greenhouse 1		
SDHW		1				
Swimming pool heater	3		1			
Small wind machine	7	2			1	
Photovoltaics	1	1				
Solar greenhouse		1				1

(Total N = 47)

[a]SDHW = solar domestic hot water system

[b]SSH = solar space heating system

[c]Three systems owned in one household—in this case, solar space heating, passive design, and a wood stove.

more than one solar feature, if wood stoves are included in the solar definition.

Compared with earlier studies, this survey shows a greater number of solar owners than previously estimated. This difference is undoubtedly due in part to misperceptions on the part of respondents in answering the item. However, it is also likely due in part to the increasing diffusion of solar technologies for residential applications in the United States—more people now own solar energy systems and homes than ever before. The rate of increase is still small, however. Yet other evidence in the survey as well as rapidly increasing utility costs suggests that the speed with which solar technologies will be incorporated in new buildings and added to existing ones is likely to quicken in the near-term future.

Energy Conservation Behavior

A 14-part item on energy conservation activities and measures was included in the survey. The sample was asked which activities or measures they had done since they had lived in their present homes. They could also indicate which items were not applicable to their situations. The list of conservation actions and the responses are presented in Table 3.33.

Mentioned most frequently as not applicable are setting thermostats at 78° or higher in summer (by 36 percent), presumably because of lack of air conditioning; car-pooling (by 30 percent), probably for several reasons, such as having only a short distance to travel; and installing equipment on fireplaces (by 38 percent), presumably because of lack of fireplaces.

Regional Differences in Energy Conservation Behavior

Each of the energy conservation items was analyzed by geographic region* to determine if statistically significant differences existed in the proportions practicing different types of conservation in various parts of the country. The sample is geographically distributed as follows: 23 percent are in the Northeast, 31 percent are in the Midwest, 32 percent are in the South, and 14 percent are in the West. Significant differences are found by region for each activity

*Regions are identical to those served by the regional solar energy centers (see Figure 3.9).

TABLE 3.33. Reported Energy Conservation Behaviors

Listed on this card are several activities or measures that conserve energy. Please tell me which, if any, you have done since you've lived here.

Response	% Have Done	% Have Not Done	% Not Applicable	Don't Know	N	Missing Data	Total N
Install weatherstripping or caulk around doors or windows	76	20	3	—	(1995)	(28)	(2023)
Set thermostat at 65° or lower in winter	64	28	7	—	(2007)	(16)	(2023)
Add insulation	61	33	4	—	(1990)	(33)	(2023)
Install storm or doublepane windows	60	32	7	—	(1982)	(41)	(2023)
Lower temperature setting on hot water heater	59	36	2	1	(1975)	(48)	(2023)
Install shutters or shades on windows	48	44	5	—	(1968)	(55)	(2023)
Drive a small car	45	44	9	—	(1992)	(31)	(2023)
Buy energy-efficient appliances	37	51	8	2	(1965)	(58)	(2023)
Set thermostat at 78° or higher in summer	35	26	36	1	(1986)	(37)	(2023)
Carpool	22	46	30	—	(1967)	(56)	(2023)
Install glass doors, heatalator, or other energy-saving device on fireplace	22	37	38	1	(1961)	(62)	(2023)
Install wood stove	18	72	7	—	(1962)	(61)	(2023)
Install timer on heating system or furnace	12	74	10	1	(1961)	(62)	(2023)
Install a heat pump	4	79	12	2	(1951)	(72)	(2023)

FIGURE 3.9. Regions of the United States

except installing a timer on a heating system, car-pooling, and installing a heat pump.

For caulking and weatherstripping, those in the Northeast (87 percent) and Midwest (86 percent) are most likely to have engaged in these practices. About the same proportion in the South and West have done so—approximately two-thirds.

Consumers in the Northeast are more likely than others to set thermostats at 65° or lower in winter (78 percent). About 14 percent of those in the South indicate this measure is not applicable to them.

Those in the Northeast are also most likely to have added insulation—71 percent report taking this action. Of those in the Midwest, 68 percent have added insulation; consumers in the South and West are least likely to have done so.

More Midwesterners (78 percent) and Northeasterners (76 percent) report installing storm or double-pane windows than do Southerners (24 percent) or Westerners (32 percent).

About two-thirds of consumers in the Northeast and 65 percent of those in the West have lowered the temperature of their hot water heaters. Over half of Southerners (54 percent) have also done so.

Those in the Northeast (58 percent) are most likely to have installed shutters or shades on windows, while those in the South (41 percent) are least likely to have done so.

Most likely to drive a small car are Westerners (56 percent); most unlikely are Southerners (42 percent).

Energy-efficient appliances are also more popular among Northeasterners (42 percent) and Midwesterners (41 percent) than among those in other parts of the country.

Southerners are the most likely group to report setting thermostats at 78° or higher in summer (48 percent). Large proportions of those in the Northeast (47 percent), West (45 percent), and Midwest (34 percent) indicate that this measure is not applicable, probably meaning that they do not have air conditioning.

Consumers in the Northeast and Midwest are most likely to indicate that installing energy-saving devices on fireplaces is not applicable to them (46 percent in each region). Otherwise, there is little difference among regions in the proportion installing these devices, ranging from 20 percent in the South to 28 percent in the West.

Least likely to have installed a wood stove are Southerners (14 percent), while 21 percent of Westerners and Northeasterners report installing a wood stove, the areas of greatest reported installation.

To summarize, for all conservation actions to conserve heating, the Northeast is significantly more likely to have taken them than have other regions of the country. Since the Northeast has a cold winter climate, dependence on fuel oil, and higher utility costs than many other parts of the country, action to conserve heat is more widespread there. The Midwest is also more likely than other regions to report more heating conservation action for two measures: (1) installation of storm or double-pane windows and doors, and (2) caulking and weather stripping. Both Northeasterners and Midwesterners are conserving electricity through use of energy-efficient appliances more than consumers in the South and West do.

Not surprisingly, Southerners are doing the least to conserve heating energy, but they are the most likely to conserve cooling energy.

The West presents an inconsistent picture on conservation activity. Westerners are more likely than those in the South and Midwest to have wood stoves and to have reduced the temperature on their hot water heaters. They, along with Southerners, are least likely to have caulked and weather-stripped, added insulation, installed storm or double-pane windows and doors, or purchased energy-efficient appliances. Westerners, however, are the most likely to drive small cars, resulting in gasoline conservation over the large distances Westerners routinely travel.

It appears, then, that the adoption of conservation activities has occurred in ways most appropriate to regional conditions. Cooling conservation is most widespread where air conditioning is most in use. Heating conservation occurs more frequently in the coldest climates and where heating costs are high. Wood stove use is highest near the forests of the West and Northeast. Gasoline conservation is practiced more in the West where the distances driven are great. These regional adaptations serve to increase the effectiveness of conservation behavior in reducing energy use by concentrating on reducing highest energy use categories. They suggest also that consumers look to reduce their energy use in areas where their pocketbooks are most affected.

With respect to potential solar energy use, the regional findings suggest that active and passive solar heating would be of most interest in the Northeast and Midwest, while active and passive solar cooling would be of interest in the South. Bioconversion applications would probably be of most interest in the Northwest and possibly the Northeast. Specific applications would be needed for the varied climates of the West, from heating needs in the Rocky Mountain region to cooling needs in the desert Southwest.

Differences in Conservation Activities by Primary Heating Fuel

Finally, primary heating source was cross-tabulated with conservation activites. The distribution of primary heating fuel is as follows: natural gas is used by 56 percent, fuel oil by 19 percent, electricity by 13 percent, wood by 8 percent, and propane by 4 percent. Those using fuel oil are more likely than others to have set their thermostats at 65° or lower in winter (77 percent). Of natural gas users, 64 percent report this practice. Those using electricity as their main heating fuel are more likely than others to have set their thermostat at 78° or higher in summer (48 percent). Of natural gas users, 40 percent report doing this.

Those using natural gas (53 percent) or fuel oil (54 percent) are more likely than others to have installed shutters or shades on windows. Propane users (20 percent) are the least likely to have done so.

Those with fuel oil (84 percent) or wood (84 percent) as their main heating fuel are more likely than others to have caulked or weather-stripped their homes. Of natural gas users, 77 percent report this activity.

The most likely group to install storm or double-pane windows are fuel oil users (73 percent), while those depending on electricity are least likely to have done so (49 percent).* Of natural gas users, 59 percent report such installations.

More likely than other groups to lower the thermostat on their hot water heaters are fuel oil users (64 percent), compared with 61 percent of natural gas users.

Wood (70 percent) and fuel oil (68 percent) users are more likely than others to have added insulation; 61 percent of natural gas users report insulating.

Those with electricity as their primary heating source (17 percent) are more likely than others to have installed a heat pump; one percent of natural gas users report this.

Those with fuel oil (15 percent), electricity (15 percent), and natural gas (13 percent) are most likely to have installed a timer on their heating system or furnace. Finally, those using natural gas are least likely (7 percent) to have installed a wood stove.

To summarize by primary heating source, seven of the conservation actions, all designed to conserve heat, have been taken signif-

---

*It is possible that most all-electric homes were built with better-than-average insulation and double-pane windows.

TABLE 3.34. Summary of "Voluntary Simplicity" Activities

Activity	Never Do	Do		
		Some	Regularly	Total
Skills for self-reliance	13%	45	42	87
Exercise for physical fitness	25	52	23	75
Change oil	30	27	43	70
Recycle	45	43	12	55
Bicycle	53	38	9	47
Buy used clothing	58	41	1	42
Contribute to ecological organizations	77	16	6	22

icantly more frequently by fuel oil users than by others. These actions are: lowering furnace and water heater thermostats, adding shutters and shades, caulking and weather stripping, using double-pane or storm windows, adding insulation, and installing a time on the heating system. Electricity users are most likely to conserve cooling and to install a heat pump. Natural gas users are most likely to use timers and shutters. Finally, wood users are more likely to caulk and add insulation. It is clear that natural gas users are among the least energy conserving of all the fuel users.

"Voluntary Simplicity" Behaviors

Respondents were asked how regularly they engaged in a set of activities thought to characterize the "voluntary simplicity" style of life. These activities, though not the same as energy conservation, relate to decreased personal consumption, generally, through a variety of life-style adaptations.*

Table 3.34 summarizes the data on voluntary simplicity behaviors reported by the nation's homeowners. Mentioned most frequently (by 87 percent) is "developing and using skills to increase self-reliance, such as in carpentry, car repair, and food preservation." Three out of four report that they "engage in exercise for physical fitness." Seventy percent "change oil in the family car at home," while a majority of 55 percent engage in recycling of newspapers, glass, or cans used at home. A plurality of 47 percent "use a bicycle for transportation or recreation," while 42 percent buy clothing at garage sales or secondhand stores. Least popular was contributing to ecologically oriented organizations (such as the Sierra Club), mentioned by 22 percent.

---

*The concept is described more fully in Chapter 4 and the detailed data are presented in Appendix D.

REGIONAL DIFFERENCES IN ATTITUDES TOWARD SOLAR ENERGY

Responses were categorized by four geographic regions of the country* and cross-tabulations were run by key study variables. Results of chi-square tests significant at $p < .025$ are reported in this section. The sample is geographically distributed as follows: Northeast, 23 percent; Midwest, 31 percent; South, 32 percent; West, 14 percent.

The Western region appears to be the least affected by the national energy situation. In response to the item on how the energy situation has impacted households, Westerners are more likely than consumers in other regions to report no effect or a few adjustments. The Northeast is most affected, where 13 percent of respondents report serious impacts on their households. This finding may be the result of the widespread dependence on fuel oil for heating in the Northeast, coupled with oil shortages and concomitant rising oil prices since 1973.

Northeasterners are somewhat more likely to view the nation's current energy situation with pessimism. About 42 percent of them rank the present energy situation as nearly the worst possible, compared with 35 percent of the sample as a whole. The other three regions are about the same in response, with the South least likely to rank the situation as the worst possible (31 percent); 1 in 5 homeowners ranks the present energy situation as nearly the best possible, with 15 percent of Northeasterners and 24 percent of Westerners responding in this way. However, Northeasterners are more likely than others to believe that the nation's energy situation will improve over the next 5 years. Westerners are slightly more likely to think the situation will stay about the same, while the South tends to be the more pessimistic region about energy.

Westerners tend to know more about solar energy and to feel that they know enough to make a decision to use solar energy in their homes more than do those in other regions. Southerners indicate that they are the least knowledgeable about solar energy of all regional groupings. Yet those in the Northeast are most likely to want more information about solar energy (63 percent), followed closely by Westerners (61 percent). About half of Southerners indicate interest in solar information. Westerners are the most likely to be aware of the federal income tax credits for solar energy (62 percent), compared

---

*Note: Regions are identical to those served by the regional solar energy centers (see Figure 3.9). See Appendix B, the homeowners interview schedule, for exact item wordings.

with 39 percent of Southerners. Consumers in the Midwest and South are more likely to be unsure of the existence of federal solar tax credits.

The South is most unsure about the feasibility and effectiveness of solar energy systems. The Midwest appears to be somewhat more confident of solar practicality than are the other regions.

When the 18 concern-about-risk items are analyzed for regional differences, most items result in significant differences with a clear pattern of results. Consumers in the Northeast are more concerned about possible problems in solar adoption than others, while Westerners are least concerned and Southerners are unsure. Specifically, Northeasterners are more concerned than others about maintenance problems, reputability of solar firms, problems in obtaining financing, solar warranties, obtaining insurance coverage, system safety, system reliability, potential for damage, laws prohibiting solar use, reducing resale value of the home, and climate. Southerners are more concerned than others about the difficulties of obtaining a system or solar home, effects of life-style, and aesthetics. Those in the Midwest, along with the Northeast, are slightly more concerned about potential problems with the utility company in solar adoption.

There are few regional differences in motivation to adopt solar energy. Among the 15 items on perceived relative advantage, only three regional differences were found. The chance to innovate or experiment with solar energy is defined as less important in the South. Reducing utility bills now is more important in the Northeast and Midwest and less important in the West. Increasing resale value of the home is also less important to Westerners than to others.

About one in four Westerners is unlikely to view a solar retrofit as a major obstacle. Southerners are more likely to be unsure about the difficulties involved in solar retrofit. Those in the Northeast and South are more likely to perceive greater difficulty, responding both that they "could not use" solar energy and "could use [it] only with major renovation."

Westerners are the most favorable to solar energy, while Southerners are the least favorable. Of Westerners, 86 percent favor the use of solar energy in homes and 70 percent favor its use in their own homes. Of Southerners, in comparison, 72 percent favor solar use in homes and 54 percent favor its use in their own homes. Southerners are more likely than those in other regions to be unsure. Those living in the Midwest and Northeast fall between the West and South in favorability. These findings suggest that Southerners associate solar energy with heating, in which they have relatively less interest than cooling.

More Westerners (7 percent) than others indicate definite plans to invest in solar energy in the near-term future, and more Westerners

(19 percent) than others may invest. Midwestern consumers are the most likely not to have considered solar adoption (73 percent), compared with 60 percent of those in the West. Southerners are the most likely to indicate that they have no interest in solar energy (25 percent), and that is the least likely region to be willing to pay more for solar energy than present monthly utility costs. About 40 percent of Westerners indicate a willingness to pay more per month for solar energy, most of them in the up-to-$25 per month category.

In summary, solar favorability and adoption plans are significantly more likely to occur in the West. Westerners are more knowledgeable about solar energy than are those in other regions, and they are less concerned about the risks associated with solar adoption. Consumers in the West are also less likely than others to perceive major obstacles to solar retrofits on existing homes. Northeasterners, while favorable to solar use, have higher concerns about the risks involved in solar adoption (including climate concerns) and they, along with Southerners, perceive greater difficulty in retrofit on existing structures. This may be associated with the fact that much of the housing stock in the Northeast is older than in other parts of the country (Annual Housing Survey 1978). The advantages of solar adoption are perceived to be the same, regardless of geographic region. The findings suggest that in the South, water heating would be the solar application of primary interest.

SUMMARY OF FINDINGS AND CONCLUSIONS

This section summarizes what has been reported in Chapter 3 on the national homeowners survey. This chapter has reported descriptive findings and a few analyses that begin to reveal how the study's variables relate to each other. The findings from this chapter are the foundation for understanding the study's multivariate analyses. The conclusions offered here may be considered descriptive of the nation's homeowners.

Attitudes toward solar energy are highly favorable. Solar energy is by far the most preferred energy supply source to meet our future energy needs; nuclear energy is least preferred. This pattern of response has been borne out by other surveys as well.

Homeowners are highly favorable to the use of solar energy in homes, and a majority favor its use on their own homes. Those using solar energy are seen as "economy-minded"—a mainstream image that is widely socially acceptable. Solar energy is not viewed solely as a favorite of iconoclasts or fringe groups. Consumers generally are not well acquainted with solar homeowners, and most have not seen any solar homes or buildings in their own neighborhoods, yet residential solar use is compatible with existing cultural norms and values.

Solar energy is not a favorite because of anticipated near-term energy shortages. A majority of homeowners are optimistic about the national energy situation five years in the future. About one in four homeowners thinks the situation will be worse in five years. So far, while most have experienced some inconvenience, the impact of the energy situation on households is not perceived as serious. This suggests that homeowners do not attribute inflation to energy price increases.

A majority of homeowners express hesitance over the present feasibility of residential solar applications. They tend to be unsure about the practicality and cost effectiveness of solar energy systems, especially as related to climate. This perspective constitutes a barrier to early adoption of solar technologies.

Most homeowners are aware of solar technologies that can be used for residential applications. Most widely known are wood stoves and windmills; least known are photovoltaics. A majority are aware of active and passive solar energy applications; yet consumers are not well informed about solar energy. Almost three out of four say they do not know enough to decide about using solar energy in their own homes. However, those who know more about solar energy tend to think it is presently feasible for use on homes. Feeling knowledgeable about solar energy is associated with the possible intention to adopt it.

A majority of homeowners would like to know more about solar energy—nearly 30 million households would like to receive such information, even if they are not actively seeking it. Kinds of information wanted are technical information, costs and savings, information on what it's like to live with solar energy, and political and business information.

Consumers are likely to use a variety of specific sources of information on solar energy. They are especially likely to turn to solar companies, state energy offices, public libraries, and federal organizations for information. Consumers want to <u>see</u> solar technologies, those in homes or for demonstrations. Reading books and articles about solar energy is also an important source. Through factor analysis, 5 important types of solar information sources are identified. The most important type is federal organizations, probably as relatively unbiased sources. The second type is media, especially print media. Third is local expertise—plumbing and heating contractors, homebuilders, and solar homeowners; solar contractors fall more within this type than any other. The fourth type is educational settings, such as seminars and workshops, which would be of particular interest to do-it-yourselfers. The fifth and final type is advocacy organizations, such as the Solar Lobby, from which consumers can learn about advantages of solar use and gain political and business-related information. Local expertise as an information source is the most persuasive type; it will very likely become more important over time, as homeowners become more aware of solar fundmentals and get closer to deciding about solar use for their own homes.

About 37 percent of homeowners are likely to request an RCS-type home energy audit and would be willing to pay the $15 audit fee. (If audits were free, this proportion would increase to about half.) From this finding we can estimate that utility companies will very likely provide between 17.8 and 20.8 million residential audits in the nation's owner-occupied homes. The audit is a particularly important information source since it provides the kinds of information on solar energy consumers may they need.

Concern about risks associated with solar adoption is widespread. Three main types of concern exist; these are concern about: (1) system safety and reliability, (2) life-style and aesthetic effects, and (3) system purchase. The most important of these is concern about system safety and reliability, including cost and difficulty of getting a system covered by homeowners' insurance, operating reliability, prohibitive local building codes, and possible damage to the solar energy system from dirt, vandalism, storms, or corrosion.

Three major types of motivation to adopt solar energy are empirically established in this survey: (1) responsible self-interest, such as reducing utility bills now and protecting against rising prices; (2) advantages of solar energy to society, such as conservation of natural resources and protecting the environment; and (3) fun, or psychological rewards from innovating or experimenting and getting attention from others. A fourth type of motivation, involving national economic interests, probably also exists, but was not measured in this survey. Motivation to adopt solar energy can be classified into four types: (1) egoistic-economic, (2) egoistic-noneconomic, (3) altruistic-economic, and (4) altruistic-noneconomic.

Underscoring the importance of these reasons to adopt solar energy, nearly one out of three homeowners indicates a willingness to pay a modest amount more than present monthly utility bills (averaging about $28/month) to own solar energy systems. Those willing to pay more per month, however, are unwilling to pay it to a utility company.

One out of three homeowners are uncertain about whether owning a solar energy system will result in sufficient utility savings to make it a good investment: one-third think solar energy is a good investment, and one-third think it is not. Almost half of those who think solar energy is a good investment are willing to pay more per month for it; the other half are not. Consumers planning to invest in solar energy tend to assess it as a good investment, while most of those who think solar energy is not a good investment have no plans to invest in it. Those to whom the initial cost of solar energy is not the most important concern tend to view solar purchase as a good investment.

A majority of homeowners (54 percent) are unaware of the 40 percent federal tax credits for solar energy. An even larger majority (75 percent) of those living in states with solar tax credits are unaware of the credits that exist. Tax credits are not encouraging the use of solar energy as effectively as they might be, owing to the fact that most consumers are unaware of them.

The most preferred form of financial incentive, the amount of benefit being equal, is low-interest loans. This is probably because home improvement loans are a long-standing, well-understood mechanism and because tax credits assist only those with sizable income taxes. Tax credits are the second most preferred form of financial incentive. Home improvement loans are the most preferred financing mechanism, as well, among those who would finance all or part of a system that they purchase.

Initial cost is a significant barrier to solar adoption, based on the following evidence:

- It is frequently mentioned as a highly salient concern.
- Homeowners are not knowledgeable about the actual costs of installing solar systems, and, if anything, they tend to underestimate costs.
- Homeowners tend to be unaware of the financial incentives that are available to them to offset system cost.

Informed consumers tend to underestimate costs for both solar domestic hot water and solar space heating systems, especially the latter.

Consumers who think solar energy is a good investment tend to underestimate or not to know system costs. Of those planning to invest in solar energy, about half know what solar domestic hot water systems cost, and a majority underestimate solar space heating costs. Those who know what systems actually cost tend not to have plans to invest in them.

Other barriers to solar adoption also exist. Consumers are generally unaware that solar products and services, other than wood stoves, are available to them locally. About half believe that solar retrofit projects would be difficult or impossible for their homes, while one in five is unsure how difficult it would be. Home energy audits, providing reliable assessments of solar retrofit potential of individual homes, will help with this problem.

Homeowners express strong support for the idea of more energy decision making at the local and individual levels. Local government can play an increasingly important leadership role in encouraging solar adoption in the future. Majorities endorse activities and policies designed to accelerate the use of solar technologies in homes, from informing consumers to mandating solar use in new home construction.

Roles are seen for all levels of government. The federal government is viewed as having primary responsibility for informing consumers, providing financial incentives, and as warrantor of last resort. State government is chosen for listing solar contractors, informing consumers, and providing financial incentives. Local government is selected to mandate the use of passive design and solar domestic hot water in new homes, to inspect systems, to list solar contractors, and to educate and inform consumers. Utilities are seen as appropriately conducting home energy audits, informing consumers, inspecting systems, and listing solar contractors.

Based on stated solar investment plans, the potential near-term market for residential solar energy systems in the next 2 to 3 years is around 18 to 19 percent of the nation's homeowners. Owing to other intervening factors, however, the actual number of adoptions is likely to be smaller. Solar domestic hot water systems appear to be the most likely solar technology to appeal to the near-term market. Other likely solar applications to be selected by the next solar buyers are, in order: solar space heating systems, small wind energy conversion systems, and passive solar homes and features.

Ownership of residential solar technologies is still quite limited in the United States. Wood stoves are the most widely owned "solar" technology, at 14.9 percent. Small wind energy systems, solar domestic hot water systems, and passive solar design features are the next most owned solar applications, by very small fractions of homeowners.

Conservation actions most frequently taken (by majorities) are caulking and weatherstripping, lowering the thermostat in winter, adding insulation, installing storm or double-pane windows, and lowering hot water heater temperatures. Those using fuel oil for their primary heating source have taken the most heating conservation action. Those using electricity for heating are most likely to conserve cooling energy and to have installed a heat pump. Natural gas users are, so far, among the least energy conserving of all the fuel users; this may change as natural gas is deregulated and prices increase markedly.

Adoption of conservation activities has occurred in ways most appropriate to regional conditions. Cooling conservation is most widespread where air conditioning is most in use. Heating conservation occurs more frequently in the coldest climates and where heating costs are high. Wood stove use is highest near the forests of the West and

Northeast; gasoline conservation is practiced more in the West, where the distances driven are great. These regional adaptations serve to increase the effectiveness of conservation behavior in reducing energy use by concentrating on highest energy-use categories.

4

DEMOGRAPHICS, VALUES, AND LIFE-STYLE PREFERENCES

Most public opinion surveys and marketing studies collect information describing the persons that they have interviewed. Generally, they collect demographic data, such as respondent age, gender, occupation, and income. Usually, however, demographic analyses are atheoretical, that is, no guiding ideas are used to define the respondent characteristics that are likely to be relevant to the topic under study.

Demographic characteristics describe persons, irrespective of the particular topic of the study. For example, people's gender remains the same regardless of their preferences for soap, leisure-time activities, or attitudes toward solar energy. The question is: does it make any difference in preference for soap or leisure-time activities if one is male or female, and, if so, what difference does it make?

In preparing for the National Study, literature was reviewed on respondent characteristics potentially relevant to solar adoption, and decisions were made on whether to include or exclude variables, based on the results. The most efficient way to achieve this was to review literature reviews; four comprehensive literature reviews were examined:*

- A review of theory and empirical research in the diffusion of innovations (Rogers and Shoemaker 1971)
- A review of the literature on environmentalism (Van Liere and Dunlap 1979)

---

*When we refer to "the literature" in this chapter, we are referring to these literature reviews.

- A review of existing analyses on public opinion about solar energy (Farhar et al. 1979b)
- A review of existing studies of solar owners (Unseld and Crews 1979)

The results of these literature reviews are summarized in this chapter, along with decisions made about variables to be included in the survey and the hypotheses that we would expect our data to support. Findings from the national homeowners survey are reported in each section. The discussion is divided into two parts, one of demographics and one on values and life-style characteristics.

DEMOGRAPHICS

The following demographic characteristics are discussed: gender, age, education, occupation, income, political orientation, religious affiliation, race, community size, opinion leadership, and innovativeness.

Gender

Based on available evidence, we predicted that there would be no significant differences between men and women in favorability to solar energy. Environmental concern is not related to gender. Two countervailing tendencies among males may tend to cancel each other out in their effect. The first is that males tend to be more politically active than females, are more involved in community issues, and have higher levels of educational achievement. They might, therefore, be more environmentally concerned than are females. However, males are more likely to be concerned about economic growth and the availability of jobs and, thus, less concerned about protecting environmental quality. They are more inclined than women to choose energy development in the energy/environment trade-off.

Women are more strongly in favor of environmental protection in energy production and of energy conservation than are men. No data on gender differences are available from existing studies of solar owners. The decision was made to include data on gender in the homeowners survey by interviewer notation, simply as a sample descriptor.

Analysis of the homeowners survey data shows, however, that there are significant differences in attitudes toward solar energy by gender. Men are significantly more likely to be favorable to solar energy, both in the abstract and in the concrete, while women are more unsure. Women are more unsure than men about solar energy being a good investment, and more men than women say it is not a good investment. About equal proportions of men and women say that solar energy is a good investment. More women than men have not considered investing in solar energy. Men are more knowledgeable about solar energy and the tax incentives than are women and are more likely to perceive solar energy as presently feasible, with women being more unsure. About a third of men feel that they know enough to make a decision about solar use, compared with 19 percent of women. Men tend to believe that the energy situation will be better in 5 years, while more women tend to believe it will be worse. Women tend to be more concerned than men about the risk of solar adoption in general. Women are significantly more concerned than men about maintenance, aesthetics, problems in acquiring a system, possible problems with the utility company, financing problems, system safety, obtaining insurance coverage, and lowering the resale value of the home. Men are slightly more concerned than women about obsolescence of present systems and warranty coverage for systems. The sexes display little difference on the perceived relative advantage of solar adoption. Women are slightly more likely than men to see solar advantages as important.

In summary, our hypothesis that gender would make no difference was not supported by the homeowner data. There are significant differences by gender in attitudes toward solar energy. Men are more favorable toward solar use, while women are more unsure and less informed. This finding suggests that men are more adventuresome and open to the risk of adopting innovations than are women.

Age

We would expect favorability toward solar energy to decrease with age. The diffusion literature shows that findings on age are mixed; in 19 percent of 228 studies reviewed, earlier adopters are younger; 48 percent showed no relationship; and 33 percent show that earlier adopters are older. Solar owners studies also result in mixed findings on age.

The evidence from public opinion surveys is also mixed, but age is negatively associated with environmental concerns, that is, as age increases, environmental concern decreases. Since younger people are more environmentally concerned than are older people, and solar energy is frequently cited as more environmentally advantageous than other energy sources, younger people could be expected to be more favorable toward it. Younger people show more consistent support than do older people for the environment when energy/environment trade-offs are posed.

As to why the young exhibit these characteristics, reasons offered are that they are more environmentally aware, they have a larger stake in environmental quality, and they are less committed to the conventional in the socioeconomic system.

Our hypothesis concerning age was supported by the homeowners data. While most consumers favor the idea of using solar energy for homes, after age 44 favorability decreases with age. Most opposed to solar energy (at 10 percent) are those 75 or older. The age group most favorable to using solar energy on their own homes is that 24 to 35 years of age (77 percent). Most opposed (at 45 percent) are those 75 or older. As age increases, the likelihood of assessing solar energy as a good investment decreases; older persons are more likely to be unsure than are younger ones. Similarly, as shown in Figure 4.1, the proportion who might invest or plan to invest in solar energy systems in the next 2 to 3 years declines with age. Conservation activities taken also decrease with age (Figure 4.1).

The most knowledgeable group is 25 to 34 years of age. Over age 35, as age increases, the feeling of being informed decreases. Feeling informed enough to make a decision to use solar energy on one's home is inversely related to age, as well. The 25 to 34 year-olds want information about solar energy more than do other age groups. High proportions (about two-thirds) of those 24 or younger and 35 to 44 also desire solar information. After age 44, desiring information about solar energy trails off. Those 25 to 44 are also most aware of the federal tax incentive. Perception of solar energy as feasible decreases with age, while being unsure increases with age.

Belief that the nation's energy situation will be worse in 5 years is higher among younger consumers (under age 35) than among older ones. Most likely to believe the situation will improve are those 75 or older (67 percent) and those 45 to 54 (61 percent). The oldest consumers do not define the energy situation as seriously impacting them. As age increases, the proportion reporting inconvenience but not serious impact, due to energy shortages, declines. At age 24 or younger, 83 percent report these effects, while at age 75 or older, 52 percent report them. In contrast, 38 percent of those 75 or older report "no effect" on their households due to energy shortages, compared with 9 percent of those 24 or younger.

The oldest consumers are the most unsure about the risks involved in solar adoption. Otherwise, perception of risk is spread evenly across age groups. The oldest group consistently defines advantages of solar adoption as less important than do other age groups. The 25 to 34 year-olds tend to define solar advantages as somewhat more important than the others. Those 24 or younger also tend to view solar advantages as slightly more important than do others.

FIGURE 4.1. Key Variables by Age

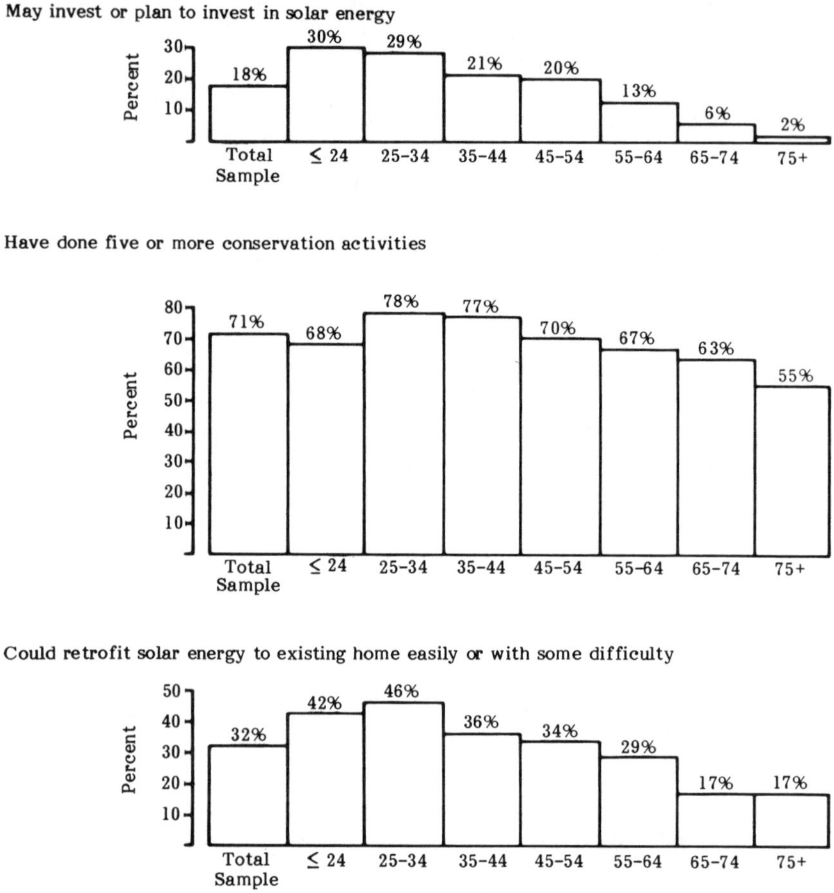

Source: Constructed by the authors.

Perception that solar retrofit would be difficult increases with age, after age 25. The unsure response is highest over age 44. As shown in Figure 4.1, perception that solar retrofits are easier is highest among those under 35 and decreases after age 35.

Most favorable toward local control of energy decision making are those 35 to 44; least favorable are those 75 or older (although almost everyone is favorable). As age increases, after age 25, preference for giving energy companies a free hand increases; most favorable to this idea are those 75 or older. Most skeptical of federal government solutions to the energy problem are those 35 to 44, while most trusting are those 75 or older. Younger groups, especially those

25 to 34, are most favorable toward conservation, while those over 75 are least favorable (although nearly everyone is favorable toward conservation).

Most likely to finance systems are those 25 to 44; as age increases, consumers are less likely to choose both paying cash cash and financing part of a system purchase. Most likely to pay cash are those 75 or older (24 percent) compared with 8 percent of those 24 or younger. Notably, about half of those 75 or older indicate that they are not interested in solar energy.

In summary, age is clearly a key variable in solar adoption. Generally, the older consumers are, the less likely they are to be informed or interested in solar energy. Solar adoption is less likely among the oldest portion of the population; yet interest remains high in age groups up to about 45 to 50.

Education

We predicted that favorability toward solar energy would increase with educational attainment. Prior research shows that education is positively associated with favorable attitudes toward solar energy and toward the environment. The diffusion literature shows that earlier adopters are more highly educated. Solar owners also tend to be highly educated.

Findings from the homeowners data bear out these predictions. Favorability toward solar energy, both in the abstract and the concrete, increases with educational level, while uncertainty decreases. The probability of adoption also increases with educational level

FIGURE 4.2. Behavioral Intention by Education

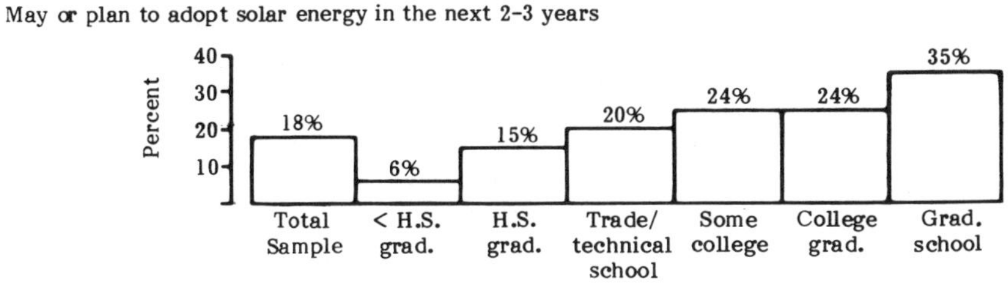

Source: Constructed by the authors.

(Figure 4.2). However, those who have college degrees and above are more likely than other educational groups to indicate that solar energy is not a good investment; high school graduates are more likely to say it is a good investment. Most energy conservation activity is reported by college graduates and trade/technical school graduates.

Knowledge about solar energy and information seeking increase with education, as does feeling informed enough to make a solar adoption decision. Awareness of federal tax credits increases with education. In general, perceived feasibility of solar energy tends to increase with educational level.

Most concerned about the risk of solar adoption are trade/technical school graduates. The highest and the lowest educated are least concerned about risk. To speculate, the more highly educated probably feel less vulnerable to risk. The lowest-educated, who tend to be unsure about risks, probably feel more unconcerned, since it is unlikely they will ever adopt solar energy.

Trade/technical school graduates also tend to define the advantages of solar energy as important more than do other educational groups. Those who have gone to graduate school are less likely to define solar advantages as important. Those with less than a high school diploma tend especially to see solar energy's advantages to society as less important. The perception of solar retrofit as reasonably easy increases with educational level. Preference for local energy decision making increases with educational attachment, while trust in the federal government decreases with education. Approval for giving energy companies a free hand tends to decrease with education; however, those with less than a high school diploma (51 percent) and college graduates (45 percent) agree most with this idea.

In general, the higher the consumers' educational level, the more favorable toward solar adoption they are. Well-educated consumers appear to want solid information about solar energy's cost effectiveness.

Occupation

We expected professionals and managers to be more favorable toward solar energy than are other occupational groups. Solar owners tend to be in professional or managerial occupations. White-collar workers are more favorable than are blue-collar workers toward environmental protection when energy costs would increase as a result. Professionals and managers are more favorable than are other workers toward solar energy, although it is favored by all occupational groups.

Occupation may affect environmentalism and energy preferences when jobs are dependent upon extracting and using natural resources. Those in jobs dependent upon these activities might be less favorable toward the use of solar energy. However, it was decided that the study could not examine whether respondent occupations were extractive, owing to space limitations.

The national homeowners data support the hypothesis that professionals and managers are the occupational groups most favorable toward solar energy. While nearly everyone is favorable, professionals, manufacturers' representatives, and managers/executives are the most favorable occupational groups. Least favorable are farm laborers. Most favorable to the idea of using solar energy in their own homes are students, manufacturers' representatives, professional workers, and business owners. Least favorable to this idea are housewives (44 percent) and the retired (37 percent). Most likely to assess solar energy as a good investment are students, manufacturers' representatives, and laborers. Least likely to do so are the retired, while housewives and farm laborers are unsure. Most energy conservation action has been taken by professionals, business owners, and those in skilled trades. Least conservation action has been taken by housewives and farm laborers. Most likely to invest in solar energy, as shown in Figure 4.3, are professional workers and manufacturers' representatives. Least likely to invest are housewives and farm laborers.

Professional workers and business owners are the most informed about solar energy. Least informed are housewives and students. As Figure 4.3 shows, professionals tend to feel that they know enough to decide about solar use, while housewives do not. The occupations that want most to learn more about solar energy are business owners, professional workers, manufacturers' representatives, and skilled workers. Less interested in learning more are students, housewives, and the retired. Most aware of federal tax credits are students, manufacturers' representatives, professionals, and managers/executives; unaware are the retired and sales workers. Professionals, farm owners, and managers/executives tend to define solar energy as feasible, while farm laborers and the retired tend to be more skeptical of solar energy's present effectiveness; housewives also tend to be unsure.

Housewives, the retired, and farm laborers report the least impact of the nation's energy situation on their households. Reporting the most impact are full-time students. Housewives are consistently unsure about the risks of solar adoption. Blue-collar workers express consistently more concern about risk than do other occupational groups. Some more concern is also expressed by clerical and sales workers, managers/executives, manufacturers' representatives, and

FIGURE 4.3. Key Variables by Occupation

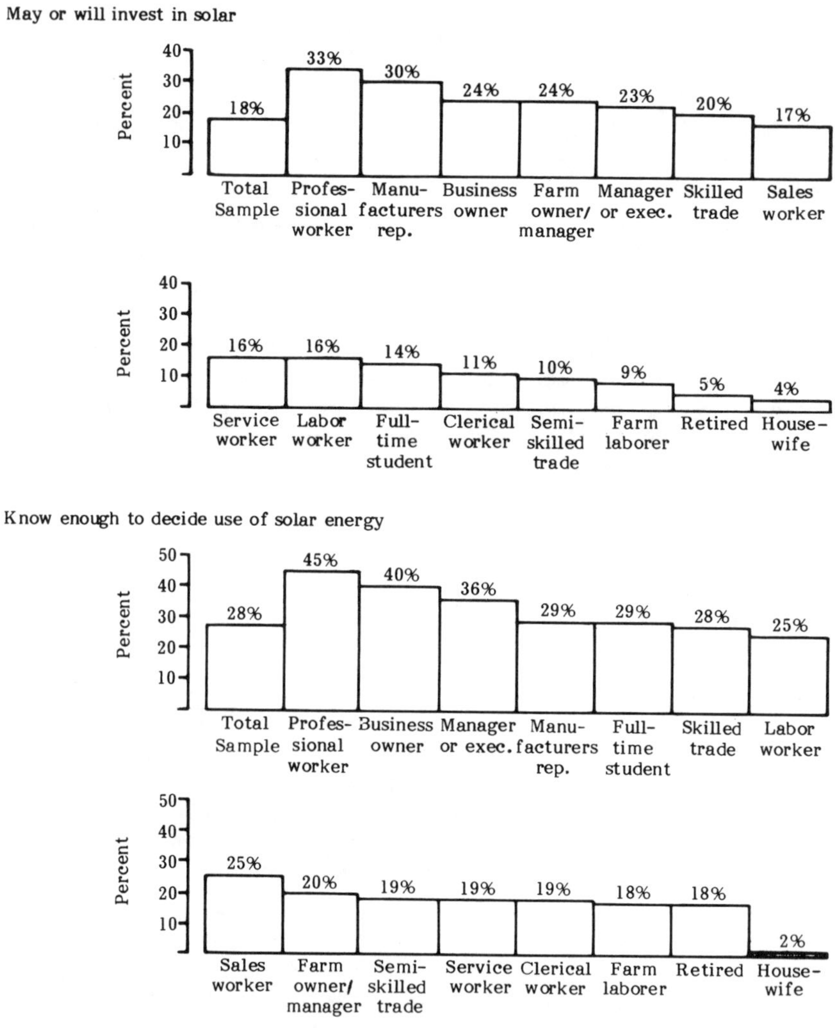

Source: Constructed by the authors.

students. Most solar advantages are defined as important more frequently by blue-collar workers than by other occupational groups. Sales workers and manufacturers' representatives also tend to define advantages of solar adoption as important.

Least difficulty in solar retrofit is perceived by manufacturers' representatives, students, business owners, farm owners, sales workers, and professionals. Most difficulty is perceived by housewives. Housewives tend to agree that we should trust the federal government to solve our energy problems, while students tend to disagree. More farm laborers and students tend to agree that we should give energy companies a "free hand" more than do others, while blue-collar workers tend to disagree.

In summary, the most favorable occupational groups, and the ones most likely to take near-term action on solar energy, are professionals, managers/executives, and business owners. Those in skilled trades may also be interested sooner than others. Business owners could also become interested in solar energy for their commercial buildings.

Income

We expected income to be an important demographic variable. It was anticipated that favorability toward solar energy would increase with income. While income is not consistently related to environmentalism, fairly consistent findings from public opinion data show that favorability toward solar energy increases with income. Also, existing solar owners tend to have higher incomes. As Maslow (1970) and Inglehart (1977) have discussed, the upper and middle social classes have solved the problem of their basic material needs and are free to focus on more "aesthetic" aspects of life, including innovating. Low income may well be a serious limiting factor in considering solar adoption, especially given initial system cost. Existing tax credits provide no financial incentive to those whose income is so low that little in taxes is paid.

These basic ideas about the importance of income as a demographic variable are supported by the homeowners data. As income increases, favorability to solar energy, both in the abstract and in the concrete, increases (Figure 4.4). Medium-income consumers are most likely to assess solar energy as a good investment. Interestingly, the number of conservation actions taken increases with income. Probability of investment in solar energy increases with income. Of high-income consumers, 22 percent may invest and 6 percent plan to invest; of medium-income consumers, 14 percent may invest and 3 percent plan to invest; of low-income consumers, 5 percent may invest and 2 percent plan to invest. Most low-income consumers have not considered solar investment (Figure 4.4).

As income increases, consumers' feelings that they know enough to decide on solar use also increase. Feeling informed about solar energy increases with income, as does wanting to know more about solar energy. Of high-income consumers, 65 percent are aware of federal tax credits; of low-income consumers, 25 percent are aware.

FIGURE 4.4. Key Variables by Income

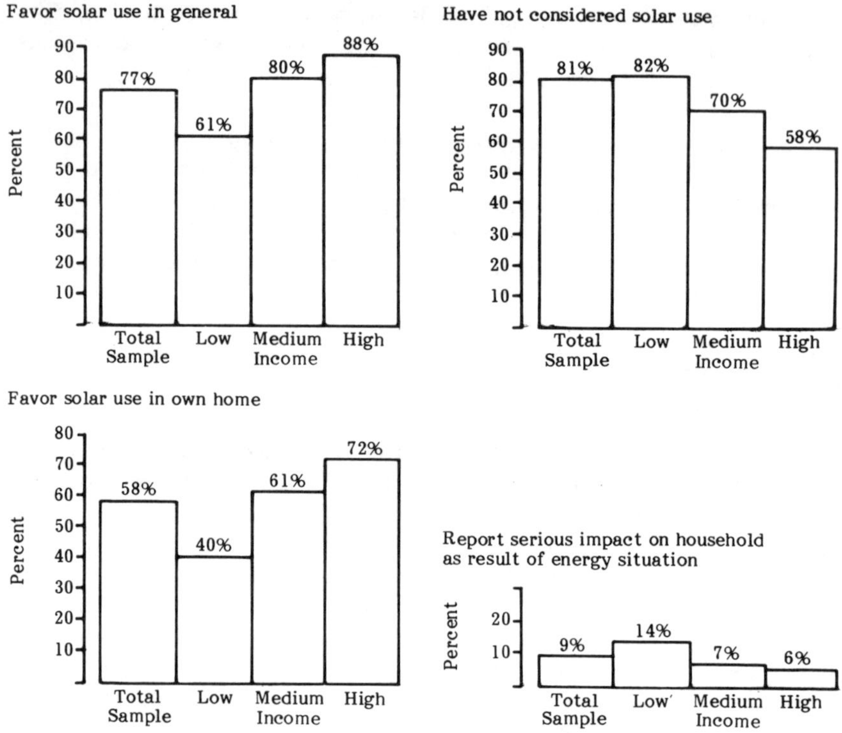

Source: Constructed by the authors.

In general, perceived feasibility of solar energy also increases with income.

Low-income consumers report more serious impacts of the energy situation on their households (Figure 4.4). Medium-income consumers are somewhat more concerned about the risk of solar adoption, in general, than are either low- or high-income consumers. Greatest concern lies in areas of system safety and purchase. High-income consumers are most concerned about purchase issues, such as warranty coverage, reputability of solar firms, and initial system cost.

Medium-income consumers tend to define as important more of the advantages of solar adoption than do other income groups. Of most interest to them are reducing utility bills now and protecting against rising costs. High-income consumers value protection against rising costs, having a more reliable source of energy, and increasing overall self-reliance. Both medium- and high-income consumers value conserving natural resources and protecting the environment. The chance to innovate, increasing the comfort of the home, and the availability of tax credits are more important to medium-income consumers than to other income groups.

As income increases, preference for local control of energy decision making increases, as does disagreement that we should trust the federal government in energy decision making.

Political Orientation

Studies have shown that partisan affiliation in unrelated to energy opinions and to environmentalism. It was expected, however, that conservative or liberal political orientation would be related to favorability toward solar energy, with liberals being more favorable than conservatives. Political liberals are found to be more favorable to the environment and to solar energy than are conservatives. Innovations and reforms are often opposed by business and industry because of the costs they involve, the changes they require, and the possibility of new governmental regulations. This helps to explain why conservatives might be less favorable toward solar energy than liberals would be.

Findings from the homeowners survey tend to support the hypothesis that liberals are more favorable, but conservatives are also favorable to solar energy. Political radicals are the most favorable (94 percent), while those unsure of their political orientation are least likely to be favorable and are more likely to be unsure (40 percent) or opposed (10 percent). Favorable to solar energy are 73 percent of strong and 80 percent of moderate conservatives (Figure 4.5). Liberals and radicals are more likely than others to favor solar use on their own homes; yet over half of conservatives favor solar use on their homes, while 26 percent are opposed to it. Those unsure politically tend to be more opposed than others and are also more unsure about solar energy. The likelihood of assessing solar energy as a good investment relates to political orientation in linear fashion: such assessment is lowest among strong conservatives and highest among radicals (Figure 4.5). No difference by political orientation was found in behavioral intention. Radicals report the most energy conservation activity, followed by strong liberals and moderate conservatives; the politically unsure report the least.

Strong liberals and radicals are more likely to be well informed, while middle-of-the-roaders tend to be slightly informed about solar energy. Politically uncommitted consumers know the least. Liberals, more than conservatives, tend to feel that they know enough to decide on solar use. Middle-of-the-roaders and those politically unsure

FIGURE 4.5. Key Variables by Political Orientation

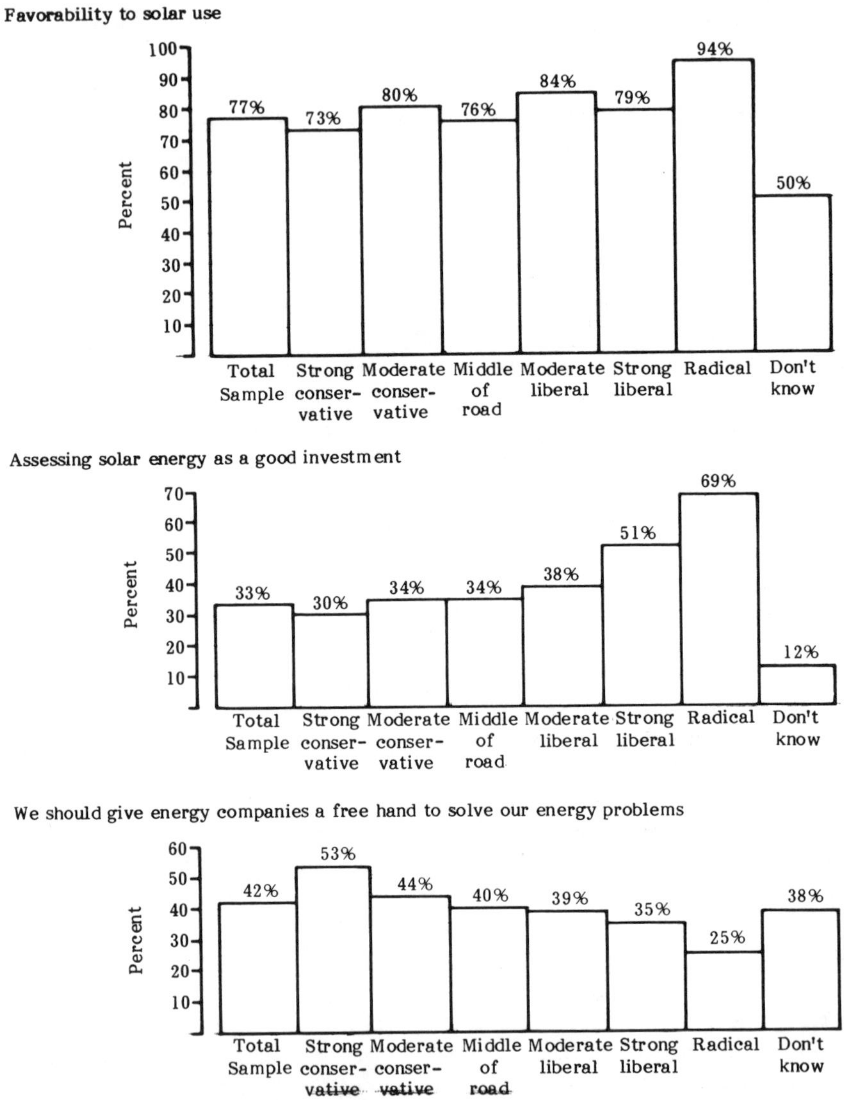

Source: Constructed by the authors.

tend not to know. Radicals want most to learn more about solar energy, followed by moderate liberals. A majority of conservatives (55 percent) also want to know more. Least interested are the noncommitted. Radicals and strong liberals tend to believe solar energy is presently practical more than others do.

With respect to the nation's energy situation, moderate liberals are the group most likely to see the situation as improving over the next five years. Those at the extreme ends of the political dimension, radicals and strong conservatives, are most likely to see it as getting worse. The politically uncommitted are most likely to report no impact of the energy situation on their households, while strong liberals and radicals are most likely to report serious effects. Moderate conservatives and strong liberals tend to define risks of solar adoption as slightly more important than do other groups; radicals tend not to define these risks as important. Those politically unsure and strong conservatives also tend to place less importance on these risks than others do.

Middle-of-the-roaders tend to define solar advantages as important more frequently than others do. Less importance is imputed to these advantages by the politically uncommitted and by radicals. Radicals, strong liberals, and moderate conservatives perceive less difficulty in solar retrofit of their homes than do others. The uncommitted and strong conservatives are more likely than other groups to indicate that they could not use solar energy on their homes.

Radicals disagree most that we should trust the federal government to handle our energy problems, followed by strong liberals and strong conservatives—the extreme ends of the political spectrum. As conservatism increases, agreement increases that we should solve energy problems by giving energy companies a free hand (Figure 4.5). Moderate conservatives to strong liberals most prefer local energy decision making.

In sum, the findings suggest that, while some differences in attitudes toward solar energy occur by political orientation, solar energy use has not become a highly politicized issue. While political liberalism is somewhat more likely to be associated with favorable attitudes toward solar energy, conservatives are also favorable toward residential solar use. The difference between liberals and conservatives is one of emphasis.

Religious Affiliation

Since there was no evidence to support the idea that religious affiliation was important in predicting attitudes toward solar energy, this variable was not included in the study.

Race

Very few data on race existed in the literature, so no prediction could be made on the relationship between race and attitudes toward

FIGURE 4.6. Key Variables by Race

Favor use of solar energy

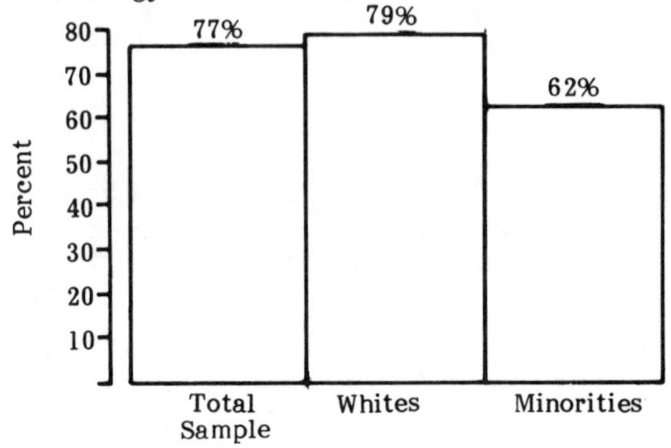

Favor use of solar energy in own home

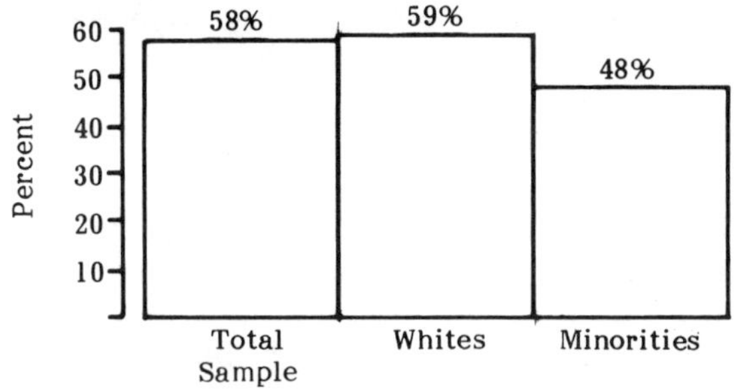

We should trust the federal government to find a solution to the energy crisis

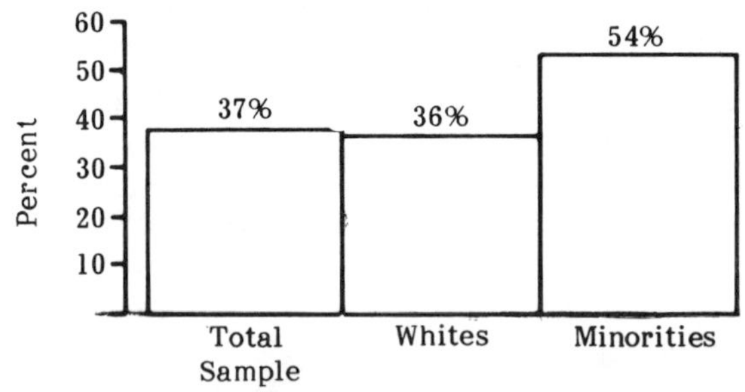

Source: Constructed by the authors.

solar energy. However, the Department of Energy has expressed interest in knowing how minorities feel about solar energy, so this variable was included in the study. Racial categories used were the standard federal statistical categories.

Data from the homeowners survey do not reveal strong differences in attitudes toward solar energy by race,* although minorities tend to be less interested than do whites. As Figure 4.6 shows, whites are more favorable toward solar use, both in the abstract and in the concrete, while minorities tend not to know. Minorities are more likely to assess solar energy as a good investment, but no significant differences occur by race in behavioral intention. Whites report engaging in more conservation activities than do minorities.

Both whites and minorities appear to be about equally informed about solar energy, with minorities expressing less confidence that they know enough to decide on solar adoption; both equally desire more solar information. Whites are more aware of the federal tax credits than are minorities and are more inclined than minorities to view solar energy as presently feasible.

According to whites, the nation's energy situation will improve, while minorities say it will worsen, over the next five years. No difference by race is reported on household impacts from the energy situation. Very little difference is found between whites and minorities in concern about the risk of solar adoption. In the few cases in which statistically significant differences occur, minorities express more concern than do whites. Similarly, little difference occurs by race in perceived relative advantage of solar adoption. Whites tend to assess solar advantages as important slightly more frequently than do minorities.

Whites tend to perceive both greater difficulty and greater ease in solar retrofit projects for their homes, while minorities tend not to know. Minorities are much more likely than whites to trust the federal government to handle our energy problems, as shown in Figure 4.6. Whites are more likely to prefer local control of energy decision making, although over 75 percent of each group express this preference.

In sum, although there is a tendency for whites to express more favorability toward and interest in solar energy, minorities are also favorable. Minority groups tend not to know enough about solar energy to evaluate its potential for their own situations. To speculate, they may feel that with limited income, they cannot realistically entertain the idea of solar retrofit on their homes and have, therefore, not informed themselves about it. However, minorities are open to receiving more information about solar energy. In certain areas of the country, such as in the San Luis Valley of Colorado, minorities have successfully applied solar technologies to save on their utility bills.

---

*All categories other than white were classified as racial or ethnic minorities for this analysis.

Community Size

Very little is known about the relationship of community size and other relevant variables. Urban residents are more likely to be environmentally concerned than are rural residents. Rural dwellers may be more likely than urbanites to favor adequate energy over environmental protection. The idea of geographic or spatial diffusion would suggest that solar adoption would spread more rapidly in neighborhoods where systems have already been installed. Thus, urban and suburban areas might be more prone to solar diffusion than rural areas. However, our field work showed solar diffusion in urban, rural, small town, and suburban settings. Information on community size was included in the national homeowners survey as a standard statistic provided by Gallup.

Community size was divided into three classifications for analysis purposes: rural (35 percent), small town (17 percent), and city (over 50,000 in population) (48 percent). Very few significant differences are found by community size, suggesting that it is not a highly salient variable, with respect to solar attitudes and knowledge. No differences exist in favorability to residential solar use by community size, either abstractly or concretely, or in assessment of solar energy as a good investment. Rural consumers (23 percent) are somewhat more likely to indicate they may invest or plan to invest in solar energy, followed by urbanites (16 percent) and small town residents (12 percent). Those in small towns are more likely than others not to have considered solar adoption. Rural residents have done the most to conserve energy, while small town residents have done the least.

No difference by community size is found in solar knowledgeability, feeling informed enough to make a solar decision, or in receptivity to solar information. Consumers in all sizes of communities assess solar feasibility in about the same ways. Little difference in concern about the risk of solar adoption occurs by community size. Differences occur in concern about solar effects on life-style and aesthetics, possible problems with the utility company and local zoning, and possible reduced resale value of home. In all these cases, rural consumers are less concerned than others. Virtually no difference occurs by community size in the perceived advantages of solar adoption.

Rural consumers are least likely to agree that we should trust the federal government to handle our energy problems, while urbanites are most likely to agree. Small town residents are most likely to feel that we should give energy companies a free hand to solve energy problems. No difference is found by community size in preference for local control of energy decision making and in preferences for financing systems. Urbanites are slightly less willing to pay more per month for solar energy, while rural residents are slightly more willing to do so. Slightly less interest in obtaining a system is expressed by small town residents.

In sum, residents of all types of communities are favorable to solar use and about equally knowledgeable. Small town and urban residents are more concerned about possible problems with their utility company if they adopt solar energy. In general, the findings suggest that small town residents may be slower than either rural or city residents to adopt solar energy in the future.

Opinion Leadership

We expected that favorability to solar energy would be higher among opinion leaders. Opinion leaders are those who tend to influence opinion in their "reference groups," whether these groups are co-workers, church members, voluntary association members, or communities.

The diffusion literature shows that 76 percent of studies found opinion leadership related to early adoption, and in a 1979 Roper study, more opinion leaders than others reported that they were considering solar adoption within the next 2 to 3 years. The relationship of opinion leadership to solar adoption is described in Chapter 7 on the "next tiers" analysis.

Innovativeness

The diffusion literature also supports the notion that earlier adopters will have higher levels of innovativeness than with others. We anticipated that favorability toward solar energy would be positively related to innovativeness. The relationship of innovativeness to solar adoption is also described in Chapter 7.

VALUES AND LIFE-STYLE CHARACTERISTICS

Values and life-style are sometimes seen as interrelated, with life-style as a daily expression of more fundamental values. In this section, values and life-style items included in the study are discussed.*

---

*Data on these items are presented in Appendix D, Demographics, Values, and Life-style Characteristics of the National Home-

Values

Some have suggested that there could be an important relationship between the values a person holds and his willingness to adopt solar energy.* Values are usually defined as enduring beliefs determining general standards of conduct, independent of specific situations. Rokeach (1973) distinguished "terminal" and "instrumental" values. "Terminal" values refer to desired end states of existence, such as freedom, security, happiness, and equality. Any person holds only 18 or so such values, and they tend to persist, changing over slowly over time. "Instrumental" values refer to the valued means to reach the desired ends. Since there are many possible routes to happiness, persons have many more instrumental than terminal values.

For the National Study, it was decided to rely on a values scale developed by Inglehart (1977) that has been used in 20 years of cross-cultural research. Inglehart found that:

> The values of Western publics have been shifting from an overwhelming emphasis on material well-being and physical security toward greater emphasis on the quality of life. The causes are that people tend to be more concerned with immediate needs or threats than with things that seem remote or nonthreatening. Thus, a desire for beauty may be more or less universal, but hungry people are more likely to seek food than aesthetic satisfaction. Today, an unprecedently large portion of Western populations have been raised under conditions of exceptional economic security. Economic and physical security continue to be valued positively, but their relative priority is lower than in the past. [p. 3]

Inglehart separated values into "materialist" and "postmaterialist" categories, based on Maslow's (1970) concept of a hierarchy of needs. Materialist values include the desire for both economic and physical security. Postmaterialist values are more expressive and include free speech, political participation, and aesthetics. Simply put, when economic times are good and such violence as war and

---

owners Sample, and in Appendix F, Demographics, Values, and Lifestyle Characteristics of the National Solar Homeowners Sample.

*For further discussion, see Appendix G, The Relevance of Values to Solar Adoption.

terrorism are not occurring, greater numbers are secure enough to pursue the "higher" things in life.

What Inglehart has found is that postmaterialist values occur more frequently in the more highly industrialized nations, seemingly as harbingers of the postindustrial era. The relevance of this to our study is that consumers likely to become solar adopters soon might tend to hold postmaterialist values. If this is the case, it has important implications for policies to increase the rate of adoption and marketing strategies that accurately address and reach the likely market. Therefore, the Inglehart values scale was included in the study.

There is some evidence that different types of solar adopters may differ in values. Leonard-Barton (1978), for example, hypothesized at least four classes of solar adopters. "Ecologists" would have purchased a system regardless of economic advantage, expressed concern about the environment, and viewed solar energy as consistent with an ideal way of life and a better future society. "Tinkerers" were intrigued by innovative technology and enjoyed working with their hands. "Comfort/convenience" adopters saw solar energy as a means of meeting their energy needs without feeling guilty about energy use. "Economy-minded" people invested in solar energy primarily to save money.

In a study of two solar subdivisions, Hamrin (1978) found differences in values and life-styles between them. One group tended to feel that their possession of solar hardware took care of their contribution to energy conservation; they consequently did not exhibit as much conservation behavior as that of homeowners in control groups in conventional housing. Homeowners in the second solar subdivision tended to view their solar home purchase as part of a life-style emphasizing energy conservation, self-sufficiency, and voluntary simplicity.

In a study of nine solar families, Zimmerman (1980) found evidence that, for at least some solar adopters, solar energy use is symbolic of a life-style focused on such values as deemphasis of material consumption, cultivation of inner resources, development of self-awareness, and appreciation of nature.

The National Study sought to gain insight into possible links between such values and solar adoption and to identify and better understand different types of adopters. The relationship between values and solar adoption is explored further in Chapter 7.

Life-styles

No evidence exists to show that stage in the family life cycle, family size and structure, and energy use characteristics affect attitudes toward solar energy (Farhar et al. 1979b). Solar owners vary widely in their family characteristics (Unseld and Crews 1979). However, it was decided to include items on family structure and size and the household's primary heating fuel, for descriptive purposes. Consumers using electricity for heating, for example, might be more interested in solar adoption than others, owing to the high cost of electricity in some areas.

Some evidence suggests that, proportionally, more mobile home owners might be interested in solar energy than those living in houses or apartments (Farhar et al. 1979a). The proportion of mobile homes in the housing market has increased over time: mobile homes accounted for less than 10 percent of housing prior to 1964 and increased to a peak of about 20 percent in the years 1969-74. The percentage thereafter declined somewhat to 13.3 percent for 1978. The figure is projected to increase in the future to 20.5 percent in 1990 (Predicasts, 78/08/02, Composites).

Because homeownership itself has also been found to be significantly related to solar adoption (Farhar et al. 1979a), it was decided to limit the National Study to homeowners.

People who move frequently might be less inclined to invest in a solar energy system for their homes. However, if they perceived that installation of a system might increase the resale value of their homes and that they could recoup their investment, it is possible that more mobile families would be interested in solar adoption. Therefore, an item on mobility/stability was included in the homeowners survey.

As measured by Leonard-Barton and Rogers (1979), the concept, "voluntary simplicity," refers to a set of behaviors expressing a life-style that strives toward decreasing material consumption. Voluntary simplicity life-styles are characterized by five basic values: (1) material simplicity, (2) human scale, (3) self-determination, (4) ecological awareness, and (5) personal growth (Elgin 1981; Lovins 1977; Mitchell and Elgin 1976).

Consuming little and producing much is the basic idea, along with recoupling production and consumption in the household. In the voluntarily simple life-style, activities include walking and bicycling instead of driving, recycling materials through source separation, engaging in residential energy conservation, and other activities possibly related to solar adoption. Voluntary simplicity is measured by asking for reports on behavior, not for agreement or disagreement with value statements. Reported behavior can then be correlated with value measures. We anticipated that solar adoption would vary positively with voluntary simplicity as a life-style.

The relationship of these life-style characteristics to solar adoption is examined in detail in Chapter 7.

SUMMARY

Consumers vary in attitudes toward solar energy by many demographic variables. All but one of the hypotheses on the relationships of demographics to solar attitudes are supported by the national homeowners data. Key differences are found by gender, age, education, occupation, and income. Political orientation, race, and community size are not found to be highly salient in attitudes toward solar energy. In general, men, adults under 45, the more highly educated, professionals and business owners, and the more affluent are more favorable to solar use, more knowledgeable, and more likely to adopt solar energy than are other groups.

The relationships of demographic variables, opinion leadership, innovativeness, values, and life-style characteristics to the likelihood of adoption are discussed in Chapter 7, as part of the "next tiers" analysis.

5

HOMEOWNER ATTITUDES TOWARD UTILITY INVOLVEMENT IN PROMOTION OF RESIDENTIAL SOLAR USE

An examination of public attitudes toward utility company promotion of residential solar use is one of the key components of this study. Questions were designed to elicit responses to such questions as, to what extent do homeowners view the utility as an appropriate entity for promoting solar use? Would a utility-backed warranty make a difference? Is the utility considered a reliable source of solar information? What role should the utility pay in home energy audits and how much is a homeowner willing to pay for this service? To what extent would consumers prefer other entities to assume these roles?

Dissatisfaction over rapidly rising utility bills and fear of increasing power shortages, among other things, have caused some consumers to view solar energy as a vehicle to personal energy independence from the utility giants. Thus, it might be expected that the public would not embrace utility marketing of solar energy and would be skeptical of utility-supplied solar information. Results from the National Study indicate this is a misperception. Consumers see key roles for utilities in promoting solar use.

Homeowners may feel that they don't have much of a choice, however, because of widespread caution toward small solar firms and a reluctance to purchase systems lacking a solid warranty. This suggests that utility company backing of solar energy systems could do much to reduce barriers to solar adoption by providing strong organizational support. There are, of course, other possibilities for such support: Sears-, Wards-, or J. C. Penney-marketed systems would probably find equal acceptance if installation were provided and guaranteed. Yet, in times of diversification, utilities may find the solar option beneficial to customer and utility alike. Utility companies could benefit from positive effects on peak-load problems, improved relations with customers and public utility commissions (PUCs), and increased cash flow. Customers could receive lower conventional utility bills, financing assistance, and easy recourse for resolving product installation and service problems.

The potential exists for a number of organizational players to enter the solar field (banks, "energy stores," architectural firms, auditing firms, and so on). Some members of the solar industry have expressed antagonism toward utility involvement in the solar field, claiming utility lack of solar expertise and unfair competition. However, utility companies have frequently subcontracted with small private firms to provide a variety of goods and services, creating a model that could be applied to a solar program. Because the possibilities for mutually beneficial arrangements among organizations dealing with solar energy are extensive, no reason currently exists to rule any one kind of organization out of the solar scene. A great deal of the outcome will depend on the creativity used to build successful cooperative ventures.

In reporting homeowner attitudes toward utility involvement, we are not recommending that utility companies "take over" solar energy. We are describing what roles are viewed as acceptable for utility companies by proportions of homeowners. In the "next tiers" analysis reported in Chapter 7, the views toward utility companies of those most likely to adopt solar energy are described. Inferences about the social acceptability of utility involvement in solar energy are drawn.

PREFERENCES FOR UTILITY ROLES IN PROMOTING SOLAR USE

Homeowners indicated which organizations—utilities, private industry, banks, and federal, state, and local governments—were viewed as suitable for carrying out the following activities:

- Conducting home energy audits
- Providing solar education and information for consumers
- Inspecting solar systems
- Compiling lists of solar contractors
- Covering warranties
- Providng financial incentives
- Mandating solar use

Results are shown in Table 3.23. Homeowners see utilities as having key roles to play, related to residential solar use. Far and away the most preferred role for utilities is conducting home energy audits, mentioned by 41 percent of homeowners. The next preferred organization to conduct audits is local government, by 27 percent.

TABLE 5.1. Preferences for Utility Roles in Promoting Solar Use by Demographics

	Proportion choosing utility company to . . .[a]					
	Provide Financial Incentives	Educate/ Inform Consumers	Conduct Home Energy Audits	Cover Warranties	List Solar Contractors	Inspect Systems
TOTAL SAMPLE	22	38	41	8	24	32
INCOME						
Low	15	25	26	6	17	26
Medium	22	40	43	7	26	31
High	28	46	52	11	28	38
EDUCATION						
H.S. grad or less	17	30	32	—	20	28
Trade or technical	26	42	38	—	22	32
Some college or more	28	47	53	—	29	37
AGE						
24 or less	22	51	47	—	32	46
25-44	26	46	51	—	30	38
45-64	20	34	36	—	21	27
65 or more	17	28	30	—	19	27
OCCUPATION						
White collar	—	44	50	10	—	36
Blue collar	—	37	38	11	—	32
Housewives	—	18	18	7	—	24
Other	—	29	29	14	—	43
GENDER						
Male	—	—	—	—	—	29
Female	—	—	—	—	—	36
REGION						
Northeast	24	43	41	10	—	—
South	18	31	40	8	—	—
Midwest	22	35	38	6	—	—
West	29	50	49	7	—	—
MAIN HEATING SOURCE						
Natural gas	—	—	—	—	—	35
Fuel oil	—	—	—	—	—	30
Electricity	—	—	—	—	—	28
Propane	—	—	—	—	—	40
Wood	—	—	—	—	—	21
Other	—	—	—	—	—	18

[a]Only statistically significant differences are reported ($p \leq .025$). Note: education, occupation, income, and age categories were collapsed to construct this table.

The second most preferred role for utilities is educating and informing consumers, mentioned by 38 percent. In comparison, while about half mention the federal government's responsibility to inform consumers, 39 and 38 percent, respectively, mention state and local governments, the same proportion as those mentioning utilities.

The third most preferred role for utilities is inspecting systems, mentioned by 32 percent. Only local government was mentioned more frequently, by 35 percent.

About one in four choose utilities for listing solar contractors; local and state governments are chosen somewhat more frequently (by up to 37 percent) as appropriately performing this role. Also, 22 percent mention utilities to provide financial incentives, such as grants and low-cost loans, although all levels of government and banks are mentioned more frequently for this role. Finally, small proportions mention utilities for covering solar warranties or mandating solar use.

Preference for utility involvement in these roles was analyzed by the following demographic characteristics: age, gender, income, region, education, occupation, and primary heating source. Relationships reported are statistically significant at $p \leq .025$, that is, the likelihood that the results are not due to chance is 97.5 percent. Relationships not discussed were not found to be statistically significant. Results of the analysis are summarized in Table 5.1.

Home Energy Audits

With respect to utility companies conducting home energy audits, differences in preference for utility involvement are found by age, income, region, education, and occupation.

Preference for utility involvement in audits decreases with age. Of those under 24, 47 percent select utilities to be involved, while 21 percent of those 75 and older do so. Preference for utility involvement increases with income; over half (52 percent) of those at the upper-income level select utilities for this role, compared with 26 percent of low-income respondents.

Westerners (49 percent) are more likely to prefer utility involvement in audits, with the rest of the country about equal (around 40 percent). Preference for utility involvement in audits increases with education level, with 58 percent of college graduates and 56 percent of those with graduate education selecting utilities, compared with 22 percent of those with less than high school and 39 percent of high school graduates.

Among occupational groups, housewives are least likely to prefer it (65 percent). A majority of managers and executives (54 percent) and of clerical workers (54 percent) prefer utility involvemen in conducting audits.

Educating and Informing Consumers

With respect to utility companies educating and informing consumers, differences in preference for utility involvement were found by age, income, region, education, and occupation.

Preference for utility involvement in educating and informing consumers decreases with age. Over half (51 percent) of those under 24 years of age select utilities for this role compared with 22 percent of those 75 or older. Of those 35 to 44, 44 percent select utilities, as do a third of those 45 to 54.

Preference for utility involvement in informing consumers increases with income. Of those with low incomes, 25 percent select utilities; of those with high incomes, 46 percent do so.

Westerners (50 percent) are most likely to prefer utility involvement in informing consumers. Southerners (31 percent) and Midwesterners (35) are less likely to select utilities for this role.

The higher the educational level, the more likely is the preference for utility involvement. While 23 percent of those with less than a high school education select utilities to inform consumers, 54 percent of those with graduate education prefer utility involvement in this role.

Among occupational groups, housewives (18 percent) are least likely to select utilities, and manufacturers' representatives (62 percent) are most likely. Managers and executives are split, fifty-fifty, on utility involvement in this role.

Inspecting Systems

With respect to utility involvement in inspecting systems to see how they are working, differences were found by all demographic variables except region. The oldest group (75 or older) is least likely to prefer utility involvement (19 percent), while those 24 and under are most likely (46 percent). However, the relationship is not linear; for example, 24 percent of those 55 to 64 select utilities for this role, as do 30 percent of those 65 to 74.

Women (36 percent) are more likely than men (29 percent) to prefer utility involvement in system inspection. Preference for utility involvement increases with income. Over a third of upper-income respondents (38 percent) select utilities for this role, compared with 26 percent of low-income respondents. Preference for utility involvement in system inspection increases with educational level. While one out of four with less than a high school education select utilities, 39 percent of those with graduate education view system inspection as an appropriate utility role.

Among occupational groups, manufacturers' representatives (47 percent) are most likely to prefer utility involvement, as are managers and executives (46 percent) and farm laborers (46 percent). Housewives (24 percent) and retired people (25 percent) are least likely to prefer utility involvement in system inspection.

Primary heating source is also found to be significantly related to preference for utility involvement in system inspection. Homeowners using propane are most likely to select utilities (40 percent); those relying on wood are least likely (21 percent).

Listing Solar Contractors

With respect to listing solar contractors, differences in preference for utility involvement are found by age, income, and education. As age increases, preference for utility involvement in this role decreases. About a third of those 24 and under select utilities for this role, compared with 13 percent of those 75 or older.

Low-income respondents are less likely to prefer utility involvement (17 percent) than are medium- and high-income groups. Those with least education are least likely to prefer utility involvement (17 percent); those with some college and above are most likely to prefer utility involvement in listing solar contractors.

Other Roles

Other roles listed in the item were covering guarantees or warranties for solar companies that go out of business and providing financial incentives such as grants and low-cost loans. While these are not roles widely held appropriate for utility companies, a few significant differences in preference for utility involvement in them were found by demographic variables.

For covering warranties, high income respondents (11 percent) are more likely than other groups to prefer utility involvement. The Northeast (10 percent) is more likely than other regions to show this preference. Among occupational groups, manufacturers' representatives (21 percent) are most likely to prefer utility involvement, as are sales workers (17 percent), students (14 percent), and managers/executives (14 percent). Farm laborers are least likely to prefer utility involvement in covering warranties.

For providing financial incentives, those aged 25 to 34 are the most likely of all age groups to prefer utility involvement (30 percent). Preference for utility involvement in these programs increases with income; 15 percent of low-income and 28 percent of high-income respondents select utilities. Those with some college and above are more likely than those with less education to prefer this involvement. Southerners are least likely to prefer this involvement; Westerners are most likely to prefer it.

UTILITIES AS A SOURCE OF INFORMATION ABOUT SOLAR ENERGY

Two items were included in this analysis. The first asked about organizational information sources generally, and the second focused on interest in home energy audits, a preferred utility role and one required through the Residential Conservation Service Program.

Preference for Utilities

Respondents were asked to indicate, for each of a list of 13 kinds of organizations, how likely they would be to contact that source for information about solar energy.* The most popular organizations for solar information were solar companies (69 percent likely to contact) and state energy offices (62 percent). Following closely were the public library (56 percent) and federal information organizations (56 percent), such as the Government Printing Office. Utility companies and federal energy agencies were selected by half the sample; thus, utility companies are a likely source of solar information for many homeowners.

Preference for the utility company as an information source was analyzed by demographic variables, with significant differences found by age, income, education, and occupation. Findings are summarized in Figure 5.1. Least likely to contact utilities for solar information are the oldest age group; likelihood of contacting the utility company decreases with age. Over two-thirds of those 24 and under prefer utilities as a solar information source, while one-third of those 75 or older prefer utilities in this role. Low-income groups are least likely to contact utilities (44 percent) for solar information. Among educational groupings, those with some college and above are

---

*Results are reported in Table 3.14.

more likely to contact utilities for information than are those with less education. Among occupational groups, sales workers (75 percent), manufacturers' representatives (65 percent), clerical workers (64 percent), and managers/executives (61 percent) are most likely to contact a utility company for solar information. Housewives (39 percent) are least likely to do so.

Likelihood of contacting the utility company for information on solar energy was cross-tabulated by survey items on knowledge and perceived solar costs, with some interesting results. Consumers significantly more likely to contact utilities for information on solar energy characterize themselves as slightly or somewhat informed (almost 40 percent of each group). Those very well informed or not at all informed are less likely to contact the utility company. However, 58 percent of the very well informed and 41 percent of the uninformed select utilities as a likely solar information source.

Of consumers likely to contact the utility company, 62 percent do not know enough about solar energy to decide on its use, while 28 percent say they do. Thus, most of those likely to be in touch with utilities are not going to know very much about solar energy.

Similarly, consumers likely to contact a utility company for solar information are unlikely to know or will underestimate the cost of solar domestic hot water (SDHW) systems. However, of those estimating SDHW costs correctly or overestimating costs, more are likely to contact the utility company for solar information. A similar pattern holds for perception of space heating costs, except that those about right in their estimates are the group most likely to contact utility companies.

Home Energy Audits

In another item on information sources, 46 percent mentioned home energy audits. As reported in Chapter 3, one of the survey items directly explored interest in RCS-type audits, which were briefly described. Over half of the sample indicates interest, with 20 percent saying they are "very interested" and 29 percent saying that they are "interested"; 2 percent say they have already received an audit. Almost half are unsure (11 percent) or not interested (38 percent).

Interested consumers and those unsure were asked what they would be willing to pay for such an audit. A third would be unwilling to pay anything, while 62 percent indicate willingness to pay from under $25 to $200 or more for an audit. Thus, approximately 37 percent of homeowners are likely to be interested in an audit and are willing to pay a $15 audit fee. (If audits were free, this proportion would increase to half.) This figure is relatively close to the RCS audit participation rate estimated by DOE for the RCS Program.

FIGURE 5.1. Utility Company as Likely Solar Information Source by Demographics

Likelihood of contacting the utility company to obtain information about solar energy

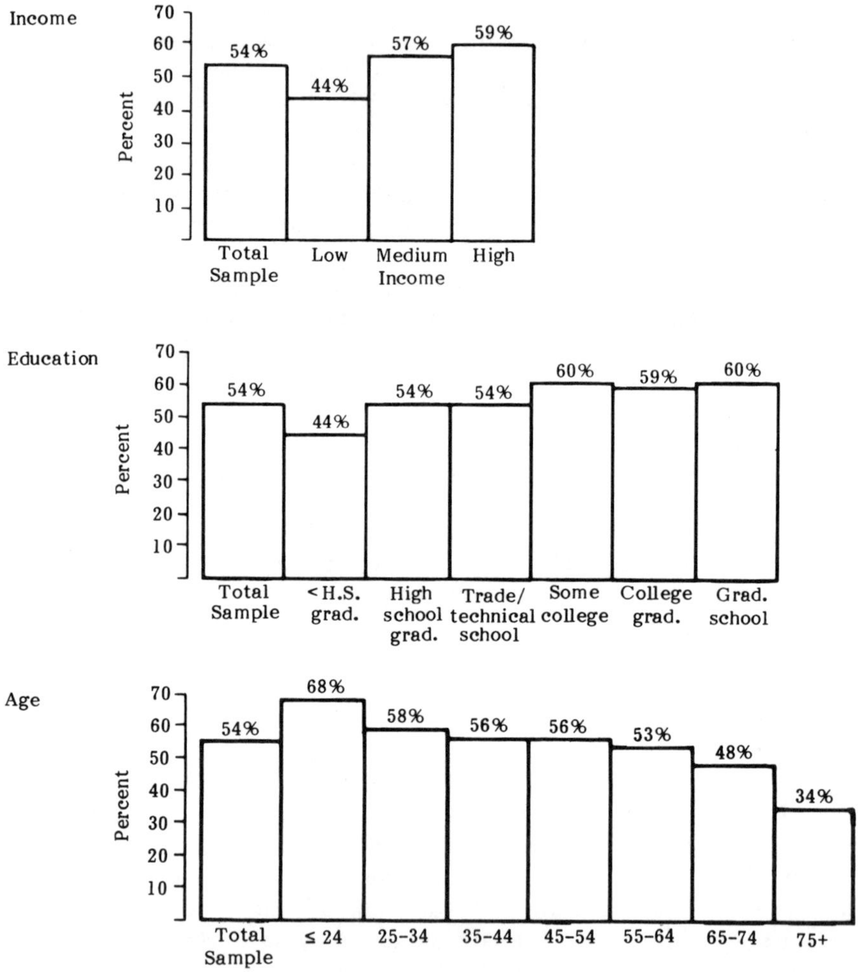

(continued)

Interest in obtaining an audit was analyzed by demographic and other key study variables. Significant differences in audit interest were found for all demographics. Results of this analysis are summarized in Figure 5.2.

FIGURE 5.1 (continued)

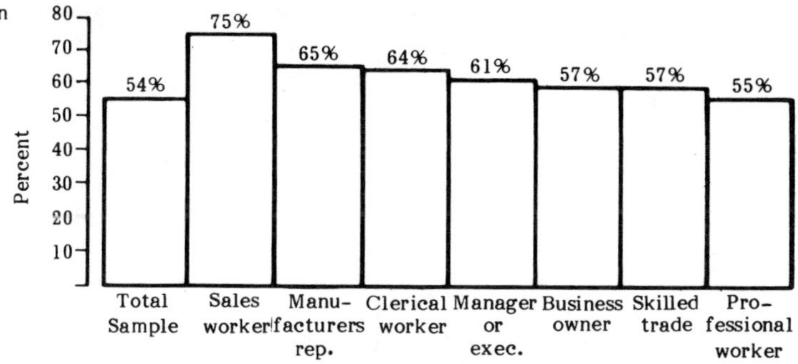

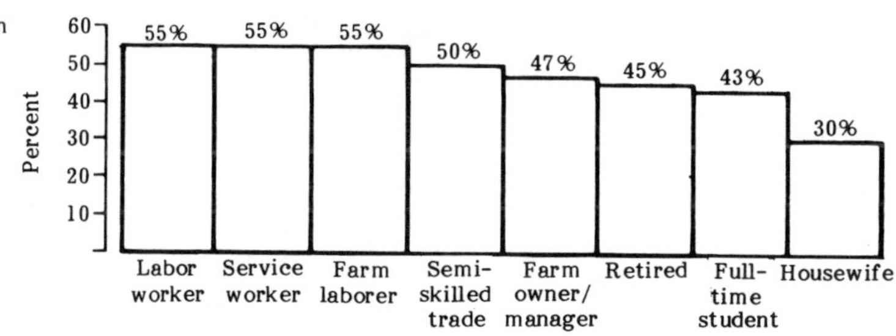

Source: Constructed by the authors.

Consumers 75 or older are least interested of all age groups in obtaining an audit (10 percent); those under 45 have the highest level of interest, for example, 74 percent of those aged 25 to 34 indicate interest in an audit. A majority of men (55 percent) are interested, while a minority of women (44 percent) are interested in an audit. Interest increases with income: of high-income respondents, 68 percent are interested compared with 28 percent of low-income respondents.

Interest in an RCS-type audit varies by region of the country. Interest is highest in the West (59 percent) and in New England (54 percent). Interest is lowest in the South (46 percent) and Midwest (48 percent).

Interest in the audit increases with education and jumps markedly among those with some college and above. Of those with less than a high school education, 28 percent express interest, while 74 percent of those with graduate education are interested.

Among occupational groups, manufacturers' representatives (81 percent), students (71 percent), and professionals (68 percent)

FIGURE 5.2. Interest in Home Energy Audits by Demographics

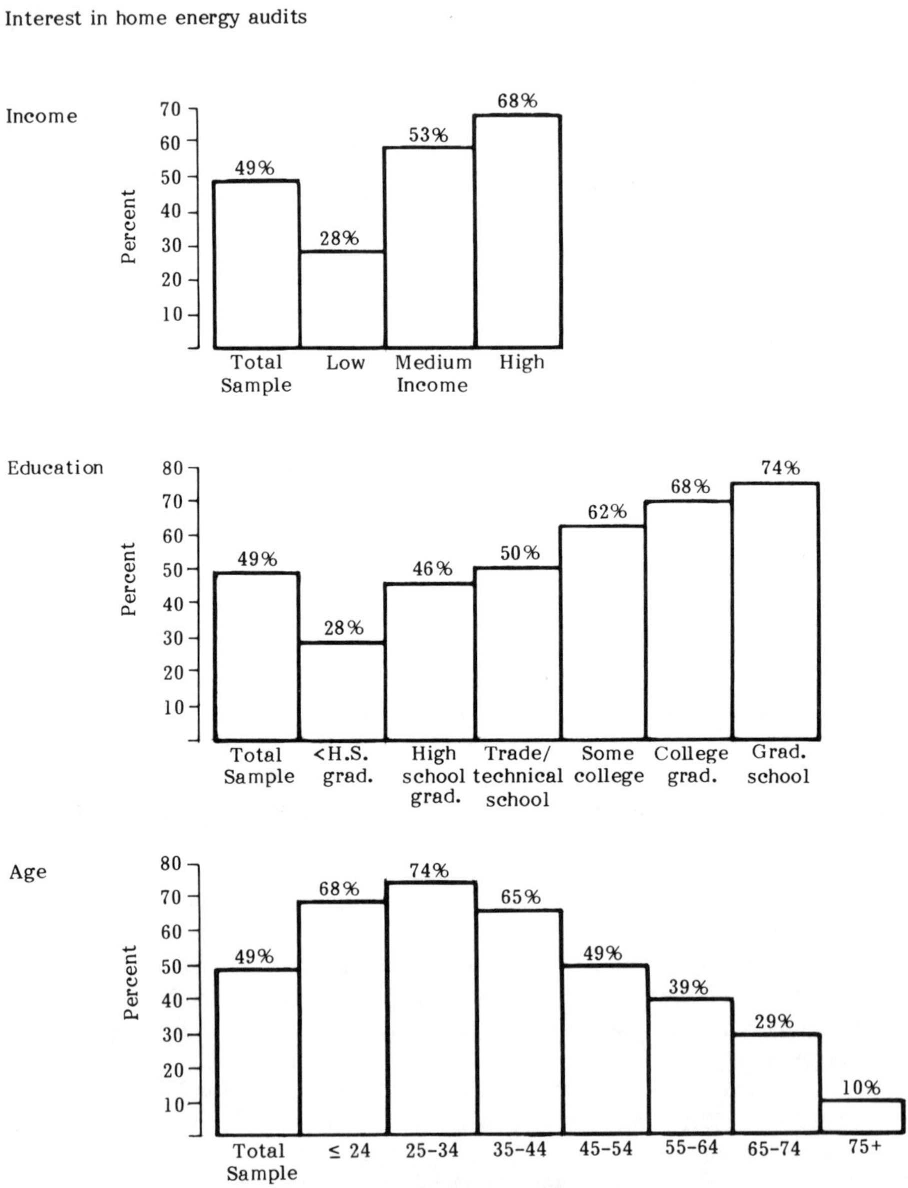

(continued)

FIGURE 5.2 (continued)

Occupation

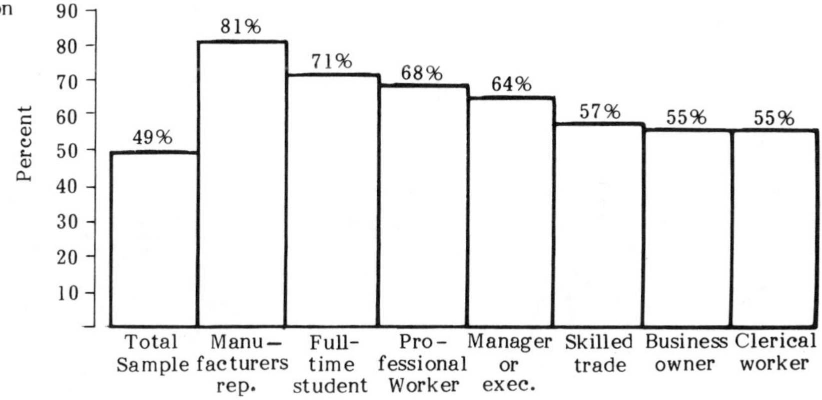

Occupation (con't)

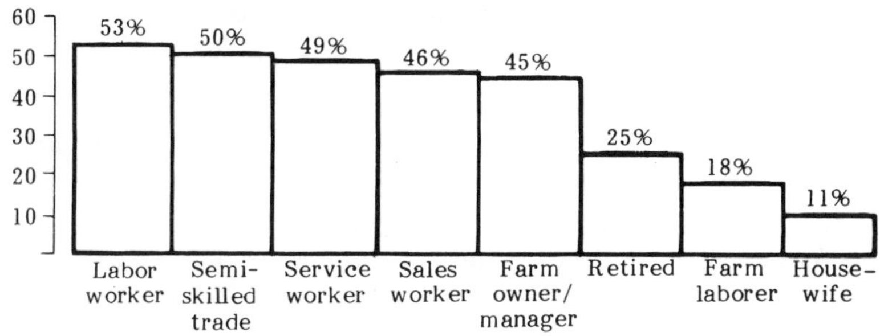

Gender

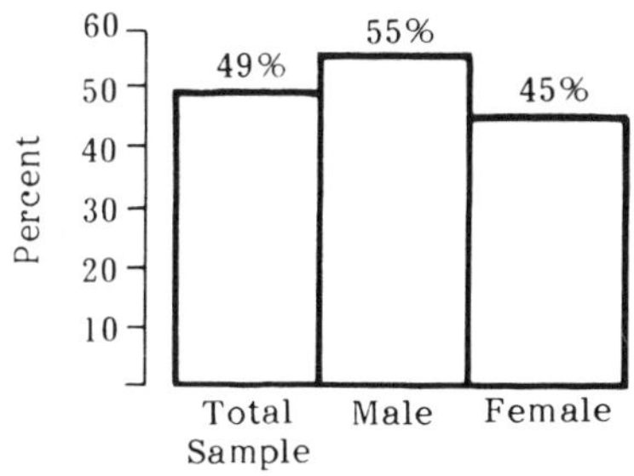

(continued)

FIGURE 5.2 (continued)

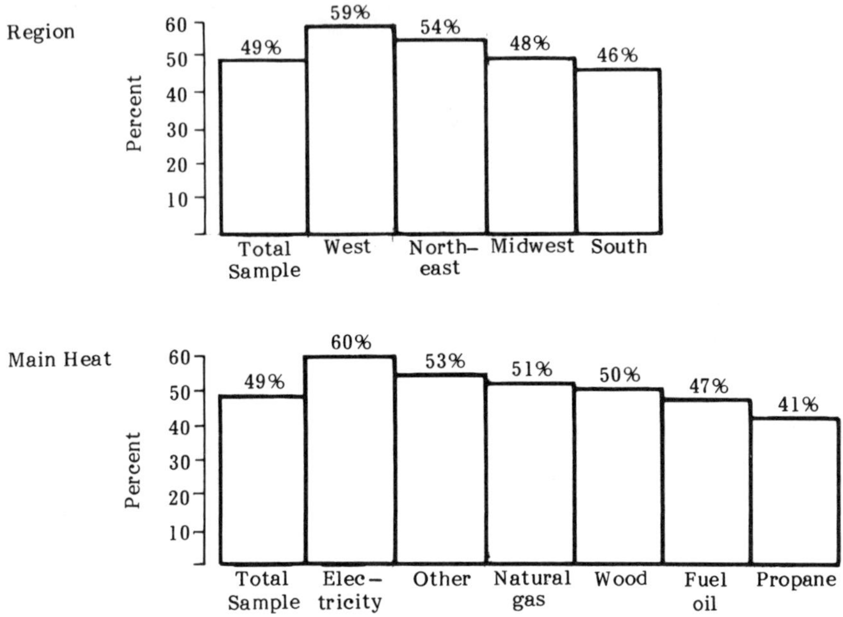

Source: Constructed by the authors.

are most likely to be interested in energy audits. Housewives (11 percent) are least likely to be interested, followed by farm laborers (18 percent). Retired persons also have relatively less interest (25 percent).

Those using electricity as their primary source of heating are most interested in an audit (60 percent); propane users are least interested (41 percent). About half of natural gas users (51 percent) are interested in an energy audit.

Interest in having a home energy audit is also significantly related to other study variables. A majority (53 percent) of those interested think utility companies should conduct home energy audits to promote the use of solar energy in homes. Additionally, of those preferring utility involvement in auditing, 65 percent are interested in having an audit done.

Of those who may invest in a solar energy system in the next 2 to 3 years, 85 percent are interested in a home energy audit, while 80 percent of those who plan to invest are also interested. This is far higher than other groups (those unsure and those who have not thought about it).

FIGURE 5.3. Interest in Audit by Favorability

Those interested in an audit favor the use of solar energy in homes.

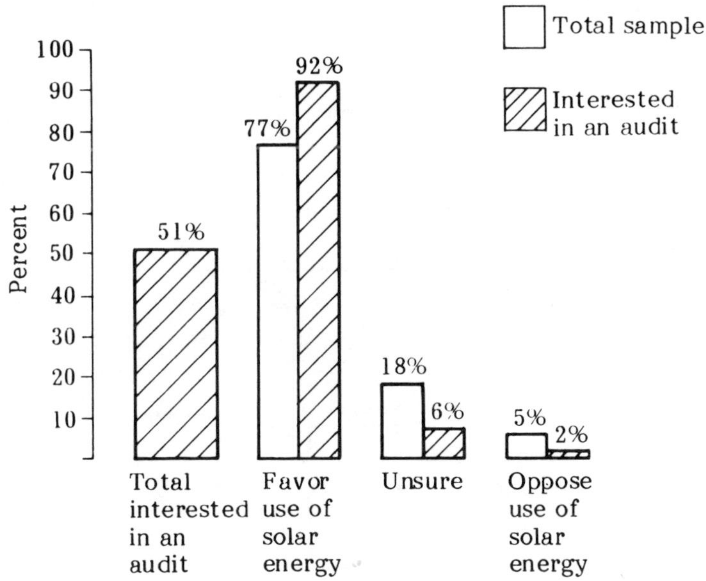

Source: Constructed by the authors.

Of consumers favoring the use of solar energy in homes, 60 percent are interested in having a home energy audit. Similarly, of those favoring the use of solar energy for their own homes, 67 percent are interested in an audit. Of consumers interested in an audit, 92 percent favor the use of solar energy in homes, and 78 percent favor the use of solar energy in their own homes (Figure 5.3).

CONCERN ABOUT THE UTILITY COMPANY

Homeowners were presented with a list of 18 things they "might see as problems with solar energy" and were asked to indicate how important a concern each item on the list would be if they were thinking about using solar energy in their own homes.* One of the items was "possible problems with the utility company." As reported in

---

*Results are presented in Table 3.17.

FIGURE 5.4. Concern about Problems with the Utility Company by Demographics

Concern about risk: possible problems with the utility company

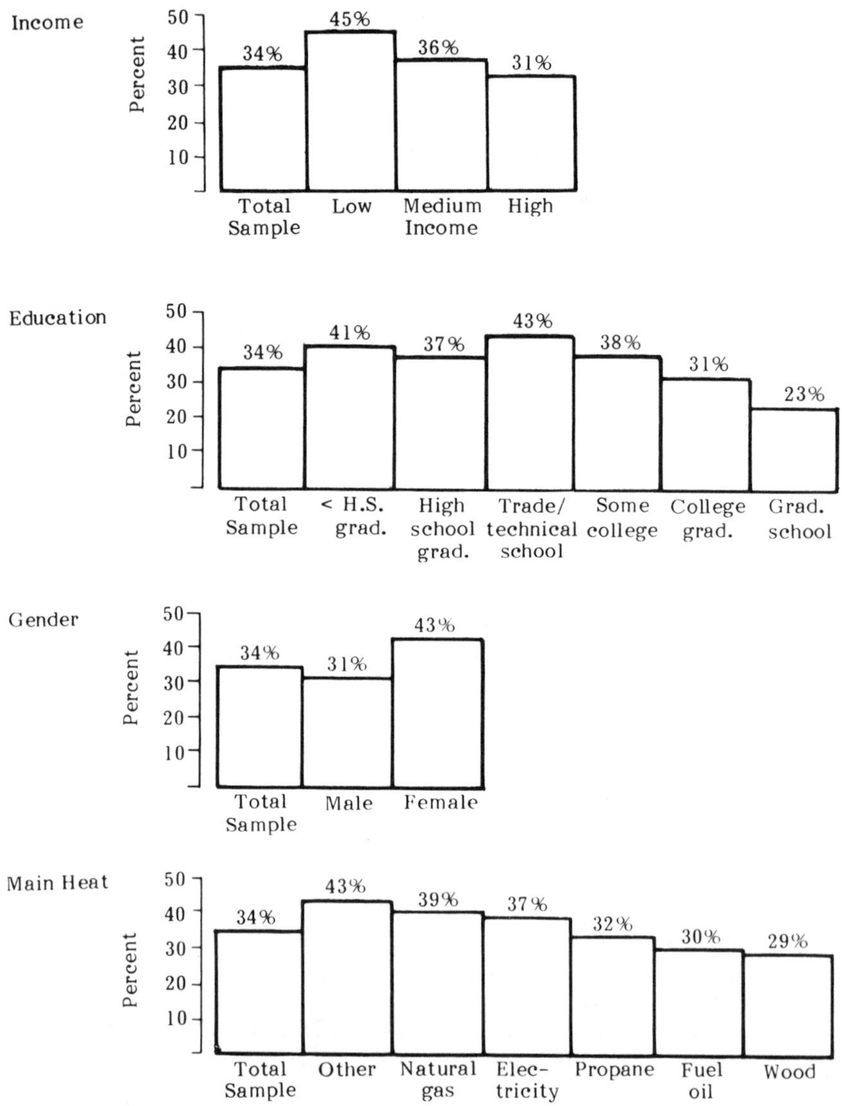

<u>Source</u>: Constructed by the authors.

Chapter 3, about a third indicate that this is an important (20 percent) or very important (14 percent) concern; 15 percent say it is somewhat important, and 49 percent say it is slightly important or not at all important. Compared with other concerns associated with solar adoption, this concern ranks low on the list.

Responses were analyzed by demographic variables, and significant differences were found in concern about utility company reaction by gender, income, education, and primary heating fuel. Results are summarized in Figure 5.4. Women (43 percent) are more concerned about possible problems with the utility company than are men (31 percent). Concern tends to decrease as income increases; 45 percent of low-income and 31 percent of high-income consumers define utility problems as an important concern.

Consumers without a college degree are more concerned about possible problems with the utility company than are college graduates and those with graduate education. For example, 43 percent of those with a trade or technical school education fear such problems, while 23 percent of those with graduate education express these fears.

More concern is expressed by consumers whose primary heating fuel is natural gas (40 percent) and electricity (37 percent) than by those using other fuel sources.

The item was cross-tabulated with a second item, asking respondents to rate in importance the dependability and reputability of solar firms. Among those defining problems with the utility company as an important barrier to solar adoption, more are also likely to define reputability of solar firms as a problem (91 percent). Among those defining reputability of solar firms as a problem, however, fewer are likely to define problems with the utility company as important (49 percent). This finding may result more from the widespread concern about solar firm reputability (mentioned by 83 percent of the sample), compared with the smaller degree of concern about utility problems (mentioned by 34 percent), than from a strong association between the two concerns. Those most concerned in general might be more likely to indicate both concerns as important.

UTILITY-RELATED MOTIVATION TO ADOPT SOLAR ENERGY

A survey item asked respondents to consider a list of 15 "possible advantages of using solar energy" that could enter into their decision about using solar energy in their homes.* Responses ranged

*Data are reported in Table 3.19.

FIGURE 5.5. Reducing Utility Bills as a Motivation to Adopt Solar Energy by Demographics

Relative advantage: reducing utility bills now

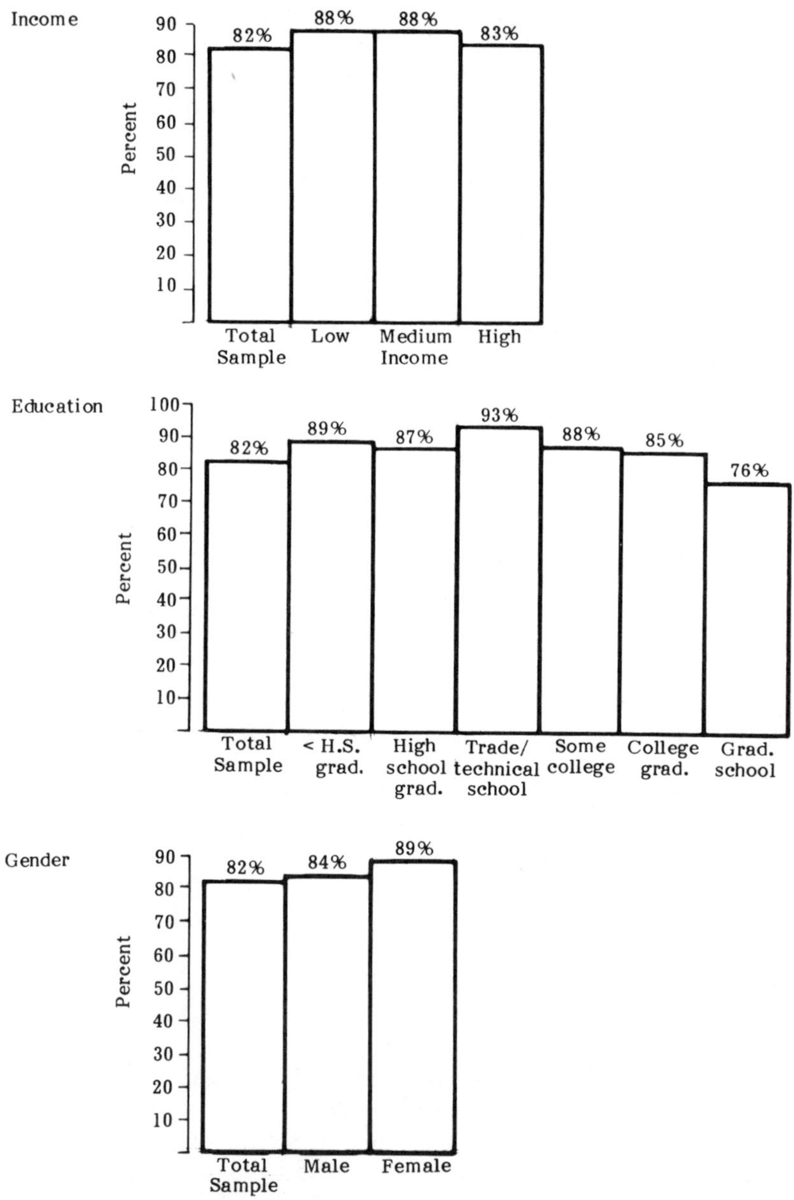

Source: Constructed by the authors.

from "not at all important" to "very important." Three items on this list are of particular interest to utility companies:

- Reducing utility bills now
- Increasing independence from utility companies
- Having a more reliable supply of energy

"Reducing utility bills now" is chosen as an important advantage of solar energy by 82 percent; only "saving money over the long term" is mentioned more frequently. "Having a more reliable supply of energy" is perceived as an important solar advantage by 79 percent, and 65 percent define "increasing independence from the utility company" as important. These items were analyzed by demographic variables; no statistically significant differences by demographics were found for defining increasing independence from utility companies from utility companies as an important solar advantage. For the other two items, significant differences were found by demographics. Results of these analyses are summarized in Figures 5.5 and 5.6.

Reducing utility bills now is more important to women (90 percent) than to men (84 percent). Low- and middle-income consumers are equally likely to see reducing utility bills as an important reason for using solar energy (88 percent); it is slightly less important for high-income respondents (83 percent).

Those with trade/technical school backgrounds are the most likely of all educational groupings to consider solar energy's potential for reducing utility bills now as an important reason for using solar energy.

Among consumers definitely planning to invest in solar energy soon, those who may invest, and those who have not considered it, reducing utility bills is an equally important advantage. Among those who have decided not to invest, it is perceived as a slightly less important advantage of solar energy.

The importance of having a more reliable source of energy in deciding to use solar energy generally decreases with age. Of those under 24, 87 percent define reliability of energy supply as important, as do 71 percent of those 75 or older.

Among occupational groups, laborers (93 percent), business owners (90 percent), and skilled workers (84 percent) are most likely to find reliability important in deciding to use solar energy. Farm laborers (67 percent) are least likely to have this concern.

Among consumers who definitely plan to install solar energy, 94 percent consider a reliable energy supply as an important reason. Of those who may invest, 88 percent consider it important. Of those who have not considered investing in solar energy or have decided

FIGURE 5.6. More Reliable Energy Supply as a Motivation to Adopt Solar Energy by Demographics

Relative advantage: having a more reliable source of energy

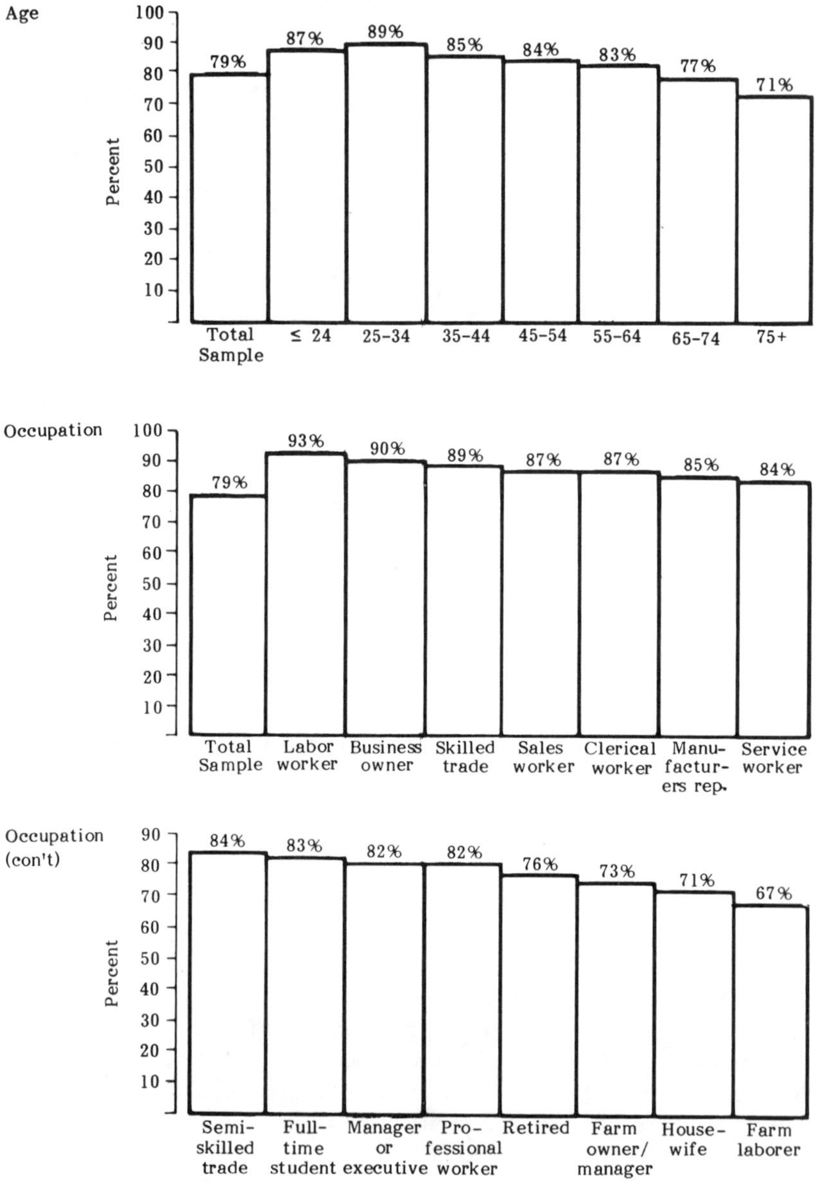

Source: Constructed by the authors.

against it, 81 percent define a reliable energy supply as an important reason for solar adoption.

UTILITY INVOLVEMENT IN FINANCING SYSTEMS

Almost two-thirds of the sample indicate that, if they were thinking of investing in a solar energy system, they would pay for the system by financing it. A list of eight alternative financing arrangements was presented, three of which were utility financing arrangements.* These were:

- Have your utility company install and maintain the system and lease it to you for a monthly rate (selected by 10 percent of the sample).
- Increase monthly payments to your utility company until the system is paid for (chosen by 6 percent).
- Have your utility pay for the system; you would repay the cost when you sell your house (selected by 5 percent).

The total proportion choosing a utility financing mechanism was 20 percent.

Respondents indicating preference for the utility financing mechanisms were divided from those not expressing this preference, and the two groups were analyzed by demographics. Significant differences were found by income and education. Results of this analysis are summarized in Figure 5.7. Preference for a utility financing scheme increases with income. Of low-income consumers, 16 percent choose a utility mechanism, while 24 percent of high income respondents do so. Those with at least some college are more likely than the lesser educated to prefer the financial involvement of utilities.

SUMMARY: PATTERNS OF DIFFERENCES IN ATTITUDES TOWARD UTILITIES BY DEMOGRAPHICS

Clear patterns showing different attitudes on the part of different groups toward utility involvement in promotion of solar use (and probably toward utilities generally) emerge from the analysis. The most significant differences were found by income, education, and age.

---

*See Table 3.26.

FIGURE 5.7. Preference for a Utility Financing Mechanism by Demographics

Preference for a utility financing mechanism for solar purchase

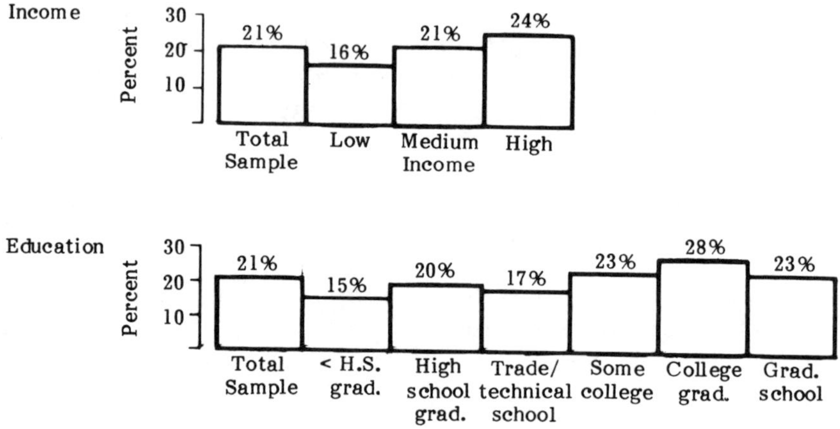

Source: Constructed by the authors.

Occupation, gender, and region had somewhat fewer significant associations, and primary heating fuel had the least.

Income

Favorability toward utility involvement in solar energy consistently increases with income:

- Interest in home energy audits increases with income.
- The lowest income group is least likely to contact a utility company for information about solar energy.
- Preference for utility involvement in financing increases with income.
- Concern about possible problems with the utility company, if solar energy is adopted, decreases with income.
- Those in lower- and middle-income brackets are most likely to regard reducing utility bills now as an important advantage of solar energy.
- Preference for utility involvement in providing financial incentives, educating and informing consumers, conducting home energy audits,

listing solar contractors, and inspecting solar energy systems increases with income.
- Preference for utility involvement in covering solar warranties if companies go out of business is highest in the high-income group.

Education

Favorability toward utility involvement in promoting the use of solar energy consistently increases with educational level:

- Interest in home energy audits increases with education, especially among those with at least some college.
- The lowest educational group is least likely to contact a utility company for information about solar energy.
- Those with some college and above are more likely to prefer utility financing mechanisms for solar adoption.
- Concern about possible problems with the utility company is highest among those without a college degree.
- Reducing utility bills now is most important to those with trade or technical school backgrounds.
- Preference for utility involvement in informing consumers, conducting home energy audits, and inspecting solar energy systems increases with educational level.
- Those with some college and above are most likely to prefer utility involvement in financial incentives for solar adoption.
- The least educated are least likely to prefer utility involvement in listing solar contractors.

Age

Younger consumers indicate consistently a more favorable attitude toward utility involvement in solar energy than do older respondents:

- Those under 45 years of age have the highest interest in home energy audits; those over 75 are not interested.
- In deciding whether to use solar energy, the importance of having a more reliable source of energy decreases with age.
- Oldest consumers are least likely to contact utilities for information about solar energy; likeliness to contact utility companies decreases with age.
- Preference for utility involvement in educating and informing consumers, listing solar contractors, and conducting home energy audits decreases with age.

- The oldest group is least likely to prefer utility involvement in inspecting systems; those under 35 are most likely to prefer this role for utilities.
- Those 25 to 34 are most likely to prefer utility involvement in providing financial incentives for solar adoption.

Occupation

Among all occupational groups, housewives are consistently the least favorable toward utility involvement in solar promotion; manufacturers' representatives are the most consistently favorable occupational group:

- Housewives are least likely to be interested in a home energy audit; manufacturers' representatives are most likely.
- Housewives are least likely to contact utility companies for solar information; sales workers are most likely.
- Having a more reliable energy supply as a reason for using solar energy is least important to farm laborers and most important to other laborers.
- Preference for utility involvement in educating and informing consumers, conducting audits, inspecting systems, and backing solar warranties is lowest among housewives and highest among manufacturers' representatives.

Gender

No clear pattern of differing favorability emerges by gender:

- Nuclear energy is least preferred by women; men tend to select oil and natural gas as least preferred more frequently than women do.
- Men are more interested in the home energy audit than women are.
- Women are more concerned about possible problems with the utility company than men are.
- Reducing utility bills now is a more important advantage of solar energy to women than to men.
- Women are more likely than men are to prefer utility involvement in inspecting systems.

Region

The findings suggest slightly more favorable attitudes toward utility involvement in the West:

- Coal is least preferred in the West; nuclear energy is least preferred in the Northeast. There is no difference by region on preference for oil and natural gas.
- Interest in audits is highest in the West, then New England, and lowest in the South and Midwest.
- The West is most likely to prefer utility involvement in educating and informing consumers; the South and Midwest are least likely.
- The South is least likely to prefer utility involvement in financial incentives for solar use; the West is most likely.
- The Northeast is most likely to prefer utility involvement in covering solar warranties.

Primary Heating Fuel

No clear pattern of difference in attitudes toward utility involvement in solar energy is found by primary heating fuel source:

- Electricity users (for heating) are most interested in home energy audits; propane users are least interested.
- More concern for possible problems with the utility company in solar adoption is expressed by those using natural gas and electricity; less concern is expressed by those using propane and wood.
- Consumers using propane are most likely to prefer utility involvement in inspecting installed systems; those using wood are least likely.

CONCLUSION

Many homeowners favor the involvement of utility companies in solar-related activities. Further, key barriers to solar adoption perceived by homeowners, such as lack of confidence in systems and solar firms and lack of relevant information, can be reduced by well-designed utility solar programs. Moreover, problems with utility companies are not viewed as a significant impediment to solar adoption by most homeowners. The data support a conclusion that utility solar programs would fill a need in society and would be socially acceptable. Most support for utility solar programs can be expected from those with higher income and educational levels and from younger persons.

6

FINDINGS FROM THE NATIONAL SOLAR HOMEOWNERS SURVEY

INTRODUCTION

Although the sample of solar homeowners included in the National Study represents the largest data base by far ever completed for this population, the sample cannot be considered to be representative of solar homeowners nationwide. Therefore, it is important to understand what solar homeowners are represented in this sample.

Most of the sample's solar owners are in their middle adult years—81 percent are 35 to 64 years of age. They are highly educated, with 80 percent having at least some college education; 41 percent are professionals; 15 percent are managers, executives, or officials; and 14 percent are retired. The sample is relatively affluent, with 12 percent making under $15,000 and about one-fifth making over $45,000. Almost three-quarters characterize themselves, politically, as middle-of-the-road to strongly conservative.

The solar owners sample is primarily white in racial composition. Over half live in nuclear families with children, and 32 percent are married with no children. About one-third use electric heat; 18 percent list solar energy as a primary home heating source.

Most of the solar owners are not solar professionals or businessmen; however, 17 percent are professionally involved with solar energy as designers, installers, sales representatives, or teachers, and 15 percent belong to solar energy organizations. This sample of solar homeowners is clearly distinct demographically from the representative sample of homeowners who participated in the National Study. To what degree it represents the nation's solar homeowners cannot be assessed.*

---

*Appendix F presents the demographics of this sample in detail.

This chapter presents descriptive findings from the mail survey of 3,809 solar homeowners. Topics covered are systems owned by the sample, the solar adoption decision itself, the purchase and installation process, and postpurchase experiences including satisfaction, problems with systems, and life-style effects.

SYSTEMS OWNED BY THE SAMPLE

The sample consists of 3,809 owners of residential solar energy systems. The questionnaire inquired in detail about the type and function of systems owned by the sample. Because of the variety of residential solar energy applications and our expectation that owners of different types of systems might have differing experiences, we asked a series of specific questions about each particular type of system.

System owners are categorized into two groups—owners of active systems and owners of passive or hybrid systems. Series of survey items were designed for each category. Owners of two active systems and/or two passive or hybrid systems could respond to each question for each system owned.

The sample consists predominantly of owners of active systems—76 percent own one active system, and an additional 11 percent own two active systems. Thus, a total of 87 percent of the sample own at least one active system. About a quarter of the sample own one passive system and 6 percent own two, for a total of 30 percent owning at least one passive system.

Cross-tabulation of active by passive system ownership indicates

TABLE 6.1. Collector Type

What type of solar panels or collectors do you own? (Asked of active system owners only.)

Response	System 1 %	System 2 %	Total %
Flat-plate collectors	80	60	77
Other	7	25	9
Evacuated tube collectors	8	9	9
"Bread box" or thermosyphon collectors	5	6	5
	100	100	100
	(3137)	(474)	(3611)
Missing data	(73)	(24)	(97)
Total N	(3210)	(498)	(3708)

TABLE 6.2. Storage Capability

Does your solar energy system have storage capability?

Response	%		N
Yes	82		(2982)
No	18		(668)
	100		(3650)
		Missing data	(159)
		Total N	(3809)

What type of storage? (CHECK ALL THAT APPLY) (Asked of those with storage.)

Response	%	N
Water	86	(2405)
Rock	16	(454)
Mass wall, floor, ceiling	15	(420)
Other	7	(194)
Phase change materials (e.g., eutectic salts)	2	(56)
Batteries	2	(44)
Don't know	1	(22)
N (2791)		
Missing data (191)		
Total N (2982)		

Approximately what is the storage capability of the system in Days? (Asked of those with storage.)

Response	%		N
0 day(s)	1		(22)
1	24		(693)
2	32		(904)
3	22		(620)
4 and over	12		(351)
Other	2		(70)
Don't know	6		(178)
	99		(2838)
		Missing data	(144)
		Total N	(2982)

$\overline{X}$ = 3.67 days
Range = 0-365 days

that many in the sample own several, and several types of, solar energy devices. Of owners of one passive system, 79 percent own at least one active solar feature, while 70 percent of those with two passive features own an active system. Of those possessing one active system, 23 percent have at least one passive feature on their homes; 3 percent of those with two active systems also own a passive feature.

Most active systems employ flat-plate collectors (77 percent), while 9 percent use evacuated tube collectors, and 5 percent use thermosyphon collectors (Table 6.1).

Of active systems, a majority employ water as a heat transfer medium, while 18 percent use chemical solutions. Air collectors are used in 15 percent of the systems.

Most systems have storage capability (Table 6.2). Water is by far the most common storage element, used by 86 percent of systems with storage. Rock is used by 16 percent, and mass walls, floors, or ceilings are used by 15 percent. Phase-change materials are employed by 2 percent, as are batteries. Most storage systems (78 percent) have a storage capability of one to three days, with a mean capability of 3.67 days.*

Nearly all systems (93 percent) employ auxiliary or backup systems. Electric furnaces or hot water heaters comprise 62 percent of backup systems, while 44 percent are wood stoves or fireplaces. Heating oil is used by 20 percent, natural gas by 17 percent, a heat pump by 10 percent, and propane by 6 percent.

A breakdown of ownership by type of system is as follows:

1 active system only	58%
1 active and 1 passive	16
1 passive system only	5
2 active systems only	5
2 active and 1 passive	3
2 active and 2 passive	3
2 passive systems only	2
2 passive and 1 active	2
Other (wind, photovoltaic, bioconversion, etc.)	7
Total	101

*The mean storage capability of 3.67 days is explained by the 12 percent of owners reporting storage of over four days. In some cases, this ranged as high as 365 days, indicating systems with annual cycle storage.

TABLE 6.3. Type of Passive Feature

What kind of passive system(s) or home do you own? (CHECK ALL THAT APPLY)

Response	System 1 %	System 2 %	Total %
House with major passive features	40	23	38
House with limited passive features	42	13	37
Attached solar greenhouse	21	29	25
Earth-bermed house with good southern exposure	9	11	10
Thermal storage wall (Trombe wall)	5	12	7
Envelope house	2	1	2
Other	12	11	12
Total N	(1099)	(291)	(1390)

Table 6.3 gives a breakdown of the type of passive systems owned by the sample. Houses with major passive features, defined as "extra south glazing, thermal mass materials in floors, walls, ceilings—brick, concrete, water containers" comprise 38 percent of passive systems owned. A nearly equal proportion of systems (37 percent) are houses with limited passive features—"broad side facing south, extra south glazing but no significant increase in thermal mass materials over conventional house."

Attached greenhouses comprise one quarter of the passive systems, 10 percent are "earth-bermed houses with good southern exposure," 7 percent are Trombe walls, and 2 percent are envelope houses.

TABLE 6.4. Active System Function

What is the function of your solar energy system(s)? (CHECK ALL THAT APPLY)

Response	System 1 %	System 2 %	Total %
Domestic water heating	70	36	65
Combined domestic water heating/ home heating	18	12	17
Home heating (separate system)	12	22	13
Swimming pool heating	5	32	8
Other	2	14	4
Don't know	--	--	--
Total N	(3210)	(498)	(3708)

Active systems are defined for this survey as "active solar energy systems or passive systems employing collectors (such as thermosyphon or breadbox systems)." Table 6.4 shows the functions of active systems owned by the sample. Domestic hot water heating is by far the most common function, for 65 percent of active systems. Combined space and water heating systems comprise 17 percent of active systems owned. Active space heating systems are 13 percent, while 8 percent are swimming pool heaters.

THE DECISION TO ADOPT SOLAR ENERGY

How did solar homeowners arrive at a decision to adopt solar energy? Retrospective data were collected on factors affecting the solar innovation-decision process. The questions asked of solar homeowners on this topic replicated (or nearly replicated) key questions asked of the national sample of homeowners. Therefore, the discussion in this section follows the general framework used in Chapter 3, where data from the homeowners survey are presented. Specifically, data are presented on:

- Favorability toward solar energy
- Social support
- Definition of the energy situation
- Perceived feasibility
- Knowledge and information sources
- Concern about risk
- Motivation
- Business and policy preferences
- Behavior

TABLE 6.5. Position on Solar Use

Based on your understanding of solar energy for homes, how do you feel about it?

Response	%		N
Strongly favor	76	96	(2800)
Favor	20		(746)
Unsure/neutral	3		(110)
Oppose	—		(9)
Strongly oppose	—		(10)
	99		(3675)
		Missing data	(134)
		Total N	(3809)

TABLE 6.6. Solar's Image

Which of the following statements, if any, would you say accurately describe people who have solar energy systems? (CHECK ALL THAT APPLY)

Response	%	(N)
Economy-minded people	73	(2678)
Do-it-yourselfers	59	(2176)
Environmentalists	50	(1851)
Socially-responsible people	44	(1634)
Upper income people	39	(1445)
Scientific types	38	(1415)
Back-to-the-land types	31	(1129)
People who enjoy being different	25	(921)
Not very different from my friends	22	(823)
Young people	19	(681)
Retired people	10	(355)
Political radicals	3	(119)
Other	7	(251)
Don't know	3	(102)
None	1	(20)
N	(3680)	
Missing data	(129)	
Total N	(3809)	

The replication of these key items between the two surveys permits the empirical verification of the "next tiers" analysis presented in Chapter 7. In this section, solar homeowner responses are described in and of themselves.

Favorability toward Solar Use

Solar owners are highly favorable toward the use of solar energy in homes—96 percent favor solar use; virtually none oppose it; data are reported in Table 6.5. This overwhelming favorability undoubtedly reflects the widespread satisfaction with their experiences, as reported by the solar owners.

Social Support

Solar imagery among solar owners is interesting—in a way, consumers are describing how they view themselves in answering the item on statements that describe people who have solar energy systems; data are presented in Table 6.6.

The most frequently occurring response, mentioned by 73 percent of the sample, is "economy-minded people." "Do-it-yourselfers" are mentioned by 59 percent, and half mention "environmentalists." More than one in five say that solar owners are "not very different from my friends." The image of the "wheat germ and granola" set is not widespread among present solar owners—3 percent mention "political radicals," 19 percent say "young people," and 31 percent choose "back-to-the-land types" as descriptive of solar owners, responses that might relate to such imagery. The economy-minded imagery suggests strong economic motivations for adopting solar energy, whether near- or long-term, a conclusion borne out by other survey results.

Those already owning solar energy systems may have experienced a favorable social milieu that encouraged their interest in solar energy. Items were included in the survey on the perceived favorability toward solar energy on the part of "significant others." Data are presented in Table 6.7.

Solar owners perceive very little governmental opposition and virtually no social opposition to the use of solar energy in homes. About two-thirds view federal and state governments as favorable, while a plurality of 44 percent indicate that their local government is favorable. Another 40 percent indicate that their local governments are mixed or neutral on solar use, while 12 percent are unsure about their local government's position.

Spouses are overwhelmingly favorable, while friends, neighbors, and co-workers are characterized as favorable by about 45 percent of solar homeowners. Mixed or neutral opinions on the part of friends, neighbors, and co-workers are perceived by about another 45 percent.

These findings show that solar owners find themselves in a social milieu that they regard as largely supportive of their decision to use solar energy. This phenomenon is conducive to solar purchase in the first place and to continued satisfaction with the purchase

TABLE 6.7. Perceived Government Favorability toward and Social Support for Solar Energy

Please indicate whether each of the following is favorable to, or opposed to, using solar energy for homes.

	Favorable	Mixed/ Neutral	Opposed	Don't know	%	N Answering
The federal government	64%	26	5	5	100	(3665)
Your state government	68	24	3	5	100	(3638)
Your local government	44	40	4	12	100	(3597)
Your neighbors and friends	43	48	1	8	100	(3616)
People you work with	47	43	1	9	100	(3453)
Your spouse/housemate	92	6	1	1	100	(3548)

decision after it has been implemented. Solar owners both affect and are affected by this social environment. They help create a supportive milieu by what they report to others concerning their solar experiences. If they are pleased and satisfied with the results of solar ownership, that is likely to be reflected in the opinions of those around them.

TABLE 6.8. Perceived Present/Future National Energy Situation

Here is a picture of a ladder. Let's suppose the top of the ladder represents the best possible energy situation for our country, and the bottom, the worst possible energy situation for our country. Please show me on which step of the ladder you think the United States is at the present time.

Just as your best guess, if things go pretty much as you now expect, where do you think the U.S. will be on the ladder, let us say, about five years from now?

Step	% Present	% Future	
10	—	2	Best
9	—	2	
8	3	8	
7	5	11	
6	9	14	
5	17	19	
4	19	15	
3	23	13	
2	14	10	
1	7	4	
0	3	2	Worst
Total	100%	100%	
	(N = 3634)	(N = 3642)	

Perceived Future Energy Situation

Recode	%	(N)
Much better	1	(25)
Better	64	(2303)
About the same	12	(424)
Worse	23	(844)
Much worse	—	(9)
	100	(3605)
	Missing data	(204)
	Total N	(3809)

TABLE 6.9. Impact of Energy Situation on Solar Households

In different parts of the nation, people have reported a variety of effects of the energy situation on their daily lives. How has the energy situation affected your family (household)? Would you say. . . .

Response	%	N
It really has had no effect on us.	12	(448)
We have had to make a few adjustments, but our lifestyle has not been affected.	52	(1975)
Our life has been less comfortable and convenient, but is is not serious.	27	(1032)
We have had to make serious changes in our daily habits.	5	(205)
It has had a positive effect on us (volunteered response).	—	(1)
Don't know.	4	(148)
Total N	100	(3809)

Definition of the Energy Situation

Solar owners, as a group, are clearly optimistic about the future energy situation in the United States; data are presented in Table 6.8. While most do not rank the present situation as particularly desirable, most respondents indicate that they expect the situation to improve over the next five years. In fact, 65 percent of solar homeowners expect the situation to be better, while almost one in four anticipates a worse future energy situation. These findings suggest that anticipated future national energy problems were not a strong motivation for solar owners to adopt solar energy. However, while not expecting shortage problems, some may have anticipated rising energy costs.

Virtually all of the solar owners report no serious impact of the national energy situation on their daily lives (Table 6.9). While 79 percent report minor to medium inconvenience due to energy shortages, 12 percent report no impact at all. Serious personal energy-related problems have obviously not been an impetus to solar adoption by these solar homeowners.

Perceived Feasibility of Solar Energy

Solar owners were asked about the feasibility of solar energy applications, by replicating three items in the homeowners survey. The first item stated, "In general, solar energy systems are technically and economically practical today for homes." About 8 out of 10 solar owners agree with this statement. The second item stated, "If they are practical at all, solar energy systems are only practical

TABLE 6.10. Perceived Feasibility

Item	Proportion resonding...						
	Strongly Agree	Agree	Total Agree	Unsure	Disagree	Strongly Disagree	Total Disagree
In general, solar energy systems are technically and economically practial today for homes.	42	37	79	11	9	1	10
If they are practical at all, solar energy systems are only practical for homes in warm, sunny climates.	5	16	21	10	45	24	69
Solar energy systems in general will perform satisfactorily over a long period of time.	33	50	83	15	2	—	2

TABLE 6.11. Awareness of Solar Technologies

Solar energy is a very broad term that includes many ways of using the sun to produce energy. Here is a list of solar features that can be used by individual homeowners to produce energy. Please indicate those you have heard of, those you have actually seen, and those you actually own. (CHECK ALL THAT APPLY)

Solar Technology	% Have Heard Of	% Have Seen	% Own	% Total
Solar panels or collectors to heat water for homes	3	13	81	97
Solar panels or collectors to heat homes	31	33	22	86
Solar swimming pool, hot tub, or spa heaters	35	40	10	85
Wood stoves to heat homes	16	39	30	85
Small wind machines (or windmills) to produce electricity or pump water	36	45	2	83
Passive solar design (such as using many south-facing windows to get heat from sun, with few windows on the north; using water containers, wall, floors, or ceilings to collect and store the sun's heat, etc.)	30	24	26	80
Bioconversion (such as gasohol and using farm wastes to produce gas for fuel)	63	15	1	79
Attached solar greenhouses to heat homes	38	30	9	77
Solar cells (photovoltaics) to produce electricity	46	29	2	77
Other	6	3	6	15

for homes in warm, sunny climates"; two-thirds of solar owners disagree with the statement. The third item stated, "Solar energy systems will perform satisfactorily over a long period of time"; 83 percent of solar owners agree with the statement, and 2 percent disagree. Data are presented in Table 6.10.

Clearly, those who own solar energy systems believe them to be practical, cost effective, and durable in a variety of climatic regimes.

Knowledge and Information

The sample exhibits a high level of awareness of all types of solar energy systems, as would be expected. Most widely known are solar domestic hot water systems, by 97 percent of the sample. Slightly less well known are attached solar greenhouses and photovoltaic systems, each mentioned by 77 percent of the sample. Data are shown in Table 6.11.

Table 6.11 also shows what kinds of solar energy systems are owned by the sample.* Domestic hot water systems are by far the most common type of solar energy system, owned by 81 percent of the sample. Wood stoves are owned by 30 percent and passive solar features by 26 percent. Active solar heating systems are owned by 22 percent, while 9 percent report ownership of solar greenhouses. Photovoltaic systems and small wind machines are owned by 2 percent each, while 1 percent own bioconversion systems.

The order of awareness and ownership of solar technologies might suggest the order in which these applications would be adopted. Solar domestic hot water systems might form an early first market, as would wood stoves and, especially in new home construction, passive solar design. Markets further down-line might be in solar greenhouses, small wind machines, and photovoltaics. Bioconversion systems are probably limited mostly to the down-line rural market.

As would be expected, solar owners judge themselves well informed about solar energy, as shown in Table 6.12. Virtually all respondents indicate that they are informed, with 47 percent assessing themselves as "very well informed" and 46 percent as "somewhat informed."

---

*These data represent percentages of owners with the various types of systems listed in this question. The data reported earlier on system ownership represent percentages of owners categorized into the "active" and "passive or hybrid" categories, used to inquire in detail about systems.

TABLE 6.12. Knowledgeability (Subjective Assessment)

How well informed would you say you are about solar energy?

Response	%	N
Very well informed	47	(1761)
Somewhat informed	46	(1718)
Slightly informed	7	(267)
Not at all informed	–	(17)
	100	(3763)
	Missing data	(46)
	Total N	(3809)

Solar owners were asked to identify the three best information sources they had found about solar energy. As data presented in Table 6.13 show, by far the most frequently chosen sources are books, journals, and reports about solar energy, by two out of three solar owners. The second best source of information, identified by about half of the solar owners, is people who have solar energy systems.

TABLE 6.13. Best Information Sources

Here is a list of possible sources of information about solar energy. Please check the three best information sources you found.

Response	Chosen in best three	
	%	N
Books, journals, and reports about solar energy	66	(2432)
People who have solar energy systems	47	(1722)
Energy fairs, exhibits, home shows	32	(1175)
Magazines and newspapers	31	(1139)
Demonstration buildings or model solar homes	24	(865)
Seminars/workshops	21	(753)
Friends, relatives, neighbors, and acquaintances	16	(601)
Local contractors (heating, plumbing, etc.)	12	(449)
Television and radio programs	9	(321)
Homebuilders, architects	9	(316)
Home energy audits	3	(117)
Adult education classes	3	(114)
Other	8	(297)
None of these	1	(46)

N	(3672)
Missing data	(137)
Total N	(3809)

About a third of the sample identify two other important sources of solar information—energy fairs, exhibits, and home shows, and the print media (magazines and newspapers). About a quarter of the sample indicate that seeing solar energy systems in demonstrations or model homes was an important source for them; one in five took seminars or workshops on solar energy and found them to be important.

Less frequently mentioned as good sources of solar information are television and radio, architects and home builders, home energy audits, and adult education classes. These may have been mentioned less frequently because information was not available from these sources when the sample's solar owners installed their systems or because the sources themselves (such as audits) were not available to them.

These findings underscore the fact that relatively unbiased publications are important in the solar decision process. Word-of-mouth communication is also of obvious importance, not only in spreading the word about solar energy but also in influencing people to purchase it. The print media are of special importance in communicating about solar energy.

Solar owners relied heavily on information from solar companies in purchasing their systems. In response to the item asking them to identify the three best organizational sources of solar information, 57 percent mentioned solar companies; data are presented in Table 6.14.

State energy offices are identified by a third of solar owners as a good source. Since many state energy offices were just coming into existence during the period when most of these solar installations occurred, their effect on this sample of solar owners is especially remarkable. Nearly a third also cite federal energy agencies or laboratories as good sources, such as the U.S. Department of Energy, the Solar Energy Research Institute, and regional solar energy centers. All of these organizations have come into existence since 1977.

A quarter of the respondents indicate that federal information organizations, such as the U.S. Government Printing Office and the National Solar Heating and Cooling Information Center were helpful to them. About one in five mention trade or professional associations as being important solar information sources for them.

Also useful to some solar owners were their public libraries, mentioned as one of the best sources by 18 percent; solar or alternative technology clubs, by 14 percent; consumer groups, by 14 percent; and utility companies, by 11 percent. Mentioned least frequently as good sources are environmental organizations, state or county agricultural extension services, lobbying organizations, and local government. Many of the latter organizations may not have had solar programs or information on solar energy to share at the time these respondents were purchasing or installing their systems. If they were to have such information in the future, potential purchasers might well turn to them for it.

Likely as a result of their generally positive experience with solar energy, solar owners are promoters of the technology (Table 6.15); 95 percent of the sample's owners have recommended solar

TABLE 6.14. Best Organizational Information Sources

Here is a list of possible sources of information about solar energy. Please check the three best information sources you found.

Response	Chosen in best three	
	%	N
Solar companies	57	(2057)
State energy or solar office	31	(1109)
Federal energy agencies or laboratories (e.g., Department of Energy, Solar Energy Research Institute, Regional Solar Energy Center, etc.)	29	(1033)
Federal information organizations (e.g., Government Printing Office, National Solar Heating and Cooling Information Center, etc.)	25	(887)
Trade or professional organizations (e.g., Solar Energy Industries Assoc.)	22	(785)
Public library	18	(638)
Solar or alternative technology clubs/associations	14	(507)
Consumer groups (e.g., Consumers Union, etc.)	14	(496)
Utility companies	11	(415)
None of these	7	(269)
Environmental organizations (e.g., Sierra Club, Wilderness Society, etc.)	6	(227)
State or county agricultural extension service agents	4	(150)
Lobbying organizations (e.g., Solar Lobby, etc.)	3	(108)
Local government	2	(74)
Other	8	(302)
N (3611)		
Missing data (198)		
Total N (3809)		

TABLE 6.15. Recommendations on Solar Energy to Others

On the basis of your experience, have you recommended solar energy to others or not?

Response	%		N
Yes, have recommended it highly	72		(2695)
Yes, with some reservations	23		(852)
Have not recommended for or against	4		(167)
No, have recommended against it	1		(45)
	100		(3759)
		Missing data	(50)
		Total N	(3809)

About how many people to whom you have recommended solar energy are using it? (Asked of those recommending solar use.)

Response	%		N
1-10	38		(1322)
11-20	4		(152)
Over 20	6		(204)
Don't know	52		(1797)
	100		(3475)
		Missing data	(72)
		Total N	(3547)

energy to others, and 72 percent have recommended it highly. This high level of recommendation suggestions that a process of "horizontal diffusion"—word-of-mouth information dissemination—about solar energy is occurring. The recommendations of present solar owners are clearly influential, based on the sample's reports. Over a third of solar owners recommending solar energy report that at least one other person has adopted solar energy after they recommended it; 10 percent of the sample making favorable recommendations report that at least 11 persons are using solar energy after they spoke with them about it.

In summary, solar homeowners are quite knowledgeable about various residential solar applications. They own solar domestic hot water systems more than any other solar application. They relied heavily on print media, people who already own solar energy systems, solar companies, and state and federal organizations for information about solar energy when they made their decision to become solar owners, and they themselves have become active promoters of solar use to other homeowners.

Concern about Risk

Homeowners were asked what problems with solar energy would be important concerns to them if they were thinking about using solar energy. Solar homeowners were asked which of the same problems were concerns to them when they were thinking about using solar energy in their homes; data are presented in Table 6.16.

The problems defined by solar owners as important and very important most frequently were those relating primarily to system purchase. Dependability and reputability of solar firms was mentioned by two-thirds of the sample as an important concern, followed by initial system cost, mentioned by 63 percent. Significantly, 37 percent of solar owners—almost 4 out of 10—did not define initial cost

TABLE 6.16. Perceived Barriers
(Concern About Risk)

Here is a list of some things you might see as problems with solar energy. Please indicate which of these were concerns to you when you were thinking about using solar energy in your home. For each item on the list, please enter the number on the scale that shows how important that concern was to you. (Response codes: Not at all important, Slightly important, Somewhat important, Important, Very important)

Response	Important %	Very Important %	Total %
Dependability and reputability of solar firms	31	38	69
Initial cost of solar home or of solar energy systems	33	30	63
Operating reliability of solar energy system (need for frequent repair and maintenance)	29	29	58
Warranty coverage for solar energy system	27	30	57
Possible damage to solar energy systems by dirt, vandalism, storms, corrosion, etc.	26	15	41
Expense of maintaining solar energy systems	23	14	37
Effects of solar energy systems on attractiveness of house/property	21	8	29
Possibility that solar energy systems now on the market will soon be obsolete and better systems will be available later	18	9	27
Safety of solar energy systems	16	11	27
Climate too cold or cloudy or not enough wind	11	7	18
Problems of obtaining a loan to finance the solar energy system	11	6	17
Codes or covenants might prohibit it	9	7	16
Cost and difficulty of getting solar energy systems covered by homeowner's insurance	11	5	16
Worry over whether getting a solar energy system would be time consuming and difficult	8	2	10
Possibility of lowering resale value of house	5	3	8
Might interfere with lifestyle	3	2	5
Possible problems with the utility company	1	2	5
What friends and neighbors would say	1	—	1

as a salient deterrent to solar ownership. This may have been due to their relatively higher income levels, as compared with the general population.

Operating reliability (mentioned by 58 percent as important) and warranty coverage (mentioned by 57 percent) were the third and fourth most widespread problems solar owners defined as important, in considering solar adoption.

Other problems on the list were defined as important by minorities of solar owners. Possible system damage by weather, vandalism, and so on was mentioned by 41 percent. Expense of maintaining the system concerned 37 percent; aesthetics was a potential barrier to 29 percent. Possible system obsolescence and system safety concerned 27 percent, in considering adoption.

The remainder of the listed problems were cited as important by fewer than 19 percent of the sample. These include concerns about climate, financing, local code barriers, insurance coverage, the difficulty of purchase and installation, lowering resale value of the house, and what friends and neighbors would say. Possible problems with their utility company were cited by 5 percent, and 5 percent mentioned fear of effects of solar energy on life-style.

For those who went forward with solar adoption plans, many of the "barriers" to solar adoption frequently mentioned by analysts simply were not important to them. Even initial cost, which is usually defined as the number one barrier to solar use, was not as important as might have been expected. The most widespread concern among solar owners was finding a solar firm that they could rely on for sound information and reliable, well-designed equipment and installation.

Motivation

Homeowners were asked what benefits of solar energy they considered to be important. Solar owners were asked how important the listed benefits were to them "in making your decision to use solar energy"; data are presented in Table 6.17.

Far and away the most important advantages solar owners anticipated when they decided to adopt solar energy were economic ones. Protecting against rising energy costs was mentioned by 87 percent, saving money over the long term by 81 percent, and reducing utility bills now by 76 percent. This set of motivations is the same as Factor 1, termed "responsible self interest," discovered in the homeowners data.*

---

*See Chapter 3, page 73.

TABLE 6.17. Perceived Benefits
(Relative Advantage)

Here is a list of factors that might have influenced your decision to adopt solar energy. For each item on the list, please enter the number on the scale that shows how important this advantage actually was to you in making your decision to use solar energy.

Response	Important %	Very Important %	Total %
Protecting against rising energy costs	30	57	87
Saving money over the long term	30	51	81
Reducing utility bills now	29	47	76
Increasing overall self-reliance	31	38	69
Easing the energy shortage	30	26	56
Conserving natural resources; protecting the environment	28	28	56
Increasing independence from utility company	23	26	49
Having a more reliable supply of energy	26	23	49
Availability of federal income tax credits	23	21	44
Increasing awareness of household energy use	26	17	43
Getting to innovate or experiment	19	16	35
Availability of state tax credits	18	17	35
Increasing comfort of home	20	14	34
Reducing the need for more large power plants	17	16	33
Increasing independence from federal government policies	12	16	28
Increasing awareness of the natural environment, for example, weather patterns	15	10	25
Increasing resale value of home	16	8	24
Increasing status, prestige, and self-esteem	6	4	10

The next most frequently mentioned advantages are items included in Factor 2 in the homeowner data, "advantages to society." These are increasing overall self-reliance, defined as important by 69 percent of solar owners; easing the energy shortage, mentioned by 56 percent; and conserving natural resources/protecting the environment, mentioned by 56 percent.

Between 43 and 49 percent of the sample define the following as important advantages of solar energy: increasing independence from the utility company, having a more reliable supply of energy, availability of federal income tax credits, and increasing awareness of household energy use.

Factor 3 in the homeowner data, "fun" or personal satisfaction, included the chance to innovate and experiment, mentioned as important by 35 percent of solar owners, and increasing status, prestige, and self-esteem, mentioned by 10 percent.

Other advantages defined as important by a third of solar owners include availability of state tax credits, increasing the comfort of the home, and reducing the need for more large power plants. A few other advantages were important considerations to about one in four solar owners: increasing independence from federal government policies, increasing awareness of the natural environment, and increasing resale value of the home.

It seems clear that most solar owners were strongly motivated by perceived near-term, but particularly long-term, financial advantages of solar ownership. This finding suggests that solar owners, when they were deciding to install a system or buy a solar home, intended to remain in their homes long enough to enjoy the anticipated financial rewards of solar ownership.

This speculation is further supported by the fact that three out of four do not define increased resale value of their homes as an important consideration; therefore, systems were probably less likely to be installed as a real estate investment for future capital gains. Direct financial benefit over the longer term appears to outweigh other factors in motivating solar adoption among those who have already done so.

Business and Policy Preferences

The items included here involve solar homeowner preferences for locus of energy decision control, such as individuals, government, and industry, and preferences for organizational roles in promoting the use of solar energy.

As noted in Chapter 3, if individuals perceive that they can do something about their own energy supply, they will be more likely to conserve energy and to adopt solar energy. They must also be willing to take action rather than fixing total responsibility on government and industry. Four items were used to assess preferred locus of energy decision control; data are presented in Table 6.18. In general, solar homeowners display a strong preference for individual and community involvement in energy decision making. They tend to disagree that energy problem solving should be left to the federal government or to private industry.

Solar owners were asked their preferences for organizational roles in promoting residential solar use, replicating the item used in the homeowners survey (Table 6.19). The proportion of solar owners implying endorsement of each listed policy is:

- Educate and inform consumers (97 percent)
- Provide financial incentives, such as grants, low-cost loans, or other kinds of financial assistance (95 percent)
- Identify and make a list available of qualified solar contractors (92 percent)
- Conduct home energy audits, in which an inspector would look over

TABLE 6.18. Preferred Locus of Decision Control

Item	Proportion responding . . .						
	Strongly Agree	Agree	Total Agree	Unsure	Disagree	Strongly Disagree	Total Disagree
People like me have a responsibility to help resolve our country's energy problems by cutting back on consumption, even if this means making some sacrifices in the way I live.	50	43	93	3	3	1	4
People like me should trust the federal government to find a solution to the energy crisis.	7	14	21	16	33	29	62
Our energy problems will be resolved when energy companies have a free hand to find the energy we need.	13	21	34	22	27	17	44
We should pay more attention to the particular energy needs of each city or town and to meeting those needs through local resources whenever possible.	34	52	86	8	4	1	5

a house and make recommendations on how energy can be saved and what kinds of solar energy could be used (86 percent)
- Inspect systems to see how they are working (81 percent)
- Require by law or ordinance passive solar design and/or solar domestic hot water in new home construction (64 percent)
- Cover guarantees or warranties for solar energy systems if a company goes out of business (59 percent)

Solar owners prefer federal and state government to have responsibility for educating and informing consumers (chosen by 63 and 62 percent, respectively); also preferred for this role are local government (selected by 46 percent), utility companies (chosen by 44 percent), and private industry (39 percent).

Most solar owners indicate that the federal government should provide solar incentives (82 percent). Two-thirds also select state government for this role. About one in five also think utility companies and banks should play a role in financial assistance for solar purchase.

Selected most frequently to list solar contractors are state and local governments, by 46 and 47 percent, respectively. Utilities are chosen by a majority of 52 percent to conduct home energy audits, a role that could also be played by private industry and local government (25 percent each).

Solar energy systems inspection is a preferred role for local government (32 percent) and private industry (30 percent).

A majority of solar owners (64 percent) selected one or more organizations to mandate solar use, with 39 percent designating local government and 36 percent selecting state government for this role; one in five thinks solar use should be federally mandated, while 30 percent are opposed to mandating solar use, the highest proportion for any policy listed.

While a majority of solar owners (59 percent) select one or more organizations to cover solar warranties, one in four thinks this should not be done, and 14 percent don't know who should be responsible for covering warranties. The plurality of response is 23 percent for federal warranty backing and 21 percent for state backing. Private industry is selected by 17 percent.

Most frequently mentioned roles for each organization are as follows: the federal government is seen as having primary responsibilities in informing consumers and providing financial incentives; state governments are chosen for the same roles as the federal government and for listing solar contractors, as well; local governments are selected primarily to list solar contractors and to inform consumers about solar energy. Primary roles selected for utility companies are conducting home energy audits and informing consumers.

TABLE 6.19. Preferences for Organizational Roles in Solar Diffusion

Here is a list of things that might be done to promote the use of solar energy in homes. For each item, please indicate which of the groups listed below, if any, should be doing these things.

Responses	Provide Financial Incentives[a]	Inform Consumers[b]	Conduct Home Energy Audits[c]	Cover Solar Warranties[d]	Mandate Passive Design or Solar DHW in New Homes[e]	List Solar Contractors[f]	Inspect Systems[g]
Federal government	82%	63	10	23	21	19	7
State government	66	62	17	21	36	46	18
Local government	38	46	25	9	39	47	32
Utility companies	23	44	52	4	3	20	19
Private industry	12	39	25	17	2	23	30
Banks	22	11	3	4	4	12	2
It doesn't matter	1	1	2	2	1	2	3
Should not be done	3	1	6	25	30	4	11
Don't know	1	1	6	14	5	2	5
N	(3739)	(3705)	(3607)	(3571)	(3627)	(3646)	(3580)

[a]Provide financial incentives, such as grants, low cost loans, or other kinds of financial assistance.

[b]Educate and inform customers.

[c]Conduct home energy audits, in which an inspector would look over a house and made recommendations on how energy can be saved and what kinds of solar energy could be used.

[d]Cover quarantees or warranties for solar energy systems if company goes out of business.

[e]Require by law or ordinance passive solar design and/or solar domestic hot water in new home construction.

[f]Identify and make a list available of qualified solar contractors.

[g]Inspect systems to see how they are working.

Private industry is chosen for informing consumers and inspecting systems. The one role mentioned most frequently for banks is to provide financial assistance, probably with reference to low-cost loans.

Behavior

Present solar owners' behavioral intentions toward solar energy aids in defining them as a potential solar market. Energy conservation and voluntary simplicity behaviors are also described in this section.

Solar owners are themselves a likely market for additional solar applications, as shown in Table 6.20. When asked how likely it is that they will add other solar features to their present homes,

TABLE 6.20. Behavioral Intention

How likely is it that you will add other solar applications to your present home?

Response	%		N
Very likely	28	48	(1049)
Likely	20		(748)
Unsure	20		(732)
Unlikely	19	31	(707)
Very unlikely	12		(454)
	99		(3690)
		Missing data	(119)
		Total N	(3809)

What kind(s) of solar energy system(s) might you invest in? (CHECK ALL THAT APPLY) (Asked of those responding very likely, likely, and unsure.)

Response	%	N
Small wind energy conversion system	39	(899)
Solar greenhouse	32	(738)
Solar cell (photovoltaic system)	32	(737)
Solar panels (or collectors) for hot water	28	(661)
Solar panels (or collectors) for home heating	28	(641)
Passive solar home features	27	(621)
Swimming pool, hot tub, or spa heater	22	(508)
Other solar energy system(s)	12	(276)
N	(2327)	
Missing data	(202)	
Total N	(2529)	

TABLE 6.21. Reported Energy Conservation Behaviors

Listed below are several activities or measures that conserve energy. Please indicate by checking the appropriate column which you have not done, which you did before acquiring your solar energy system or house and which you have done since acquiring your solar energy system or house.

Response	% Have not Done	% Did Before Using Solar Energy	% Did After Using Solar Energy	% Not Applicable	% Total Did
Drive a small car	21	57	14	8	71
Lower temperature setting on hot water heater	18	46	24	12	70
Install weatherstripping or caulk around doors or windows	14	52	14	20	66
Add insulation	18	44	17	21	61
Install storm or doublepane windows	19	44	14	23	58
Install shutters or shades on windows	26	40	18	16	58
Buy energy-efficient appliances	30	37	20	13	57
Set thermostat at 60° or lower in winter	21	41	11	27	52
Set thermostat at 78° or higher in summer	11	28	9	52	37
Install glass doors, heatalator, or other energy-saving device on fireplace	22	22	14	42	36
Install wood stove	41	20	14	25	34
Carpool	42	22	6	30	28
Install timer on heating system or furnace	48	10	8	34	18
Install a heat pump	52	8	5	35	13

nearly half say it is likely, while 31 percent indicate it is unlikely; 20 percent are unsure.

While present solar owners are interested in a variety of additional solar features, small wind energy conversion systems are mentioned most frequently, by 39 percent of those who might purchase additional solar energy systems. Interest in solar greenhouses and photovoltaics is expressed by about a third of potential purchasers. About a quarter of them mention interest in solar domestic hot water and space heating systems and in passive solar design.

The 14-part item on conservation behavior in the homeowner survey was replicated in the solar homeowners survey. Solar owners were asked whether they engaged in specific conservation activities before or after becoming a solar owner. We anticipated that some solar owners might be prompted by their experiences with solar energy to conserve energy in other ways. Table 6.21 presents data on this item.

Solar owners are very likely to drive gas-efficient cars—71 percent report that they drive small cars. Those who have lowered the temperature setting on their hot water heaters constitute 70 percent; 24 percent did this after becoming solar owners. Two-thirds report weather stripping and caulking their homes, while 61 percent have added insulation. A majority have also installed storm or double-pane windows, installed shutters and shades, bought energy-efficient appliances, and lowered their thermostats in winter.

About a third report having installed a wood stove; 14 percent did so after becoming solar owners. Not applicable to 52 percent was raising the thermostat in summer; to 42 percent, were fireplace modifications; to 35 percent, heat pumps; to 34 percent, installing a timer on their heating system; and to 30 percent, car-pooling.

Most solar owners engage in both heating and transportation conservation. Significant numbers of conservation actions have been taken after the onset of solar ownership, notably, lowering the temperature of the hot water heater and buying energy-efficient appliances. However, while the data suggest that solar ownership may heighten conservation awareness and activity, the evidence for this is not conclusive. Solar owners may have continued to increase their conservation action, whether or not they purchased a solar energy system or home. One clear conclusion is possible, though; adoption of solar energy does not appear to result in cessation of energy conservation behavior.

As were homeowners, solar owners were asked how regularly they engage in a set of activities associated with the "voluntary simplicity" life-style. Generally, these activities relate to decreased

TABLE 6.22. Summary of Voluntary Simplicity Activities

Activity	Never Do	Do: Some	Do: Regularly	Do: Total
Skills for self-reliance	4%	45	50	95
Exercise for physical fitness	7	58	35	93
Recycle	16	70	14	84
Change oil in family car(s)	26	27	46	73
Bicycle	32	59	10	69
Contribute to ecological organization	47	31	21	52
Buy used clothing	56	44	1	45

personal consumption and to providing some of one's own goods and services.*

Table 6.22 summarizes data on voluntary simplicity behaviors reported by solar owners. Mentioned most frequently (by 95 percent) are "developing and using skills to increase self-reliance, such as in carpentry, car repair, and food preservation." Virtually everyone (93 percent) reports engaging in "exercise for physical fitness." Most solar owners (84 percent) also recycle newspapers, glass, or cans used at home; three out of four change the oil in their family cars, and two-thirds report that they bicycle for transportation or recreation. A majority of 52 percent contribute to ecologically oriented organizations, such as the Sierra Club. Finally, 45 percent report that they buy some of the household's clothing at garage sales or secondhand stores.

Since such large proportions of solar homeowners report engaging in activities generally associated with the voluntarily simple life-style, these findings lend support to the idea that solar ownership for many owners may be associated with this preferred style of life.

Summary

To summarize what we found about solar homeowners' actual decision process, solar homeowners are highly favorable toward residential solar use. The long-term financial advantage of solar

---

*Voluntary simplicity is described more fully in Chapter 1 and detailed data are presented in Appendix F, Demographics, Values, and Life-style Characteristics of the National Solar Homeowners Sample.

ownership seems to be their single most important motivation in adoption. They tend to view solar owners as "economy-minded" and perceive themselves to be in a social milieu supportive of solar adoption. Past impacts of energy shortages and anticipated future national energy problems did not motivate solar adoption.

The sample's solar homeowners tend to view solar energy systems as currently feasible and cost effective. Their most widespread concern in adoption was locating a solar firm that they could rely on for sound information, equipment, and installation. Most widely owned by this sample are solar domestic hot water systems. Wood stoves and passive solar design are also widely owned by the sample.

Solar homeowners strongly prefer individual and community involvement in energy decision making. They do not display a strong anti-utility bias. Solar owners are themselves a likely market for additional solar applications; nearly half indicate interest in additional solar features.

PURCHASE AND INSTALLATION OF SOLAR ENERGY SYSTEMS

Who supplied, designed, built, and installed the systems owned by the sample of solar homeowners? Did solar owners "shop around" before buying their systems? Did they buy them locally? Were systems built primarily in new homes or were they added to existing homes? Answers to these questions are discussed separately for active and passive systems; then, system costs (for all types of systems) are described. Solar homeowner use of tax credits and other financial incentives is also described.

TABLE 6.23. Active System Installation

Was the system(s)....?

Response	System 1 %	System 2 %	Total %
Added to an existing house	70	60	69
Built into the house when originally constructed	27	31	27
Part of a room or wing addition to an existing house	3	9	4
	100	100	100
N	(3204)	(497)	(3701)
Missing data	(6)	(1)	(7)
Total N	(3210)	(498)	(3708)

TABLE 6.24. Active System or Component Purchase

Was your solar energy system(s)....?

Response		System 1 %	System 2 %	Total %
Bought as complete "packaged" system and assembled on site		54	30	51
Bought as components and assembled on site		28	33	29
Built on site		15	34	18
Other		2	2	2
		99	99	100
	N	(3200)	(497)	(3697)
	Missing data	(10)	(1)	(11)
	Total N	(3210)	(498)	(3708)

Of active systems, 85 percent were purchased, as opposed to being owner built. Of passive systems, 60 percent were purchased, suggesting that passive systems are a more attractive option for the do-it-yourself market. Data regarding choice of a firm (type of firm, number contacted, reasons for choice) are reported, with percentages based on the subsamples of active and passive owners who purchased systems.

Active Systems

As data in Table 6.23 show, two-thirds of active systems were retrofitted into existing houses; one-third were part of new construction. About a quarter were built into a new house and 4 percent were

TABLE 6.25. Installer of Active Systems

Who installed your (active) solar energy system?

Response		System 1 %	System 2 %	Total %
A contractor		64	49	62
Yourself		27	44	29
Other		8	6	8
Both yourself and a contractor		1	1	1
		100	100	100
	N	(3210)	(497)	(3707)
	Missing data	(--)	(1)	(1)
	Total N	(3210)	(498)	(3708)

TABLE 6.26. Suppliers of Purchased Active Systems

From whom did you purchase your system(s) or parts and materials for it? (CHECK ALL THAT APPLY)

Response	System 1 %	System 2 %	Total %
Local solar company	70	48	67
Local plumbing, heating, ventilating or air conditioning store	17	20	17
Local hardware, lumber, etc. supply house	12	22	13
Other	11	18	12
Mail order from a nonlocal solar manufacturer	7	13	7
Local building contractor	6	9	6
Local outlet of a chain store	2	4	3
Local architect	1	3	2
Nonlocal solar company	—	1	—
Don't know	—	1	—
Total N	(2721)	(415)	(3136)

included in a new addition to an existing house. Most active systems were purchased as complete packaged systems (51 percent); 29 percent were purchased as components; 18 percent were site-built systems (Table 6.24). The majority of active systems (62 percent) were installed by a contractor, while 29 percent were installed by the owner, as shown in Table 6.25. Local sources were the predominant suppliers of active solar energy systems, as illustrated in Table 6.26. Local solar companies were by far the major suppliers, supplying over half of the active systems, while local plumbing or HVAC suppliers and hardware or lumber companies provided nearly one-fourth of systems.

TABLE 6.27. Manufacturers of Active Systems

What kind of firm manufactured your (active) solar energy system(s) or components?

Response	System 1 %	System 2 %	Total %
Large national firm	31	26	31
Small local firm	29	29	29
Small national firm	21	21	21
Other	8	15	9
Myself (volunteered response)	8	8	8
Foreign firm	2	1	2
	99	100	100
N	(2721)	(407)	(3128)
Missing data	(--)	(8)	(8)
Total N	(2721)	(415)	(3136)

TABLE 6.28. Number of Firms Contacted before Purchase of Active Systems

About how many firms did you contact before purchasing your (active) system(s) or components?

Responses	System 1 %	System 2 %	Total %
0	9	25	11
1	30	34	31
2	15	14	15
3	20	11	19
4	9	5	8
5	6	2	5
6	3	1	3
Over 7	7	6	7
Other	—	1	—
Don't know	1	1	1
	100	100	100
N	(2650)	(415)	(3065)
Missing data	(71)	(--)	(71)
Total N	(2721)	(415)	(3136)

System 1

$\bar{X}$ = 3.23
Range = 0 - 130

System 2

$\bar{X}$ = 2.71
Range = 0 - 100

The solar market has been dominated by domestic manufacturers. Neither small nor large firms yet have the edge in active system manufacture, judging from these data. About 30 percent of the active systems were manufactured by large national firms, while another 30 percent came from small local firms. Small national firms supplied about one-fifth of the systems, 8 percent of active systems were built by the owner, while foreign firms supplied 2 percent of the systems. Data are presented in Table 6.27.

Most active system owners shopped around prior to purchasing their systems. The mean number of firms contacted per system was slightly over three. However, nearly a third of systems were purchased after contact with only one other firm, and another 11 percent were purchased without any shopping around (Table 6.28). By the time they purchased their second systems, solar owners apparently felt less need to shop around—over half of second systems were bought with only one other contact or even no other contacts than the firm from whom the original system was purchased. Perhaps owners felt more confidence in their own knowledge and in the firms they had dealt with previously by the time of the second system purchase.

Why were particular firms chosen as suppliers of active solar energy systems? An open-ended inquiry yielded insight for this question (Table 6.29). The two most often cited reasons for choosing a firm indicate that solar shoppers were concerned with the quality of the system and of the firm marketing it; 30 percent mention system quality or such system attributes as ease of maintenance and reliability. The firm's reputation is mentioned by one-fourth of the sample.

System price was a key attraction, according to nearly a quarter of the sample. Solar features are probably best marketed locally—19 percent mention the availability or convenience provided by dealing with a local firm. Also, 10 percent mention prior experience with the firm as a factor affecting their choice.

TABLE 6.29. Factors Affecting Choice of Firm (Active Systems)

Why did you choose the firm(s) you did? (OPEN-ENDED ITEM)

Response	%	N
Quality of system (and materials); maintenance simple, reliable, etc.	30	(618)
Firm's reputation (long time in business, etc.)	25	(524)
Price was advantageous	22	(457)
Local, nearby, available, convenient	19	(388)
Knowledgeable, competent	16	(320)
Prior experience with the firm or person; special connections to firm	10	(212)
Firm recommended by friends, neighbors, etc.	7	(143)
Approved or rated or licensed by federal agency or state, Better Business Bureau, etc.	7	(140)
Offered warranty, guarantee	5	(111)
Service availability	4	(85)
Firms recommended by architect, HVAC, other building-related professionals	3	(70)
Not applicable	3	(65)
Firm recommended by other owners	3	(64)
Saw the system (on display, on a home, etc.)	3	(53)
Other	2	(36)
Advertisement	2	(31)
No answer	--	(7)
Aesthetics	--	(4)

Total N* 2063

(*Based on those purchasing active systems or components.)

Evidence of knowledgeability or competence displayed by the firm was a factor in persuading 16 percent of the sample to choose a particular firm; 7 percent cite recommendations of friends or neighbors; 3 percent mention recommendations of other owners; and another 3 percent mention recommendations of professionals as factors.

Agency licensing or rating of a firm helped persuade 7 percent to choose a particular firm. For 5 percent, warranties or guarantees, and for 4 percent, service availability, influenced the choice of firm.

Advertising was a factor for 3 percent. This small proportion might be further evidence that solar energy systems and products are not well advertised. Advertising alone will probably not sell systems, but it serves to attract potential customers to a firm. The firm's attributes and the performance of its personnel will determine whether or not a sale is made.

Passive Systems

Installation as part of new construction accounted for three-quarters of the passive systems, with 68 percent built into a new house and 7 percent as part of an addition to an existing house (Table 6.30). About one in four passive systems were retrofits. The ratio of passive retrofits to new construction is roughly the opposite of that found for active systems. A majority of 80 percent of passive systems were built on site, 11 percent were purchased as components and assembled on site, and 7 percent were bought as packaged systems and assembled on site. Data are shown in Table 6.31. As shown in Table 6.32, the majority (52 percent) of passive systems were

TABLE 6.30. Passive System Installation

Was the system(s)...?

Response	System 1 %	System 2 %	Total %
Built into house when originally constructed	70	63	68
Added to an existing house	24	26	24
Part of a room or wing addition to an existing house	6	11	7
	100	100	99
Total N	(1099)	(291)	(1390)

TABLE 6.31. Passive System or Component Purchase

Was your passive system(s)...?

Response	System 1 %	System 2 %	Total %
Built on site	80	79	80
Bought as components and assembled on site	10	13	11
Bought as a complete "packaged" system and assembled on site	8	7	7
Other	2	1	2
	100	100	100
N	(1076)	(282)	(1358)
Missing data	(23)	(9)	(32)
Total N	(1099)	(291)	(1390)

designed by their owners.* Architects designed 20 percent of the systems, while contractors designed 13 percent. The levels of skill among the passive solar owners in this sample must be reasonably high for them to attempt passive system design. Contractors, however, were responsible for installing nearly half of the passive systems, while 44 percent were installed by the owners (Table 6.33). The kinds of firms that installed purchased passive systems are shown in Table 6.34. General building contractors were responsible for building 43 percent of purchased passive systems, while solar building companies built 21 percent; thus, general contractors are apparently learning the skills involved in solar construction. They built more of the passive systems owned by the sample than did more specialized solar building companies.

Table 6.35 shows the suppliers of purchased passive systems or components. Local supply houses (such as for hardware and lumber) were the sources of 43 percent of passive systems or components. Local building contractors accounted for 28 percent of these purchases, while local solar companies supplied 21 percent. All other sources accounted for fewer than 15 percent of systems, each. The passive solar market seems to be dominated by local firms, based on these findings. The do-it-yourself market in passive systems is particularly strong.

---

*This may help explain why print media (books, journals, and reports on solar energy) were among the best information sources solar owners found. They probably referred to these sources for how-to-do-it information on passive features.

TABLE 6.32. Passive System Designer

Who designed your passive system(s)?

Response	System 1 %	System 2 %	Total %
Yourself	51	56	52
An architect	21	18	20
A contractor	14	9	13
Other	11	13	11
Acquired plans	3	4	3
	100	100	99
N	(1083)	(282)	(1365)
Missing data	(16)	(9)	(25)
Total N	(1099)	(291)	(1390)

TABLE 6.33. Passive System Installer

Who installed your passive system(s)?

Response	System 1 %	System 2 %	Total %
A contractor	51	40	48
Yourself	42	51	44
Other	5	7	6
Both yourself and contractor	2	2	2
	100	100	100
N	(1082)	(281)	(1363)
Missing data	(17)	(10)	(27)
Total N	(1099)	(291)	(1390)

TABLE 6.34. Builder of Passive Systems

What kind of firm built your (passive) system(s)? (Asked only of those who purchased systems)

Response	System 1 %	System 2 %	Total %
General building contractor	44	38	43
Solar building company	20	22	21
Myself (volunteered response)	3	3	3
Other	31	34	31
Don't know	2	3	2
	100	100	100
Total N	(640)	(191)	(831)

TABLE 6.35. Suppliers of Purchased Passive Solar Energy System

From whom did you purchase your (passive) system(s) or parts and materials for it? (CHECK ALL THAT APPLY) (Asked only of those who purchased systems)

Response	System 1 %	System 2 %	Total %
Local hardware, lumber, etc. supply house	44	42	43
Local building contractor	29	24	28
Local solar company	21	19	21
Local heating, plumbing, ventilating or air conditioning company	13	10	12
Mail order from a non-local solar manufacturer	10	14	11
Local outlet of a chain store	4	5	4
Local architect	4	3	4
Non-local solar company	—	1	1
Other	15	17	15
Don't know	—	1	—
Total N	(640)	(191)	(831)

TABLE 6.36. Number of Firms Contacted before Purchase of Passive Systems

How many firms or architects, etc. did you contact before purchasing your (passive) system(s)?

Response	System 1 %	System 2 %	Total %
0	24	40	27
1	38	34	37
2	14	9	12
3	9	7	9
4	5	3	5
5	3	1	3
6	1	1	1
7 and over	4	4	4
Other	1	—	1
Don't know	1	1	1
	100	100	100
N	(619)	(178)	(797)
Missing data	(21)	(13)	(34)
Total N	(640)	(191)	(831)

System 1

$\bar{X} = 2.27$
Range = 0-111

System 2

$\bar{X} = 2.17$
Range = 0-100

Most passive system owners shopped around prior to purchasing their systems or components for it (Table 6.36). The mean number of firms contacted per system was 2.27, fewer contacts than for active system purchases. However, 37 percent of passive systems were purchased after contact with only one other firm and another 27 percent were purchased without any shopping around. By the time owners bought their second systems, they did less shopping around—40 percent of second systems were purchased after contact with only one firm.

Passive system owners, when asked to explain why a particular firm was chosen as the supplier of their systems, mention reasons similar to active system owners. Firm's reputation (17 percent) and advantageous prices (16 percent) are mentioned most often (Table 6.37); but 18 percent report the question is not applicable, likely

TABLE 6.37. Factors Affecting Choice of Firm (Passive Systems)

Why did you choose the firm(s) you did? (OPEN-ENDED ITEM)

Response	%	N
Not applicable	18	85
Firm's reputation (long time in business, etc.)	17	82
Price was advantageous	16	77
Quality of system (and materials); maintenance simple, reliable, etc.	16	76
Local, nearby, available, convenient	15	71
Prior experience with the firm or person; special connections to firm	12	59
Knowledgeable, competent	9	42
Firm recommended by architect, HVAC, other building-related professionals	6	28
Firm recommended by friends, neighbors, etc.	4	20
Other	4	17
Approved or rated or licensed by federal agency or state, Better Business Bureau, etc.	3	16
Saw the system (on display, on a home, etc.)	3	16
Service availability	2	10
Aesthetics	2	9
No answer	2	8
Firm recommended by other owners	1	4
Advertisement	1	3
Offered warranty, guarantee	--	1

Total N* 479

(*Based on those purchasing passive systems or components.)

TABLE 6.38. System Cost

What was the total installed cost of your solar energy system(s). NOTE: If you own a solar home, give cost of solar portion.

Response	%	N
$0-999	12	(425)
1000-1999	15	(533)
2000-2999	20	(729)
3000-3999	20	(725)
4000-4999	10	(351)
5000-5999	4	(160)
6000-6999	3	(105)
7000-7999	2	(73)
8000-8999	2	(70)
9000-9999	1	(49)
10,000 & over	9	(346)
Don't know	2	(90)
	100	(3656)
	Missing data	(153)
	Total N	(3809)

$\bar{X}$= $4,507.00
Range = $1,000 - 94,300

because do-it-yourselfers among passive system owners who purchased parts of materials for their systems did not find the choices for response relevant to their purchase process.

Convenience of a local firm was a factor affecting choice of firm for 15 percent; prior experience with the firm was a factor for 12 percent.

Recommendations of others, whether professionals, friends, or other owners, are considerably less important as a factor among passive system purchasers than among active system purchasers. The greater extent to which passive systems are custom designed for particular dwellings rather than purchased as packaged systems probably explains this difference.

System Costs

The majority (67 percent) of solar energy systems owned by the sample cost less than $4,000. A minority of very expensive systems raised the mean cost of systems to $4,507. Data on costs are presented in Table 6.38. Most of the solar homeowners in the sample received an income tax credit for their system purchase (Table 6.39). Of those who were potentially able to take advantage of tax credits, that is, those whose systems were acquired after credits were available, 80 percent received such a credit. Of those who received income tax credits, 99 percent received federal credits and 54 percent received state credits. Of those receiving credits, one quarter had between 20 and 29 percent of system costs covered by credits; another quarter fell into the 30 to 39 percent of the system cost range; 16 percent received credits covering less than 20 percent of system costs.

TABLE 6.39. Use of Tax Credits

Did you receive an income tax credit for your solar energy system or home?

Response	%	N
Yes	74	(2758)
No	19	(706)
Not applicable, no credits existed when you acquired system	7	(261)
	100	(3725)
	Missing data	(84)
	Total N	(3809)

Which tax credits? (CHECK ALL THAT APPLY) (Asked only of those who received a tax credit.)

Response	%	N
Federal	99	(2627)
State	54	(1426)
N (2655)		
Missing data (103)		
Total N (2758)		

Subsidized Portion of System Cost

What portion of the cost of the solar energy system(s) did your tax credit(s) cover? (Asked only of those who received a tax credit.)

Response	%	N
0-9%	4	(97)
10-19	12	(318)
20-29	25	(657)
30-39	25	(653)
40-49	10	(260)
50-59	12	(307)
60-69	1	(33)
70-79	1	(19)
80-89	—	(6)
90-99	—	(3)
100 & over	2	(49)
Don't know	8	(194)
	100	(2596)
	Missing data	(2213)
	Total N	(3809)

$\bar{X}$ = 28.91%
Range = 0 - 100%

TABLE 6.40. Other Aid

Aside from income tax credits, have you received any government or industry aid in connection with your solar energy system(s) or home?

Response	%		N
Yes	27		(1013)
No	72		(2659)
Don't know	1		(28)
	100		(3700)
		Missing data	(109)
		Total N	(3809)

What kind of aid have you received? (CHECK ALL THAT APPLY) (Asked only of those who received other aid.)

Response	%	N
HUD Solar Hot Water Demonstration Program Grant	67	(608)
Local property tax exemption	33	(304)
State sales tax credit or exemption	29	(261)
Other aid	23	(211)
State property tax exemption	10	(88)
HUD Solar Heating and Cooling Demonstration Program Grant	8	(74)
State Low income loan	6	(58)
Local sales tax credit or exemption	5	(48)
Rebate from industry	4	(38)
N	(914)	
Missing data	(99)	
Total N	(1013)	

The average portion of system costs offset by combined federal and state tax credits, for this sample, was 28.91 percent.* Thus, the average system cost to owners was $3,204. In addition, solar homeowners received other aid. Table 6.40 shows that 27 percent of the sample received additional government or industry aid. The most commonly received, by 67 percent of those receiving additional aid, was a HUD Solar Hot Water Demonstration Grant.† Most solar homeowners in the sample spent around $3,000 for their systems, after

*It should be noted that federal tax credits have increased since these owners bought their systems to 40 percent of the first $10,000 in cost for many system types. In addition, many states have tax credits (see Appendix E).

†U.S. Department of Housing and Urban Development (HUD).

availing themselves of existing tax credits. About one in four found additional aid to reduce costs of solar adoption still further.

In summary, most active systems owned by the sample were added to existing houses, while most passive systems were part of new construction. Both active and passive markets, judging from these data, are dominated by domestic manufacturerers, with systems provided by local companies. Solar owners shopped around only a modest amount prior to purchasing their systems, contacting a total of about three firms, on the average. After availing themselves of existing tax credits, which most did, solar owners spent an average of about $3,000 to purchase their systems.

SATISFACTION WITH SOLAR ENERGY SYSTEMS

As described in Chapter 1, two stages of the solar decision process could only be examined in the study of solar homeowners. These postpurchase stages were termed "observation of effects" and the "continuance/discontinuance decision." We anticipated that, after installing solar energy, homeowners would observe the positive and negative consequences of their decision and would eventually arrive at a point where they are satisfied or dissatisfied with it. At that point, they would make a decision to continue use of solar energy—and full adoption of the innovation can be said to have occurred—or they would reject solar energy, in which case they would have made a discontinuance decision.

Of the solar homeowners sample, 97 percent were solar owners at the time of the survey while 3 percent had discontinued solar use (Table 6.41). It should be noted that former solar owners may have discontinued use owing to factors other than rejection of the innovation, such as a job relocation that necessitated a move.

TABLE 6.41. Solar Ownership

Do you presently own (or did you ever own) a solar home or a house with solar features? ("Solar" includes: collectors or panels, passive features such as many south-facing windows and skylights installed specifically for solar benefit, specially constructed walls or floors for heat storage, solar greenhouses, photovoltaic or solar cells, small wind energy conversion systems or windmills. Solar energy does not include wood stoves for the purpose of this question.)

Response	%	N
Yes, presently own	97	(3710)
Yes, once owned but no longer do	3	(99)
Total N	100	(3809)

TABLE 6.42. Owner Satisfaction

Altogether, how satisfied are you with your experiences as a solar owner?

Response	%	N
Very satisfied	59	(2264)
Satisfied	30	(1136)
Unsure, no opinion yet	5	(188)
Dissatisfied	5	(172)
Very dissatisfied	1	(49)
Total N	100	(3809)

Therefore, most solar homeowners in the sample appear to have arrived at a continuance decision or to be moving in that direction. This finding constitutes one piece of evidence that solar owners are, by and large, satisfied with their decision to use solar energy and will fully complete the solar adoption process. The satisfaction of current solar owners will contribute to other consumers adopting solar energy in the future.

A related item in the survey asked about satisfaction directly: "Altogether, how satisfied are you with your experiences as a solar owner?" Table 6.42 presents the responses. Almost 90 percent of solar owners report satisfaction, with 59 percent "very satisfied" and 30 percent "satisfied"; 6 percent report dissatisfaction and 5 percent are unsure, having not yet arrived at an opinion. Other indications of owner satisfaction found in the survey and reported in previous sections include owners' reports of recommending solar energy to others, their intentions to purchase additional systems, and their high degree of favorability toward residential solar use.

Thus, the data confirm that, for the most part, solar owners are satisfied with their solar experiences. One limitation of the sample should be kept in mind in evaluating this evidence, however. It is possible that the more satisfied solar owners were more willing to participate in the survey than were the less satisfied ones. Some evidence to this effect is found in the reasons offered for refusal to complete the survey. Of all refusals, 9 percent mentioned that they were unhappy or dissatisfied with their solar experience. Some of the refusers indicated that they did not wish to compound their problems by filling out a questionnaire. It is therefore possible that the frequency of satisfaction among solar homeowners is slightly less widespread than the survey findings would suggest. Nevertheless, the overwhelming positive response suggests that solar owners, in general, are much more likely to be satisfied than not satisfied with their solar purchases. Further, the more meaningful question is whether solar owners evidence a higher degree of dissatisfaction than do adopters of other innovations, or than, for example, new car buyers (when innovation itself is not an issue). While we have no strictly comparable data, observation and anecdotal information suggest that solar owners, in general, are probably more satisfied with their solar purchase decisions than are buyers of many consumer products today.

OBSERVATION OF EFFECTS

Having positive and negative experiences after moving in with solar energy, or having it "move in" with the family, will obviously affect the way household members feel about it. In this section, we describe the factors that can contribute to the overall esteem—or lack of it—in which solar energy systems are held by their owners. First, problems solar owners have experienced with active systems, passive systems, wood stoves, and small wind energy systems are explored. Costs of these problems, repairs, and warranty coverage are then described. Solar homeowners' expectations for system performance are compared with their actual experiences, in terms of system condition, utility savings, performance claims, and durability. Finally, effects of solar ownership on life-style are examined.

Negative Experiences

Solar homeonwers were given the opportunity to describe what they considered to be the "most negative experiences associated with owning a solar energy system." Responses to this open-ended question were content analyzed; the results are shown in Table 6.43.

Over one quarter of the sample gave no answer to this question, while 21 percent answered that they have had no negative experiences. Nearly half of the sample's solar owners were apparently hard pressed to think of any serious negative experiences at all. Conversely, most were able to report positive experiences, as will be discussed later.

Negative experiences mentioned are each reported by fewer than 10 percent of the sample. Mentioned most frequently are system performance (8 percent) and initial cost (7 percent). These complaints probably relate to the economic reasons motivating system purchase. To speculate, system purchase probably did not pay off as rapidly as these owners expected that it would.

Examination of the list of negative experiences, and the relatively few owners mentioning them, indicates that some of the anecdotal reporting of solar homeowner problems can give a misleading impression that such problems are widespread.

TABLE 6.43. Negative Experiences

Please describe what you consider to be the most negative experiences associated with owning a solar energy system. (OPEN-ENDED ITEM)

Response	%	N
No answer	26	(973)
None	21	(798)
System inefficient, not performing as stated	8	(291)
Initial cost	7	(274)
Difficulties caused by cloudy weather	6	(221)
Difficulties arising from improper system design	6	(217)
Maintenance problems (daily)	6	(217)
Installation problems	5	(185)
Lack of clear, reliable information about solar energy	4	(144)
Company went out of business, failed to honor warranty, guarantee	4	(142)
Aesthetic disadvantage (e.g., noise, glare)	4	(136)
Other	3	(113)
Finding a qualified contractor, architect or builder	3	(108)
Failure of utility companies or federal agencies to provide support	3	(108)
Repairs	3	(108)
Lack of knowledge/information on how to repair/maintain system	3	(96)
Difficulty finding qualified service personnel	2	(85)
Educating others about solar, conserving energy, etc.	2	(75)
Minor problems, now fixed	2	(74)
Difficulties resulting from leaks in the system (e.g., inconvenience, damage to the roof)	2	(70)
Difficulty finding parts and equipment for initial purchase or repair	2	(63)
Difficulty in evaluating and choosing among competing systems	2	(57)
House has become less comfortable	1	(54)
Lifestyle disadvantages (educating the family, etc.)	1	(49)
Hot water shortages	1	(42)
Conflicts with local building codes and other local regulations	1	(41)
Loss of privacy, inquiries	1	(39)
Federal forms, this questionnaire	1	(39)
Damage due to weather (e.g., storms, lightening, etc.)	1	(38)
Overheating	1	(35)
Difficulty in obtaining financing	1	(25)
Obstruction of your solar energy system by vegetation, new buildings, etc.	1	(19)
Increased property taxes	--	(16)
Utility raised rates for solar owners	--	(13)
Resale value of house has decreased	--	(10)
Vandalism to the solar system	--	(9)
Opposition of neighbors	--	(9)
Difficulty obtaining solar access rights	--	(1)

Total N 3759

System Problems

This section presents data on problems with solar energy systems as reported by owners. Self-reports by owners of systems cannot be considered completely valid measures of performance in the absence of more objective indicators, such as readings from system instrumentation; nevertheless, perceived performance affects satisfaction and deserves attention in any analysis of postpurchase experience.

We asked solar owners a general question regarding problems: "As a solar owner, which of the following problems have you actually experienced?" Responses are presented in Table 6.44. The list of potential problems was derived from literature reviews and field work.

The plurality (41 percent) of owners report having experienced none of the listed problems, while no problem has been experienced by more than 17 percent of the sample. The low percentages of the

TABLE 6.44. Problems Experienced by Solar Owners

As a solar owner, which of the following problems have you actually experienced? (CHECK ALL THAT APPLY)

Response	%	N
None of these	41	(1463)
Maintenance problems	17	(628)
Installation problems	17	(620)
Difficulty in evaluating and choosing among competing systems	17	(610)
Lack of clear, reliable information	16	(563)
Difficulties arising from improper system design	12	(425)
Difficulty finding parts and equipment for initial purchase or repair	11	(382)
Finding a qualified contractor, architect or builder	9	(327)
Overheating	8	(289)
Conflicts with local building codes	7	(250)
Obstruction of your solar energy system by vegetation, new buildings, etc.	7	(246)
Increased property taxes	6	(210)
Difficulty in obtaining financing	5	(187)
Utility raised rates for solar owners	3	(119)
Opposition of neighbors	2	(58)
Vandalism to the solar energy system	1	(52)
House has become less comfortable	1	(50)
Difficulty obtaining solar access rights	1	(36)
Resale value of house decreased	1	(35)
Don't know	—	(17)
N		(3589)
Missing data		(220)
Total N		(3809)

TABLE 6.45. Active System Problems with Pumps and Fans

Have you ever had any of the following problems with your pumps or fans? (CHECK ALL THAT APPLY)

Response	System 1 %	System 2 %	Total %
No problems with pumps or fans	53	39	51
Pump or fan has been replaced	10	5	9
Pump or fan has been repaired	8	4	8
Pump operates on warm days when the sun is not shining	5	3	5
Pump or fan operates noisily	5	4	5
Pump or fan operates all the time	3	2	3
Pump or fan seldom operates	1	1	1
Pump or fan never operates	1	1	1
Other	6	6	6
Don't know	2	2	2
Total N	(3210)	(498)	(3708)

sample reporting such problems as overheating, increased property taxes, and vandalism give some indication that anecdotal reports of problems with solar homes may tend to overestimate their actual occurrence and to present as typical problems that are likely to be rare. More detailed analysis of problems, however, can be expected to yield information of value to policymakers as well as to the solar energy industry. Given the difference in technologies employed by active and passive or hybrid systems, we asked a series of detailed questions on problems from subsets of the sample, categorized by system type.

TABLE 6.46. Active System Controller Problems

Have you ever had any of the following problems with your controller? (CHECK ALL THAT APPLY)

Response	System 1 %	System 2 %	Total %
No problems with controller	44	30	42
Controller has been replaced	13	7	12
Controller has been repaired	8	6	8
Controller does not activate the pump or fan to provide efficient operation of the system	6	3	6
Controller fails to initiate freeze protection for the collector	2	3	2
Other	5	5	5
Don't know	5	4	5
Total N	(3210)	(498)	(3708)

TABLE 6.47. Active System Collector Problems

Have you ever had any of the following problems with your collectors? (CHECK ALL THAT APPLY)

Response	System 1 %	System 2 %	Total %
No problems with collectors	54	41	53
Water had condensed inside the collector	11	6	10
Collector has frozen or ruptured	6	6	6
Collector glass or cover has separated from the collector box	4	3	4
Collector has corroded	2	1	2
Transparent part of collector has corroded	2	1	2
Collector was mounted loosely	1	2	1
Other	11	10	11
Don't know	2	1	2
Total N	(3210)	(498)	(3708)

Active System Problems

Problems with active systems are categorized by system components, such as pumps, fans, and collectors. As Table 6.45 indicates, of those systems with pumps or fans, 51 percent have none of the problems listed. All other problems asked about occur in fewer than 10 percent of systems. Of systems with controllers, 42 percent are reported to be free of any of the problems listed in Table 6.46;

TABLE 6.48. General Active System Problems

Have you ever had any of the following problems? (CHECK ALL THAT APPLY)

Response	System 1 %	System 2 %	Total %
None of these	47	37	46
Insufficient insulation anywhere in system	12	7	11
Roof penetration leaked	6	4	6
Storage tank has leaked	6	3	6
Storage tank was undersized	4	2	4
House was damaged during installation of the solar energy system	3	2	3
Piping or pipe joints have corroded	3	3	3
Pipes were mounted loosely	2	1	2
Panel mounting failed	1	2	1
Health hazards from anti-freeze mixture or chemicals in system	—	1	—
Other	11	9	11
Don't know	1	2	1
Total N	(3210)	(498)	(3708)

TABLE 6.49. Passive System Problems

Have you ever had any of the following problems? (CHECK ALL THAT APPLY)

Response	%	(N)
None of these	40	(513)
Uneveness of temperature throughout the house	19	(266)
Overheating	16	(221)
Excessive condensation	9	(125)
Excessive glare	4	(60)
Mechanical failure of fans	4	(57)
Mechanical failure of movable insulation or shutters	3	(39)
Total N (1390)		

12 percent of controllers have been replaced; 8 percent have been repaired. Of systems employing collectors, 53 percent have none of the problems listed in Table 6.47. All other problems listed are reported as having occurred in 10 percent or fewer of the systems. The most commonly reported problem is water condensation, for 10 percent of the collectors.

A final question asked of active system owners concerned nine miscellaneous problems often cited as potential results of poor design or installation; 46 percent of the systems have experienced none of the problems listed in Table 6.48. A total of 11 percent of the systems have insufficient insulation somewhere in the system; all other problems are reported as having occurred in 6 percent or fewer of the systems.

Passive System Problems

Table 6.49 indicates that 40 percent of owners of passive or hybrid systems have experienced none of the listed six problems. The two problems cited most often relate to thermal properties of the passive system or house—unevenness of temperature throughout the house (19 percent) and overheating (16 percent). Excessive condensation (9 percent), excessive glare (4 percent), failure of fans (4 percent), and failure of shutters or moveable insulation (3 percent) are reported far less often.

Problems with Wood Stoves

Table 6.50 portrays results of a question asked of 1,151 owners of wood stoves in the sample. No problems are reported by 49 percent, while fewer than 20 percent have experienced any particular listed problem. Most frequently mentioned are the rising cost of wood (17 percent), local inspection required (11 percent), and problems with indoor air quality (10 percent).

TABLE 6.50. Problems with Wood Stoves

Have any of the following ever occurred? (CHECK ALL THAT APPLY) (Asked of wood stove owners.)

Response	%	N
None of these	49	(564)
Experienced wood price increases of over 50% since installation	17	(196)
Local inspection required	11	(132)
Problems with indoor air quality	10	(115)
Had difficulty in cleaning and maintaining system	9	(99)
Chimney fire	8	(95)
Had difficulty obtaining wood	8	(88)
Problems with installation of wood stove	7	(85)
Increase in cost of homeowner's insurance	2	(25)
Other	8	(89)
Don't know	1	(15)
N (1151)		

Problems with Small Wind Energy Systems

Of the sample, 80 solar owners own small wind machines. No problems are reported by about a quarter of wind owners (Table 6.51). Nearly half (48 percent) report lack of sufficient wind as a problem, suggesting that proper siting and resource assessment are important

TABLE 6.51. Problems with Small Wind Machines

Have any of the following ever occurred? (CHECK ALL THAT APPLY) Asked of wind system owners.)

Response	%	N
Lack of sufficient wind	48	(38)
Excessive vibration	24	(19)
None of these	24	(19)
Blade throw	16	(13)
Excessive noise	15	(12)
Hazards in servicing the machine	9	(7)
Leakage of battery acid	8	(6)
Problems hooking up with utility grid	6	(5)
Birds flying into the system	5	(4)
Increase in homeowners insurance	5	(4)
Tower collapse	4	(3)
"Attractive nuisance" problems with people	4	(3)
Signal interference with TV	1	(1)
Hazard to utility repairmen	1	(1)
Other	23	(18)
Don't know	1	(1)
N (80)		

TABLE 6.52. Active System Failures

Approximately how many incidents of failure or "down time" have you had with your (active solar energy systems(s))?

Responses	System 1 %	System 2 %	Total %
None	53	72	55
1	17	8	15
2	12	6	12
3	7	3	6
4	2	2	2
5	2	1	2
6	1	1	1
7 and over	3	3	3
Other	1	—	1
Not applicable	2	3	2
Don't know	—	—	—
	100	99	99
N	(2726)	(437)	(3163)
Missing data	(484)	(61)	(554)
Total N	(3210)	(498)	(3717)

How many occurred . . . ? (Asked of those with incidents of failure.)

Response	System 1 %	System 2 %	Total %
Within 3 months of installation	26	27	27
Within 3-6 months of installation	15	15	15
Within 6 months to 1 year of installation	21	22	21
Within 1-2 years of installation	21	19	21
After 2 years of installation	16	17	16
	99	100	100
Incidents, Total N	(3717)	(477)	(4194)

issues for residential wind energy systems. Excessive vibration is mentioned by 24 percent, blade throw by 16 percent, and excessive noise by 15 percent. Nearly a quarter of wind owners mention that they have problems other than those on the list. Problems of wind system owners should be investigated more thoroughly in future research.

Costs, Repairs, and Warranty Coverage

Incidents of system failure are another measure of performance. Both active and passive system owners were asked about incidents of system failures to assess how frequently systems had to be repaired, how soon repairs were necessary after the systems were acquired, how much these repairs cost, and who paid for them.

Active Systems

Active system owners were asked about incidents of failure or "down time." Over half of systems have had no such incidents. Slightly more than one quarter (27 percent) have had one or two incidents; 6 percent, three incidents; and 8 percent, four or more incidents (Table 6.52).

Over one quarter (27 percent) of these incidents occurred within the first three months of installation; 63 percent occurred in

TABLE 6.53. Cost of Active System Failures

Taken together, how much have these problems cost you so far?

Response	System 1 %	System 2 %	Total %
Nothing	48	44	47
$1-99	24	27	24
100-199	9	6	9
200-299	4	4	4
300-399	4	—	3
400-499	1	2	1
500-599	2	3	2
600-699	1	—	1
700-799	—	—	—
800-899	1	1	1
900-999	—	4	1
1000-1999	2	4	3
2000 and over	1	2	1
Don't know	2	1	2
	99	98	99
N	(1188)	(135)	(1323)
Missing data	(31)	(33)	(64)
Respondents mentioning incidents, Total N	(1219)	(168)	(1387)

TABLE 6.54. Active System Problem Costs Covered by Warranty, Insurance, and System Owners

What percent of the total costs involved were paid by warranty?

Response	System 1 %	System 2 %	Total %
25%	3	2	3
25-49%	6	—	5
50-74%	3	4	3
75-99%	5	32	7
100%	83	62	82
	100	100	100
N	(584)	(53)	(637)
Missing data	(37)	(22)	(59)
Respondents mentioning costs, Total N	(621)	(75)	(696)

What percent of the total costs involved were paid by insurance?

Response	System 1 %	System 2 %	Total %
25%	13	33	16
25-49%	15	—	13
50-74%	10	17	11
75-99%	23	33	24
100%	31	17	29
Don't know	8	—	7
	100	100	100
N	(39)	(6)	(45)
Missing data	(582)	(69)	(651)
Respondents mentioning costs, Total N	(621)	(75)	(696)

What percent of the total costs involved were paid by yourselves?

Response	System 1 %	System 2 %	Total %
25%	7	21	8
25-49%	3	1	3
50-74%	6	3	6
75-99%	3	1	2
100%	81	72	80
Don't know	—	1	1
	100	99	100
N	(547)	(71)	(618)
Missing data	(74)	(4)	(78)
Respondents mentioning costs, Total N	(621)	(75)	(696)

the first year after installation, another 21 percent occurred within one to two years of installation; and 16 percent occurred after two years.

Nearly half (47 percent) of the failures in active systems cost owners nothing, and another 24 percent of failures cost under $100 to repair (Table 6.53). Of active system owners who incurred costs

TABLE 6.55. Passive System Failures

Approximately how may incidents of failure or "down time" have you had with you (passive) solar energy system(s)?

Responses	System 1 %	System 2 %	Total %
None	74	76	74
1	7	6	7
2	4	2	3
3	2	1	2
4	—	1	—
5	1	—	1
6	—	—	—
7 and over	1	1	2
Other	—	—	—
Not applicable	9	13	10
Don't know	—	—	—
	98	100	99
N	(974)	(267)	(1241)
Missing data	(125)	(24)	(149)
Total N	(1099)	(291)	(1390)

How many occurred...? (Asked of those with failure incidents.)

Response	System 1 %	System 2 %	Total %
Within three months of installation	24	53	29
Within 3-6 months of installation	13	12	12
Within 6 months to 1 year of installation	18	19	18
Within 1-2 years of installation	21	9	19
After 2 years of installation	24	7	22
	100	100	100
Incidents, Total N	(413)	(75)	(488)

due to system failures, 92 percent received warranty payments; 6 percent, insurance payments; and 89 percent, some cost was borne by themselves. Of those problems covered by warranty payments, 82 percent received total coverage. Insurance coverage of costs due to problems was lower (Table 6.54).

Passive Systems

Table 6.55 indicates that 74 percent of passive or hybrid systems have had no such incidents; another 5 percent, two or three incidents; and 3 percent, more than five incidents. Passive owners who considered this question not applicable were 10 percent, most likely because the nature of the technology employed was such that mechanical failure or down time were irrelevant concepts.

Of those incidents that have occurred in passive systems, 29 percent were within the first three months of installation. Another 30 percent occurred within one year; 19 percent, within one to two years; and 22 percent, after two years.

As Table 6.56 indicates, 40 percent of these problems with passive systems cost nothing to repair; 25 percent cost under $100; another 13 percent, under $200; and another 7 percent, under $300; 12 percent involved costs of $300 or over.

Table 6.57 shows that, of those problems covered by warranty, 71 percent received total coverage. For those problems for which some costs were borne by the owner, 79 percent were paid for in total by the owner.

Over a third of the sample with costs have some warranty coverage; 2 percent have some insurance coverage; and 60 percent have no coverage.

The fact of having experienced problems with one's system would not necessarily be expected to result in dissatisfaction. The information presented on costs of repairs and responsibility for

TABLE 6.56. Cost of Passive System Failures

Taken together, how much have these problems cost you so far?

Response	System 1 %	System 2 %	Total %
Nothing	39	44	40
$1-99	26	20	25
100-199	13	16	13
200-299	7	8	7
300-399	3	—	3
400-499	—	—	—
500-599	1	4	2
600-699	—	—	—
700-799	1	—	1
800-899	2	4	2
900-999	—	—	—
1000-1999	3	—	3
2000 and over	1	—	1
Not applicable	1	—	1
Don't know	3	4	3
	100	100	101
N	(152)	(25)	(177)
Missing data	(9)	(4)	(13)
Respondents mentioning incidents, Total N	(161)	(29)	(190)

TABLE 6.57. Passive System Problem Costs Coverage

What percent of the total costs involved were paid by warranty?

Response	System 1 %	System 2 %	Total %
25%	2	11	3
25-49%	2	—	2
50-74%	9	11	9
75-99%	12	33	15
100%	75	44	71
	100	99	100
N	(57)	(9)	(66)
Missing data	(95)	(16)	(111)
Respondents mentioning costs, Total N	(152)	(25)	(177)

What percent of the total costs involved were paid by yourselves?

Response	System 1 %	System 2 %	Total %
25%	10	25	13
25-49%	1	—	1
50-74%	7	5	7
75-99%	—	—	—
100%	81	70	79
	99	100	100
N	(86)	(20)	(106)
Missing data	(66)	(5)	(71)
Respondents mentioning costs, Total N	(152)	(25)	(177)

paying these costs indicates that most system failures did not result in significant financial burdens to owners. Still, it might be expected that the inconvenience of having a malfunctioning system could detract from owner satisfaction. To investigate this possibility, a survey item was included to elicit a judgment of the quality of repair service provided, as compared with that provided for conventional heating or hot water systems.

Table 6.58 illustrates the response for this comparison. Claiming their systems have never needed repair are 22 percent; 16 percent state they do their own repairs; 7 percent have no opinion; and 6 percent find the question not applicable to their systems. Table 6.58 presents findings for those who actually made a direct comparison. When considering only these respondents, one-third have found solar repair service better than for conventional systems, 45 percent about the same, 10 percent have found their solar repair service less reliable but adequate, and 12 percent have found solar repair service inadequate. Thus, more than three out of four solar owners with service rate their solar repair service as good as or better than conventional service. It appears that repair service for the sample experiencing problems was mostly satisfactory. This probably contributes to owner satisfaction.

Almost two out of three solar owners received a written warranty with their solar energy system when they purchased it (Table 6.59). Of these warranties, most cover service and maintenance

TABLE 6.58. Satisfaction with Solar Repair Service

Relative to the repair service for conventional heating or hot water systems, how would you rate the quality of your solar repair service?

Response	%	N
About the same	22	(782)
Never needed repair	22	(772)
Better	16	(575)
I do my own repairs	16	(569)
No opinion/don't know	7	(260)
Not applicable to my system	6	(229)
Inadequate	6	(200)
Less reliable but adequate	5	(165)
	100	(3552)
	Missing data	(257)
	Total N	(3809)

Solar Repair Service Compared to Service for Conventional Systems[a]

Recalculation	%	N
Better	33	575
About the same	45	782
Less reliable but adequate	10	164
Inadequate	12	200
	100	(1722)

[a]Considering only those making a direct comparison and omitting the following responses: never needed repair, I do my own repairs, no opinion/don't know, and not applicable to my system.

TABLE 6.59. System Warranty Coverage

Did you receive a written warranty for your solar energy system(s)?

Response	%		N
Yes	64		(2321)
No	19		(693)
Don't know	4		(147)
Not applicable	13		(481)
	100		(3642)
		Missing data	(167)
		Total N	(3809)

Does your warranty cover...? (CHECK ALL THAT APPLY) (Asked of those receiving a warranty.)

Response	%		N
Service and maintenance calls	64		(1307)
On-site repairs and replacement of parts	78		(1593)
Don't know	12		(242)
			(2032)
		Missing data	(289)
		Total N	(2321)

How satisfied are you with your warranty coverage? (Asked of those receiving a warranty.)

Response	%		N
Very satisfied	30		(696)
Satisfied	42		(959)
Don't know/didn't use	20		(471)
Dissatisfied	4		(101)
Very dissatisfied	4		(83)
	100		(2310)
		Missing data	(11)
		Total N	(2321)

calls (64 percent) and/or on-site repairs and replacement of parts (78 percent). Most solar owners (72 percent) are satisfied or very satisfied, while 8 percent are dissatisfied or very dissatisfied, with their warranties. Having satisfactory warranty coverage is another reason to be satisfied with a solar energy purchase decision.

Assessment of System Condition

Active system owners assess their systems as being in very good working order (Table 6.60). Six percent of the systems are

TABLE 6.60. Overall Assessment of Active Solar Energy System Condition

What overall condition describes your solar energy system(s) right now?

Response	System 1 %	System 2 %	Total %
Excellent condition, never were any problems	33	32	33
Very good, the few problems have been solved	45	43	45
Good, some minor problems remain	16	18	16
Fair, system problems difficult to resolve	3	5	3
Poor, system has chronic problems	3	2	3
	100	100	100
N	(3192)	(490)	(3682)
Missing data	(18)	(8)	(26)
Total N	(3210)	(498)	(3708)

judged as "fair" (problems difficult to resolve) or "poor" (chronic problems). About one-third have never had problems; nearly one-half (45 percent) have had their few problems resolved; and 16 percent have minor problems remaining.

Similarly, most passive systems are assessed to be in good to excellent condition (Table 6.61). A plurality of 43 percent of systems are reported to be in excellent condition, while 32 percent are rated very good. One system in four still has problems; however, 21 percent of systems are rated in good condition with minor problems; 4 percent of the systems are rated fair to poor.

The vast majority of systems are performing well as determined by owner report, contributing to the high levels of satisfaction

TABLE 6.61. Overall Assessment of Passive Solar Energy System Condition

What overall condition best describes your solar energy system(s) right now?

Response	System 1 %	System 2 %	Total %
Excellent condition, never were any problems	43	44	43
Very good, the few problems have been solved	32	32	32
Good, some minor problems remain	21	21	21
Fair, system problems difficult to resolve	3	2	3
Poor, system has chronic problems	1	1	1
	100	100	100
N	(1096)	(286)	(1382)
Missing data	(3)	(5)	(8)
Total N	(1099)	(291)	(1390)

reported. There is evidence from other studies, however, that owners of solar energy systems do not always have accurate knowledge of how well their systems are performing and, in fact, may report satisfaction with systems that are not operating at all (Unseld and Crews 1979).

To judge better the accuracy of reported performance, owners were asked if they had a way of determining how well their systems are performing. As Tables 6.62 and 6.63 indicate, a large majority of systems are owned by people claiming to have such a means of determining performance; 84 percent of active systems are owned by consumers who claim to have such knowledge. Decreases in utility bills are by far the most common means of establishing that a system is working, reported for 75 percent of systems. Meters or other instrumentation provide information for 45 percent of systems; owners' inspections of equipment, for 32 percent; owners' feelings of personal comfort, for 30 percent; and professional inspection, for 9 percent.

TABLE 6.62. Source of Information on Active System Performance

Do you have a way of knowing how well your (active) solar energy system(s) is working?

Responses	System 1 %	System 2 %	Total %
Yes	84	83	84
Unsure	10	10	10
No	6	7	6
	100	100	100
N	(2734)	(363)	(3097)
Missing data	(476)	(135)	(611)
Total N	(3210)	(498)	(3708)

How can you tell? (CHECK ALL THAT APPLY) (Asked of those with a way of knowing.)

Response	System 1 %	System 2 %	Total %
From reduced utility bills/energy consumption	78	55	75
From meters or instruments	45	48	45
From your inspection of the equipment	32	30	32
From personal comfort	28	42	30
From professional inspection	9	11	9
Other	8	10	9
Total N	(2294)	(300)	(2594)

TABLE 6.63. Source of Information on Passive System Performance

Do you have a way of knowing how well your (passive) solar energy system(s) is working?

Responses	System 1 %	System 2 %	Total %
Yes	78	79	78
Unsure	12	12	12
No	10	9	10
	100	100	100
N	(937)	(247)	(1184)
Missing data	(162)	(44)	(206)
Total N	(1099)	(291)	(1390)

How can you tell? (CHECK ALL THAT APPLY) (Asked of those with a way of knowing.)

Response	System 1 %	System 2 %	Total %
From reduced utility bills/energy consumption	74	67	72
From personal comfort	61	70	63
From meters or instruments	35	39	36
From your inspection of equipment	21	22	21
From professional inspection	10	8	9
Other	6	6	6
Total N	(733)	(195)	(928)

Over three-quarters of passive systems are owned by consumers claiming such knowledge. As with active systems, the most common means of feedback is from reduced utility bills (for 72 percent of systems). Personal comfort is reported as an assessment method for 63 percent of passive systems. Probably because of the nature of the technology, fewer passive systems (36 percent) possess meters or other instrumentation.

Reported system performance has more credibility for this sample than in previous research, in which no check for objective measures was made. Based on the evidence, this sample owns solar energy systems, the performance of which is, for the most part, quite good.

Expectations for and Experience with System Performance

Expectations of solar owners play an important role in determining satisfaction, as discussed in Chapter 1. Briefly, the more realistic an owner's expectations are, the more likely they are to be fulfilled. Expectations that are fulfilled or surpassed result in buyer satisfaction.

A series of questions was included in the survey on expectations for system performance and on the extent to which these expectations have been borne out through actual experience. As Table 6.64 indicates, expectations regarding cost savings appear quite realistic.

TABLE 6.64. Expectations for Cost Savings

When you bought or built your solar energy system(s), approximately what percent of your total monthly utility costs did you expect to save?

Response	%	N
I did not expect to save at all	3	(114)
Between 1 and 10%	8	(285)
Between 10 and 20%	18	(678)
Between 20 and 30%	19	(726)
Between 30 and 40%	16	(591)
Between 40 and 50%	13	(491)
Between 50 and 60%	9	(323)
Between 60 and 70%	4	(147)
Between 70 and 80%	3	(105)
Over 80%	2	(90)
Don't know	5	(199)
	100	(3749)
	Missing data	(60)
	Total N	(3809)

On what source of information did you base your estimate of savings? (CHECK ALL THAT APPLY)

Response	%	N
Own calculations	56	(1971)
Installer/builder	30	(1067)
Books, magazines	27	(951)
Seller of system/home	27	(939)
Advertisements	15	(523)
Other	11	(402)
Architect/engineer	10	(360)
N	(3508)	
Missing data	(241)	
Total N	(3749)	

TABLE 6.65. Experience with Cost Savings

Since you acquired your solar energy system, what percent of your total monthly utility costs do you think you have actually saved?

Response	%	N
I have not saved at all	4	(131)
Between 1 and 10%	9	(334)
Between 10 and 20%	17	(633)
Between 20 and 30%	19	(700)
Between 30 and 40%	15	(543)
Between 40 and 50%	10	(363)
Between 50 and 60%	6	(240)
Between 60 and 70%	5	(169)
Between 70 and 80%	3	(107)
Over 80%	3	(127)
Don't know	9	(319)
	100	(3666)
	Missing data	(143)
	Total N	(3809)

About two-thirds of the sample expected to save from 10 to 50 percent on their total monthly utility bills as a result of their solar investment. Thus, in making their decisions, owners had fairly realistic expectations about how much they might save. A majority of solar homeowners, 56 percent, based these expectations on their own calculations. They also relied heavily on the system salesman, builder, or installer and on books and magazines for information on what to expect.

As Table 6.65 shows, solar owners' expectations have essentially been met; two-thirds of those sampled feel they are saving between 10 and 50 percent of their total monthly utility bills. Fulfilled expectations on utility cost savings very likely contribute to their satisfaction with owning a solar energy system.

Present savings on utility bills are one form of financial advantage of owning solar energy systems. A second financial advantage is "pay-back"—how many years it takes for a system to pay for itself. The heat, hot water, or electricity the system provides after the pay-back period is essentially free, except for system maintenance. Thus, the perceived longer-term financial advantage that we find to be a strong motivation to adopt solar energy is probably an important factor in owner satisfaction. We asked solar owners whether they feel their systems will pay back and, if so, how long it will take. Data are presented in Table 6.66.

Most solar owners (83 percent) feel their systems will eventually pay for themselves through savings on utility bills. About a third expect pay-back in four to five years; 13 percent expect pay-back in

TABLE 6.66. Expectations for Pay-back

Do you think that your solar energy system(s) will eventually pay for itself in savings on utility bills?

Response	%	N
Yes	83	(3049)
Unsure	10	(383)
No	7	(257)
	100	(3689)
	Missing data	(120)
	Total N	(3809)

How many years from the time of installation do you think that it will take to pay for itself? (Asked of those who think their systems will pay for themselves.)

Response	%	N
0 year(s)	—	(21)
1 year	2	(45)
2 years	3	(89)
3 years	8	(224)
4 years	10	(295)
5 years	20	(605)
6 years	9	(258)
7 years	9	(262)
8 years	8	(233)
9 years	2	(70)
10 years	16	(472)
Over 10	11	(331)
Don't know	2	(68)
	100	(2973)
	Missing data	(76)
	Total N	(3049)

$\bar{X}$ = 7.16 years
Range = 0-100 years

three years or less; the balance expect pay-back in a period ranging from six years to over ten years. An expectation of pay-back in under four years may or may not be realistic, depending on the system or solar home and the occupants' life-style. However, to the degree that expectations for pay-back are not met, owner satisfaction could decline in the years ahead. With rising utility costs, however, many owners' estimates of pay-back may be surpassed, resulting in even greater satisfaction with the solar adoption decision.

We were interested in learning whether present solar owners felt that they had been misled about system performance by those selling them their systems. Data on these questions are reported in Tables 6.67 and 6.68. Active system owners report that, for 59 percent of systems, in their experience, their systems performed

TABLE 6.67. Perceived Accuracy of Claims for Active Systems

In light of your experience since your purchase, in general, how accurate do you think the claims regarding your (active) solar energy system(s) were?

Response	System 1 %	System 2 %	Total %
System performed better than claimed	13	13	13
System performed about as claimed	61	50	59
System performed worse than claimed	11	8	11
Don't know	3	4	3
Not applicable, no claims were made	12	24	14
	100	99	100
N	(2731)	(364)	(3095)
Missing data	(479)	(134)	(613)
Total N	(3210)	(498)	(3708)

"about as claimed." Performance "better than claimed" is reported for 13 percent of active systems, while 11 percent of active systems do not perform as well as it was claimed they would. Active system owners report that they received no claims for 14 percent of systems, and for the remaining 3 percent, they are still unsure whether performance claims are being met.

Passive system owners report that, for 42 percent of systems, their systems have performed about as claimed. Performance better than claimed is reported for 15 percent of passive systems, while 7 percent of passive systems do not perform as well as claimed. Apparently fewer claims are made for passive than for active system

TABLE 6.68. Perceived Accuracy of Claims for Passive Systems

In light of your experience since your purchase, in general, how accurate do you think the claims regarding your (passive) solar energy system(s) were?

Response	System 1 %	System 2 %	Total %
System performed better than claimed	15	17	15
System performed about as claimed	44	38	42
System performed worse than claimed	7	4	7
Don't know	4	3	4
Not applicable, no claims were made	30	38	32
	100	100	100
N	(930)	(250)	(1180)
Missing data	(169)	(41)	(210)
Total N	(1099)	(291)	(1390)

TABLE 6.69. Expected Active System Durability

How long do you expect your (active) solar energy system(s) to last?

Response	System 1 %	System 2 %	Total %
0-9 years	5	20	7
10-19	39	27	37
20-29	39	30	37
30-39	5	5	5
40-49	2	3	2
50-59	2	2	2
60 years and over	1	5	2
As long as the house	3	4	3
Don't know	4	4	4
	100	100	99
N	(2652)	(424)	(3076)
Missing data	(558)	(74)	(632)
Total N	(3210)	(498)	(3708)

performance—no claims are reported for about a third of passive systems. However, this may be, in part, because they were owner designed and built.

In general, then, the majority of solar homeowners in the sample do not feel they have been misled by sellers' claims for system performance. About one in ten systems is assessed by owners as not performing as well as it was claimed it would.

A final measure of expectations in the surveyed concerned

TABLE 6.70. Expected Passive System Durability

How long do you expect your (passive) solar energy system(s) to last?

Response	System 1 %	System 2 %	Total %
0-9 years	5	13	7
10-19	19	13	18
20-29	28	26	27
30-39	9	8	9
40-49	4	4	4
50-59	8	4	7
60 years and over	9	13	10
As long as the house	15	15	15
Don't know	3	4	3
	100	100	100
N	(867)	(252)	(1119)
Missing data	(232)	(39)	(271)
Respondents mentioning incidents, Total N	(1099)	(291)	(1390)

system durability (Tables 6.69 and 6.70). The majority (64 percent) of active systems are expected by their owners to last from 10 to 29 years. These appear to be reasonable estimates, in light of reported durability of solar energy systems.

Nearly half of the passive systems are expected to last from 10 to 29 years; 7 percent, not that long; 30 percent, longer than 29 years. Many passive systems are integral, structural components of houses. Reflecting this, 15 percent of systems are expected to last as long as the house.

Fulfillment of expectations regarding pay-back and system durability could not be measured, given the relatively short time that these systems have been owned by the sample. To the extent that these expectations are borne out, satisfaction levels will remain high. Should they prove to be unrealistic, satisfaction with systems may diminish with time.

Life-style Effects

Frequently mentioned as a potential barrier to solar energy adoption is the unwillingness of people to put up with the "hassle" of owning a solar energy system. Preliminary field work indicated that, while nonowners sometimes viewed this as a concern, few owners reported disruptive or negative life-style changes. An in-depth longitudinal investigation of life-styles of solar owners, conducted as part of the National Study, supported this lack of negative life-style impacts, and in fact suggested positive life-style changes associated with solar adoption (Zimmerman 1980).

Two questions on the incidence of life-style change, and on solar owners' feelings about these changes, were included in the survey.

We were aware that solar energy systems may require more personal interaction or attention on the part of owners than do more conventional systems. Questions were included in the survey on what maintenance activities owners of active and passive systems, respectively, have and how they feel about these activities.

Active systems, for the most part, do not require more than occasional maintenance (Table 6.71). The most frequently mentioned activities are turning auxiliary systems on and off, cleaning collectors, and maintaining system components, such as pumps. These activities are performed occasionally or seasonally. Most active system owners (63 percent) find these maintenance activities "not at all inconvenient," while 11 percent enjoy them. About one-fourth of active system owners find maintenance activities somewhat or very inconvenient.

TABLE 6.71. Frequency of Maintenance Activity for Active Systems

Many solar energy systems require owners to engage in activities to operate or maintain them properly. For each of the following activities you engage in, please indicate how often you do so.

Response	Daily	Weekly	Monthly	Occasionally	Seasonally	Yearly	Never	N/A
Drain down collectors or other freeze protection activities	1%	—	—	4	4	7	46	37
Clean collectors	—	1	2	16	17	16	43	14
Maintenance of pumps, etc.	—	—	2	15	6	19	41	16
Change liquids in system	—	—	—	4	1	10	49	35
Turn auxiliary system on/off	4	1	1	26	17	2	30	19

How do you feel about these activities? Are they...? (Asked of those reporting maintenance activities.)

Response	%	N
Enjoyable	11	(273)
Not at all inconvenient	63	(1613)
Somewhat inconvenient	23	(636)
Very inconvenient	2	(58)
	99	(2580)
Not applicable or missing data		(630)
Total N		(3210)

TABLE 6.72. Frequency of Maintenance Activity for Passive Systems

Many solar energy systems require owners to engage in activities to operate or maintain them properly. For each of the following activities you engage in, please indicate how often you do so. (CHECK ONE RESPONSE FOR EACH ACTIVITY)

Response	Daily	Weekly	Monthly	Occasionally	Seasonally	Yearly	Never	N/A
Open/close drapes or shutters on passive house	64%	1	—	6	5	1	8	15
Open/close vents/ skylights	20	1	—	9	17	1	11	40
Open/close flues or dampers	19	1	—	9	16	1	11	43
Control insects in greenhouse	2	6	4	15	5	1	11	56

How do you feel about these activities? Are they...? (Asked of those reporting maintenance activities.)

Response	%	N
Enjoyable	14	(137)
Not at all inconvenient	63	(621)
Somewhat inconvenient	22	(221)
Very inconvenient	—	(5)
	99	(984)
Not applicable or missing data		(406)
Total N		(1390)

A majority of passive systems require daily minor maintenance activities, such as opening and closing drapes or shutters (Table 6.72). Passive solar owners, in general, find these maintenance activities reasonably convenient or even enjoyable; about one in five report that they are somewhat inconvenienced by maintenance activities. Virtually no one defines maintenance of passive systems as a major disruption.

Another question relating to maintenance inquired about instructions provided to the owner (Table 6.73). Over half (53 percent) received operation and maintenance instructions for their systems upon purchase. Most of those receiving such instructions (81 percent) consider them helpful or extremely helpful.

Contrary to common misperceptions, solar energy systems are not viewed as disruptive of family routines or living patterns by those living with them (Table 6.74). Over half of solar owners report no life-style change; 41 percent report moderate change; and 5 percent report significant effects on life-style. The kinds of life-style changes reported by the same are illustrated in Table 6.74. The responses derive from content analysis of an open-ended question.

TABLE 6.73. Operation and Maintenance Instructions

Did you receive written operation and maintenance instructions for your solar energy system(s)?

Response	%		N
Yes	53		(1944)
No	30		(1100)
Don't know	2		(89)
Not applicable	15		(551)
	100		(3684)
		Missing data	(125)
		Total N	(3809)

How helpful were the instructions to you? (Asked of those receiving instructions.)

Response	%		N
Extremely helpful	24		(474)
Helpful	57		(1105)
Don't know/didn't use	13		(247)
Not helpful	6		(108)
Harmed instead of helped	—		(5)
	100		(1939)
		Missing data	(5)
		Total N	(1944)

TABLE 6.74. Life-style Changes

Has living in a solar home changed your life-style in terms of family routines or living patterns, such as times for showers, use of rooms, etc.? Would you say your family routines changed a great deal, changed somewhat, or did not change?

Response	%	N
No, unchanged	53	(1975)
Yes, changed somewhat	41	(1525)
Yes, changed a great deal	5	(199)
	99	(3699)
	Missing data	(110)
	Total N	(3809)

In general, how do you feel about these changes? (Asked of those reporting changes.)

Response	%	N
Positive	76	(1345)
Neutral/mixed	23	(403)
Negative	1	(24)
	100	Total N (1772)

How have your routines changed? (OPEN-ENDED ITEM)

Response	%	N
Scheduling of laundry	45	620
Scheduling of bathing/showers	41	555
Scheduling of dishwashing	33	454
Adopted additional energy conserving practices/ measures	15	199
Closing/opening shutters, other systems maintenance activities	12	166
More hot water available; don't worry about shortage	10	135
Change use of space	7	100
Increased consciousness of energy use	6	80
Increased enjoyable activities (e.g., greenhouse gardening, swimming longer in year)	5	74
Cutting firewood, maintaining fire	4	60

(continued)

TABLE 6.74 (continued)

Response	%	N
Increased awareness of natural environment (e.g., weather, etc.)	4	59
Dress more or less warmly	3	39
Positive changes in social interactions (more time with family and friends, fighting less, etc.)	3	32
Other	2	28
Scheduling of appliance use	2	26
Increases space available	1	15
None	1	13
Increased use of energy other than hot water	1	8
No answer	1	8
Negative changes in social interactions (difficulty getting family to cooperate, etc.)	—	1
Total N		1369

The three most frequently mentioned changes involve scheduling of hot water use—laundry (45 percent), bathing (41 percent), and dishwashing (33 percent). Since nearly all solar domestic hot water systems have some form of backup system, scheduling of hot water use is not necessary but can be done to increase the solar contribution to hot water needs. Decreased concern about use of hot water, due to its greater availability, is mentioned by 10 percent.

For some homeowners, solar energy contributes to an increased consciousness about and attention to energy use: 15 percent mention more adoption of energy-conserving practices after installing solar energy and 6 percent report greater awareness about energy use; 4 percent cite an increased awareness of the natural environment, such as weather patterns. System maintenance activities are mentioned by 12 percent, and 7 percent cite changes in use of space in the home. Increases in enjoyable activities, such as swimming or greenhouse gardening, are mentioned by 5 percent.

Those reporting that they have experienced change (46 percent of the sample) were asked how they felt about these changes; three out of four report positive feelings about them, while one out of four are mixed or neutral (23 percent) or report negative feelings (3 percent).

Positive Experiences

Solar homeowners were asked to describe in an open-ended fashion "the most important advantages that have actually occurred since you became a solar owner." These responses were content analyzed; results are shown in Table 6.75.

Examination of these data reinforces many of the conclusions drawn from more specific questions. Further evidence for owner satisfaction with the solar purchase is that 85 percent volunteer descriptions of positive experiences they have had.

Clearly, economic advantages, an important motivating factor, have been realized by most solar owners. Reduced utility bills or saving money are reported by 62 percent as one of the most important occurrences. Protection from rising energy costs is mentioned as an advantage by 19 percent, while 16 percent mention long-term economic savings. Increase in the resale value of their home is cited by 15 percent. Availability of federal tax credits and state credits are mentioned by 14 and 12 percent, respectively.

Satisfaction of altruistic motives is also reported: 28 percent feel they are actually conserving natural resources, 14 percent cite easing the energy shortage, 5 percent mention protecting the environment, and 4 percent mention reducing the need for more large power plants.

The desire of some solar homeowners to become more self-reliant or independent is evidenced by the proportions citing the following advantages:

- Having a more reliable supply of energy (24 percent)
- Increasing overall self-reliance (17 percent)
- Increasing independence from utility company (8 percent)
- Independence from energy industry policies (4 percent)

Owners report other advantages that result from the experience of living in a solar home. These range from physical changes (increased comfort of home, 11 percent) to more psychological or emotional advantages. Owners mention increased self-esteem or happiness (14 percent), the opportunity to innovate or experiment (11 percent), the opportunity to educate and help others (8 percent), and increased status or prestige (5 percent).

Other actual advantages experienced by solar owners yield further evidence that owning a system can lead to changes in awareness or consciousness. Increased awareness of energy use is mentioned by 9 percent, learning experiences by 8 percent, and increased awareness of the natural environment by 5 percent.

TABLE 6.75. Positive Experiences

Please identify the give most importange advantages that have actually occurred since you became a solar owner. (OPEN-ENDED ITEM)

Response	%	N
Reducing utility bills now, saving money	62	(2339)
Conserving natural resources, saving resources	28	(1069)
Having a more reliable supply of energy	24	(909)
Protecting against rising energy costs	19	(698)
Increasing overall self-reliance	17	(633)
Saving money over the long term	16	(605)
Increasing resale value of home	15	(565)
Availability of federal income tax credits	14	(541)
Increasing self-esteem, happiness	14	(535)
Easing energy shortage	14	(513)
No answer	13	(480)
Availability of state tax credits	12	(447)
Increasing comfort of home	11	(430)
Getting to innovate of experiment	11	(402)
Increasing awareness of household energy use	9	(322)
Increasing independence from the utility company	8	(319)
Learning experience for self, family members on energy problems, costs, solutions	8	(305)
Personal role in educating/helping others	8	(293)
Increasing awareness of natural environment, for example, weather patterns	5	(201)
Increasing status, prestige	5	(196)
Protecting the environment	5	(194)
Increased enjoyable activities (e.g., greenhouse gardening, swimming longer in year)	5	(182)
Independence from energy industry policies especially oil companies	4	(137)
Reducing need for more large power plants	4	(135)
None	2	(91)
Led to business opportunity	2	(68)
Increased beauty, aesthetic pleasure in home	2	(68)
Behavior change--more energy conserving in general, will do more in the future	2	(63)
Other	2	(63)
Simplified life-style, in harmony with nature, etc.	1	(55)
Availability of grants	1	(44)
Increasing independence from federal government policies	1	(44)
Increased sense of safety, security in home	1	(31)
Positive effect on family relationships	--	(8)

Total N 3768

This investigation of both positive and negative experiences of solar consumers suggests that solar homeownership can mean different things to different people. For most, the solar energy system or home is not simply a substitute for a conventional energy system. Solar energy systems represent unique advantages and disadvantages to owners. The weight of the evidence suggests that the advantages far outweigh the disadvantages. To the extent that the experiences of this sample of innovators and early adopters affects the future decisions of potential solar homeowners, the rate of adoption will be positively influenced.

SUMMARY

This section summarizes the findings reported in this chapter on the solar homeowners survey. The descriptive findings, while based on the most comprehensive survey of solar homeowners in the nation, cannot be considered representative of all solar homeowners. Conclusions about solar homeowners generally based on findings from this sample can be drawn only with great care. Discussion of "solar owners" refers to this sample only, not necessarily to solar owners everywhere.

Solar owners are highly favorable to the use of solar energy in homes—96 percent favor solar use; virtually none oppose it.

Solar owners tend to view other solar owners as "economy-minded," "do-it-yourselfers," "environmentalists," and "not very different from my friends." The fringe-group image sometimes associated with solar use is not shared by solar owners. Owners appear to have experienced a favorable and supportive milieu that encouraged their interest in solar energy.

Solar owners perceive very little government opposition and virtually no social opposition to the use of solar energy in homes. This perceived general favorability to solar use is conducive to solar purchase in the first place and to continued satisfaction with the purchase decision.

Solar owners, as a group, are clearly optimistic about the future energy situation in the United States. While most of them do not rank the present situation as particularly desirable, most expect it to improve over the next five years. Virtually none of the sample report serious impacts of the energy situation on their daily lives.

The solar owners overwhelmingly consider solar energy systems to be practical, cost effective, and durable in a variety of climates.

The sample exhibits a high level of awareness of all types of solar energy systems. Solar owners judge themselves to be well informed about solar energy. Books, journals, and reports were considered the best sources of information on solar energy, and other solar owners were considered the next best source when these owners were making their purchase decisions. Of organizational sources, solar companies were considered the best by far, followed by state and federal offices or agencies.

Solar owners are also promoters of the technology—95 percent have recommended it to others.

Issues that concerned solar consumers prior to making their purchases related primarily to the purchase process. Dependability and reputability of solar firms was the foremost concern, followed by initial system cost, operating reliability, and warranty coverage.

Economic considerations, such as savings on utility bills and protection against rising energy costs, motivated solar purchases. Other important motives were altruistic, such as easing the energy shortage, conserving resources, and increasing self-reliance.

Solar homeowners generally display strong preferences for individual and community involvement in energy decision making. They tend to disagree that energy problem solving should be left to the federal government and to private industry.

Solar homeowners are themselves a likely market for additional solar applications with nearly half indicating it is likely that they will make additional purchases. Small wind systems, greenhouses, and photovoltaics are likely future purchases.

Most solar homeowners engage in heating and transportation conservation. Large proportions report engaging in activities associated with the voluntarily simple life-style.

Most active systems owned by the sample were added to existing houses, while most passive systems were part of new construction. Both active and passive markets, judging from this study, are dominated by domestic manufacturers, with systems provided by local companies. Solar owners shopped around only a modest amount prior to purchasing their systems, contacting a total of about three firms, on the average. After availing themselves of existing tax credits, which most did, solar owners spent an average of about $3,000 to purchase their systems.

Most solar homeowners in the sample are satisfied with their decisions to adopt solar energy. Most systems have performed well, as determined by the self-reports of the owners. While systems have not been trouble-free, most problems experienced appear to have been minor and fairly easily resolved, at no great expense to owners. Repair service is judged to be good. At time of purchase, expectations regarding cost savings, pay-back, performance, and durability were high but reasonable and have been borne out, for the most part. Life-style changes associated with solar ownership are generally reported as minor and positive in nature.

7

CONSUMER PROFILE OF LIKELY RESIDENTIAL SOLAR BUYERS

Who are the next likely users of solar energy and how many of them are there? We briefly described the ideas guiding the "next tiers" analysis in Chapter 1. This analysis permits an assessment of the size of the potential solar market and a description of its characteristics. It also defines what needs to be done to encourage consumers close to solar adoption to take action in the near future. For example, if a person is ready to adopt solar energy in every way except that he sees initial cost as a significant barrier, a financial incentive might cause him to act. If finances are not the problem, but a lack of specific information is holding him back, then carefully designed consumer information programs will hasten a decision. If he is skeptical about current solar feasibility, then demonstrations, exhibits, and model solar homes could serve to educate him on the practicality and cost-effectiveness of present solar applications.

In this chapter, we will describe the solar market segmentation analysis performed, using the national homeowner data. First, we describe how we conducted the analysis and how the market segments relate to favorability and behavioral intention; then, a consumer profile is presented for each market segment, including its demographic characteristics. Next, the location of the likely solar markets is identified. The relationship of the study's independent variables to market segments is also examined. Demographics, values, and lifestyle characteristics are related to likelihood of adoption. Finally, the way in which market segments view utility involvement in promotion of solar use is discussed.

HOW THE ANALYSIS WAS PERFORMED

The idea of the solar innovation-decision process was used to guide the analysis. To recapitulate what was described in Chapter 1, using solar energy is a new idea to most residents of the United States. Buying a solar energy system means willingness to adopt something new and relatively untried. Adoption of innovations is conceived as a process, with stages over time. An individual is, at first, unaware that the new technology exists. He then hears about it, but doesn't know much in detail about its operation or benefits. If his interest is piqued, he may take the time to become more informed about the new technology and to evaluate its potential application to his own situation.

As he becomes more knowledgeable about solar energy, for example, he may become interested enough to plan to purchase a system. Once he has decided to act, he begins shopping around, buys a system, and evaluates how well he likes it. If he is satisfied, the process of adoption of the new technology is complete. If he is dissatisfied, adoption is discontinued, as the outcome of the process. These stages in the adoption-decision process are, in summary form: (1) Ignorance, (2) Initial Awareness, (3) Knowledge-Evaluation, (4) Decision-Intention, and (5) Action.

At any point in time, such as when we conduct a survey, any given individual will fall somewhere along the continuum from ignorance of residential solar applications to ownership. The next likely solar purchasers would be those in the Decision-Intention stage of the adoption process. The least likely solar purchasers would be those still relatively unaware of the new solar technologies.

To perform the analysis, the independent variables—responses on attitudes, knowledge, and preferences regarding solar energy—were subjected to multivariate analyses.* The data were reduced to nine pieces of information on each respondent. These nine variables were:

- How concerned is the person about the risk of solar adoption?
- How advantageous does the person see solar adoption to be?
- How likely is it that the person is seeking information about solar energy?
- To what degree does the person hold an "intellectual" imagery about solar users (related to environmentalism, back-to-the-land types, and scientists)?
- To what degree does the person experience support for solar use in his own social world (including perceptions of government favorability)?
- How feasible does the person think solar energy presently is?

---

*The "boiling down" process involved several multivariate analyses, including factor and scale analyses.

TABLE 7.1. Results of the Cluster Analysis

Cluster	Concern about Risk Scale	Relative Advantage Scale	Factor 1	Factor 2	Factor 3	Factor 4	Factor 5	Factor 6	Factor 7	Size %
			Seeking Information	Preference for Government Involvement	"Intellectual" Imagery	Social Governmental Support	Perceived Feasibility	Preference for Industry Involvement	Knowledge	
1	Avg. .080	Avg.+ .366	Very high 1.159	Very high -1.388	Very high 1.342	Avg.- -.184	High .871	Very high -1.799	High -.801	8
2	Avg. .014	Avg.- -.300	Low -.813	Low .574	Avg.- -.141	Mixed/neutral 1.456	Low -.653	Avg. -.055	Avg. .080	11
3	Low .947	Very low -1.075	Low -.720	Low .546	Low -.833	Avg.- -.264	Avg. -.086	Low .563	Avg.- .411	13
4	Low .798	Avg.+ .165	High .805	Avg.- .182	Avg.+ .360	Avg.- -.163	Very high 1.297	Avg. .011	Very high -1.298	10
5	Avg.- .114	Avg. .091	High avg. .491	Avg.- .259	High .902	Avg. -.013	Low -.554	High -.647	Avg.+ -.348	12
6	High -.678	Avg.+ .162	Very low -1.033	Avg.- .327	Low -.797	Avg.- -.145	Low -.845	Low .728	Very low 1.020	16
7	High avg. -.471	High .622	High .695	Very high -1.377	Avg.+ .413	Avg.- -.388	Avg. .037	Avg.+ -.183	Avg.- .162	12
8	High avg. -.478	High .573	Avg.+ .188	Avg.- .243	Avg.- -.383	Low -.504	Avg.+ .378	Avg.- .395	Avg.- .213	18

±.5 = Avg.±
.5 to 1 = High or low
1.0 = Very high or low

- To what degree does the person favor government involvement in promoting solar use?
- To what degree does the person favor private industry (including utility) involvement in promoting solar use?
- How knowledgeable is the person about solar energy?

Can these nine pieces of information actually distinguish homeowners into an interpretable number of groups? The answer is yes, using a cluster analysis technique. Results of the cluster analysis are shown in Table 7.1. Cluster analysis groups together consumers with similar characteristics and distinguishes among groups with different combinations of characteristics; eight distinct types of consumers or market segments, were produced using the nine variables. The question remained—how did these eight groups arrange themselves along the adoption-decision continuum? The groups nearest adoption would be the next likely solar market, while the groups farthest away from it would be an extremely unlikely solar market.

RELATIONSHIP OF CLUSTERS TO DEPENDENT VARIABLES

To answer this question, the groups were analyzed by the study's dependent variables—favorability to solar use and behavioral intention. Tables 7.2, 7.3, and 7.4 show how each group, labeled now as types, relates to the dependent variables. Clearly, Types A and B are the most favorable toward solar use and the most likely to be planning a near-term solar purchase. Conversely, Types G and H are least favorable to solar use and least likely to have any purchase plans. The other types fall along a continuum between these two extremes. This analysis permitted us to arrange the groups into types, approximating stages of the adoption-decision process, and to determine the size of the market segments at each stage. The market segments were then analyzed demographically.

Figure 7.1 shows how the market segments are distributed across stages of the solar adoption-decision process and the proportion of homeowners located at each stage of the process. The "next tier" is about 18 percent of homeowners and is actually composed of two distinct market segments that are described below. This "tier" is at the Decision-Intention stage. At the Knowledge-Evaluation stage are later potential solar adopters, consumers who will take more time to make up their minds. About one in four homeowners falls into this stage; again, two distinct market segments are involved. At the Initial Awareness stage are consumers who ultimately may or may not adopt solar energy. They would be later adopters, if they adopt at all. Two

TABLE 7.2. Behavioral Intention by Cluster

	Will or may invest	Have not considered	Will not invest
Likely Solar Adopters			
Type A (Cluster 4)	48%	40	12
Type B (Cluster 1)	43	44	13
Later Solar Adopters			
Type C (Cluster 7)	21	72	7
Type D (Cluster 5)	17	65	18
May or May Not Adopt			
Type E (Cluster 8)	17	72	11
Type F (Cluster 2)	8	74	18
Will Not Adopt			
Type G (Cluster 3)	6	77	17
Type H (Cluster 6)	1	89	10

Chi-square test, d.f. = 14, $p \leq .0000$.

types form this market segment, which amounts to 29 percent of homeowners. Finally, about 29 percent of homeowners are very unlikely ever to adopt solar energy.

LIMITATIONS OF THE ANALYSIS

As would be expected, behavioral intention and evaluation are highly correlated ($p \leq .0000$). If we compare the responses of the "next tier" on behavioral intention (Table 7.2) to their responses on specific evaluation (Table 7.4), an interesting fact emerges. While 12 percent of Type As have decided against solar use, 6 percent—half that proportion—oppose its use on their own homes. This means that the other 6 percent have probably decided against solar use for some other reason, such as the need to move or financial constraints. Similarly, 13 percent of Type Bs have decided against solar use, while 3 percent oppose its use on their homes. Therefore, 10 percent have likely decided against it for some other reason. The implication is that, for the next tier, the impediments to solar use are not attitudinal but have more to do with such other factors as unsuitability of their housing for solar retrofit, plans to relocate, and so forth.

A discriminant analysis was run for the entire sample, using the nine pieces of information used in the cluster analysis, with behavioral intention as the dependent variable. Results are presented in Table 7.5. This analysis indicates that the nine pieces of information accurately predict indicated behavioral intention for 75.3 percent of the homeowners. Of those who indicate they will or may adopt, 79 percent are predicted and 21 percent are not predicted. Of those who say they will not adopt, 75 percent are predicted and 25 percent are not predicted.

Although more refinement of this analysis is possible, we conclude that the predictive scheme used to form the clusters is sufficiently accurate to characterize the residential solar markets adequately and to make policy and marketing recommendations on the basis of this analysis.

TABLE 7.3. General Evaluation of Solar Energy by Cluster

How do you feel about the use of solar energy on homes?

	Favor %	Unsure %	Oppose %
Next Likely Solar Users			
Type A (Cluster 4)	98	2	—
Type B (Cluster 1)	96	3	1
Later Solar Adopters			
Type C (Cluster 7)	93	7	—
Type D (Cluster 5)	89	9	3
May or May Not Adopt			
Type E (Cluster 8)	87	12	1
Type F (Cluster 2)	63	30	7
Will Not Adopt			
Type G (Cluster 3)	66	26	8
Type H (Cluster 6)	41	46	13

Chi-square test, d.f. = 14, $p \leq .0000$.

TABLE 7.4. Specific Evaluation of Solar Energy by Cluster

How do you feel about using solar energy on your own home?

	Favor %	Unsure %	Oppose %
Likely Solar Adopters			
Type A (Cluster 4)	90	4	6
Type B (Cluster 1)	89	8	3
Later Solar Adopters			
Type C (Cluster 7)	76	16	8
Type D (Cluster 5)	69	18	13
May or May Not Adopt			
Type E (Cluster 8)	69	20	11
Type F (Cluster 2)	33	33	33
Will Not Adopt			
Type G (Cluster 3)	42	28	30
Type H (Cluster 6)	21	39	40

Chi-square test, d.f. = 14, $p \leq .0000$.

FIGURE 7.1. Distribution of Homeowners Across the Stages of the Solar Decision Process

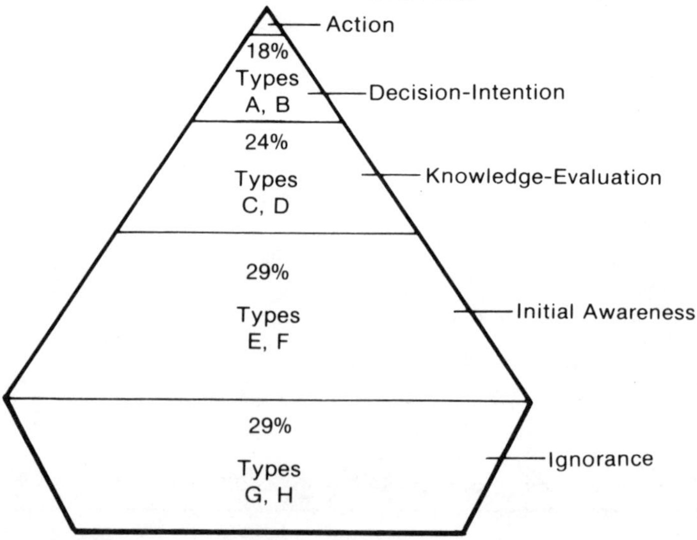

TABLE 7.5. Discriminant Analysis, Predictor Variables by Behavioral Intention, Total Sample

Indicated Behavioral Intention	Predicted Behavioral Intention	
	Will or may adopt	No plans to adopt
Will or may adopt	78.8%	21.2
No plans to adopt	25.5	74.5

(N = 2023)
75.3 percent of cases correctly classified
Chi-square = 514.805

$p \leq .0000$

DESCRIPTION OF THE TIERS

Tables 7.6 through 7.9 characterize each of the four tiers. At the top of each table, each market segment is described, with respect to the nine pieces of information used in the "next tiers" analysis. This portrays how the market segment feels about solar energy. The bottom of each table profiles the demographic characteristics of the market segments.

Type A: Successful Young Families

This group, representing about 10 percent of homeowners, is in good economic condition, relatively well educated, more settled, measured, and establish than Type B consumers (Table 7.6). They appear to be serious, hardworking, motivated, responsible, and

TABLE 7.6. Next Tier: Next Solar Adopters (18%)

	Type A 10% Successful Young Families	Type B 8% Young Activist Homeowners
Seeking information on solar	High	Very high
Knowledge	Very high	High
Perceived feasibility	Very high	High
Concern about risk	Low	Average
Relative advantage	Average	Average
Intellectual imagery	Average	Very high
Preference for government involvement	Average	Very high
Preference for industry involvement	Average	Very high
Social/governmental support	Average	Average
Gender	Male (74%)	About equal
Age	3rd youngest; slightly more 25-35	Youngest (but range 25-54)
Income	High (4th highest)	High (2nd highest)
Education	High average	Highest
Occupation	More professionals; businessmen	More professionals; managers
Political orientation	Strong political opinions; tend toward strong conservative; also more strong liberals	Moderately liberal
Innovativeness	High (2nd highest)	High (highest)
Opinion leadership	High	High
Region	All over the country	Slightly more in West and Midwest
Stability	Low average (23% likely to move)	Low average (27% likely to move)
Community	More rural; half urban	More urban
Family type	Nuclear; more kids<13	More nuclear; kids < 13; some extended families
Main heat	More likely electricity and wood	More likely fuel oil and electricity
Race	White (slightly more Asian and Indian minorities; fewer Hispanics/ Chicanos)	White (some Asian/Pacific Islanders)

capable, yet they are on the innovative side. Primarily male, they are definitely leaders in their communities and work places. These households tend to represent young families with preteenage children. A highly opinionated group, they are strong conservatives or strong liberals in political orientation. They are more likely to use electricity and wood for home heating than other homeowners; they have more rural households than other groups. They are not limited geographically but may be found all over the country, working as professionals and businessmen, on the way up.

They tend to read magazines more frequently than other households, with their favorites being Reader's Digest, National Geographic, and Better Homes and Gardens. They, along with Type B, read Time and Newsweek more frequently than other households. Type A consumers are more likely to read Popular Mechanics, Popular Science, U.S. News and World Report, Playboy, and Mother Earth News than other groups.

They have studied solar energy and evaluated it, yet their information seeking is still high—they want to know more. Many of them have already made a decision to adopt solar energy, while others have not yet seriously considered it. Nevertheless, virtually all of them are favorable to solar use. Their primary interests in solar applications are solar domestic hot water and solar space heating. Those who have not thought about it need to be "sold" on solar energy.

Type B: Young Activist Homeowners

These are younger people on the way up—fashionable and trendy, comprising about 8 percent of all households (Table 7.6). They appear to be positive thinkers, idealistic with a great deal of energy, people who make things happen. They are the most well-educated group with the highest incomes at their age level and the second highest of all households.

They are professionals and managers located around the country, but with slightly more of them in the West and Midwest. Both men and women are represented about equally in this group of urbanites. About one in four is likely to move in the next two to three years. They tend to live in nuclear families with young children, and a few live in extended families. This group is comprised of opinion leaders, tending to be moderately liberal in political orientation. They are more likely to use fuel oil and electricity for home heating than do other groups.

Type B consumers also read magazines more frequently than other groups. Their favorites are the same as Type A consumers; they read Consumer Reports more frequently than any other group.

This group is actively seeking information on solar energy, with less knowledge overall than Type A. While they perceive solar energy to be feasible, they need more information to reduce concern about risks involved in solar adoption and about the financial advantages—particularly long-term financial advantages—of solar adoption. Because they may be somewhat trendy, this group may have a higher potential to adopt solar energy quickly and then reject it. Their primary interests in solar applications are in solar domestic hot water and in solar space heating. They have a more limited interest in passive solar design; however, those buying new homes in the near future would almost certainly be interested in passive design.

Type C: Mainstream Younger Families

These consumers are getting along fairly well financially (Table 7.7). They tend to be younger "silent majority" types, middle-of-the-road politically, yet on the innovative side. More skilled laborers may be found in this group, with some do-it-yourselfers. Divided equally between men and women, this is the second youngest group, ranging in age from 25 to 45. They tend to live in nuclear families with children both under and over age 13. They are more likely to read People magazine than other groups. They are somewhat more likely than other groups to use natural gas for home heating.

This group, while seeking information on solar energy, does not know very much about it yet. They are worried about the costs involved in a solar purchase, as well as other potential problems, but they appear to be highly motivated to adopt. They have strong preferences for government involvement in promoting solar use. Along with Type D consumers, they are more likely than other groups to view solar retrofit as a "major renovation," although they do not rule it out as impossible.

Type D: Older Solid Families

Representing about 12 percent of owner-occupied households, these homeowners have the highest income of all groups (Table 7.7). About equally represented by men and women, they tend to be successful professionals, managers, and businessmen. They evoke the image of three-piece suits and wing tip shoes. While they tend to be leaders, they are moderately conservative and not inclined to be innovative. More of them are located in the Northeast urban centers than are other groups. Their children tend to be older, while they themselves are middle-aged, with more over age 55 in this group. Well-educated, their magazine preferences are similar to Types A and B; they are more likely than other groups to read Family Circle, Woman's Day, and Redbook.

Type D consumers tend to be skeptical of solar feasibility, but they don't know much about solar energy. They are somewhat active in seeking more information about it and would be most amenable to change through sound information. They have some concern about cost effectiveness, yet they have the financial wherewithal to adopt

TABLE 7.7. Second Next Tier: Later Solar Adopters (24%)

	Type C 12% Mainstream Younger Families	Type D 12% Older Solid Families
Seeking information on solar	High	High average
Knowledge	Average	Average
Perceived feasibility	Average	Low
Concern about risk	High average	Average
Relative advantage	High	Average
Intellectual imagery	Average	High
Preference for government involvement	Very high	Average
Preference for industry involvement	Average	High
Social/governmental support	Average	Average
Gender	Equal	Almost equal
Age	2nd youngest (range 25-45)	Older middle (more 55+)
Income	High	High (highest)
Education	Average	High (2nd highest)
Occupation	More skilled labor	More professional, managers, businessmen
Political orientation	Middle of the road	Moderately conservative
Innovativeness	High	Average
Opinion leadership	High average	High
Region	Average (slightly North-east and Midwest)	More in Northeast
Stability	Average (19% likely to move)	Average (22% likely to move)
Community	Average (slightly more urban)	Average (slightly more urban)
Family type	Nuclear; more kids<>13	Average (slightly more nuclear kids >13)
Main heat	More likely natural gas; less likely electricity	Average distribution
Race	Average distribution	Average distribution; more Whites

solar energy if they become convinced it is a viable investment. They tend to associate "intellectual" imagery with solar use, which for them may be a barrier to more active consideration.

Type E: Mainstream Midlife Families

About 18 percent of homeowners fall into this category (Table 7.8). These are the older members of the "silent majority"—not leaders and not particularly innovative. They tend to be blue-collar

TABLE 7.8. Third Next Tier: May or May Not Adopt (29%)

	Type E 18% Mainstream Midlife Families	Type F 11% Uninterested Group
Seeking information on solar	Average	Low
Knowledge	Average	Average
Perceived feasibility	Average	Low
Concern about risk	High average	Average
Relative advantage	High	Average
Intellectual imagery	Average	Average
Preference for government involvement	Average	Low
Preference for industry involvement	Average	Average
Social/governmental support	Have opinions, mostly favorable	Mixed, no opinion
Gender	Equal	More female (51%)
Age	Average distribution	Older middle (55-75)
Income	Average	Average
Education	Low average	Average
Occupation	More blue collar	More retired; average distribution
Political orientation	Middle of the road	Most conservative
Innovativeness	Average	Low
Opinion leadership	Average	Average
Region	Average in distribution	More in South
Stability	Average (15% likely to move)	High (10% likely to move)
Community	Average (slightly more small town)	Average distribution
Family type	Older families, kids > 13	Average (more live alone)
Main heat	Average distribution (slightly more wood)	Average (slightly more propane)
Race	White (slightly more Asian, Hispanic/Chicanos)	Average distribution

workers with older families (children over 13). They are not particularly well educated and their income level is about average for all homeowners. There are some do-it-yourselfers in this group. They read Field and Stream magazine more frequently than other groups do.

As would be expected, Type E consumers are not particularly well informed about solar energy, nor are they very actively seeking information on it. They exhibit some concern about risks of solar adoption, one worry clearly being cost. They have high motivation to adopt solar energy, particularly if it would aid them financially, and they experience a highly supportive social environment for solar use.

Type F: Uninterested Group

About one in ten homeowners just doesn't care about solar energy (Table 7.8). This group is bored with the idea and may be bored with life generally. These are very conservative, very settled consumers who are not leaders nor are they innovative. About average in educational attainment, they tend to be older and middle-aged, with slightly more women in the group. Most of them are retired and live alone than in other groups. They are somewhat more likely to be found in the South.

They know very little about solar energy and don't want to know more. Type F consumers do not believe solar energy is currently cost-effective or reliable, and they are not inclined to view government involvement in it favorably. They experience a social milieu that is unsure or mixed in opinion about solar use rather than a supportive one, which most others perceive.

Type G: Resigned Group

About 13 percent of homeowners are not in very good shape (Table 7.9). Relatively uneducated, they "got lost in the shuffle" somewhere along the line, and they've given up hope for the future. Basically, they are old and poor, with older couples, widows, and widowers overrepresented in this group. They are not leaders and not at all innovative. They tend to be settled types, more likely to be found in small towns, especially in the South. More of them are likely to use wood for heating and to be retired, with a few farmers in the group. They may be somewhat active in local church activities or have some ongoing social life in their communities.

Type G consumers know very little about solar energy and are uninterested in knowing more. They see no advantage to it, nor are they bothered by risks of adopting it since they are very unlikely to do so. They don't associate solar use with intellectualism, and they don't care whether government or industry supports solar energy or not. In general, it is quite unlikely that Type G consumers will ever adopt solar energy.

Type H: Frightened Group

About 16 percent of homeowners make up the oldest and poorest group (Table 7.9). More of them women, they are concerned about survival, and they are averse to taking risks. They tend to be loners and isolates, although some are still living with one other person.

TABLE 7.9. Final Tier: Will Not Be Solar Adopters (29%)

	Type G 13% Resigned Group	Type H 16% Frightened Group
Seeking information on solar	Low	Very low
Knowledge	Low average	Very Low
Perceived feasibility	Average	Low
Concern about risk	Low	High
Relative advantage	Very low	Average
Intellectual imagery	Low	Low
Preference for government involvement	Low	Average
Preference for industry involvement	Low	Low
Social/governmental support	Average	Average
Gender	More male (60%)	More female (60%)
Age	2nd oldest	Oldest
Income	Low (2nd lowest)	Low (lowest)
Education	Low average (2nd lowest)	Lowest
Occupation	More retired; slightly more farmers	More retired; more housewives
Political orientation	Don.t know; some strong conservatism	Average in distribution
Innovativeness	Low	Low
Opinion leadership	Low average	Low
Region	More in South	More in Midwest, slightly more in South
Stability	High (10% likely to move)	Highest (10% likely to move)
Community	More small town	Average distribution
Family type	More live with one other adult	More live alone or with one other adult
Main heat	More wood	More natural gas; less electricity
Race	Average (slightly more Black, Asian)	Mostly White, some Black

There are more retirees and housewives in this group than in others; their educational level is the lowest of all groups. They tend to be quite settled, with somewhat more likelihood of being located in the Midwest and South. They are somewhat more likely than other groups to use natural gas for heating and less likely to use electricity. Minority groups are slightly more likely to be Type G and H consumers.

Type H consumers are basically ignorant of solar energy and actively resistant to knowing more. They are frightened of it. They are slightly interested in the government being involved, if there is anything to it at all. These consumers are unlikely ever to use solar energy unless there is a government program to install solar energy systems on low-income homes.

LOCATION OF THE LIKELY SOLAR MARKET

The solar market segments were analyzed by geographic location to determine whether or not they are concentrated in one part of the country. Results are presented in Table 7.10. To summarize the findings:

- Type A (likely solar adopters) may be found all over the country, with slightly more in the South and West.
- Type B (likely solar adopters) are also located everywhere, but there are more of them in the West and Midwest, and fewer in the South.
- Type C (later solar adopters) are more likely to be found in the Northeast and slightly more likely in the Midwest.
- Type D (later solar adopters) are located more in the Northeast and less in the South.
- Type E (may or may not adopt) are located all around the country, with about average geographic distribution.
- Type F (may or may not adopt) are more likely to be Southerners and less likely to be Westerners.
- Type G (will not adopt) are more likely to be found in the South and less likely in the Northeast.
- Type H (will not adopt) are least likely to be in the West and more likely to be in the South and Midwest.

TABLE 7.10. Regional Distribution of Solar Market Segments

	Total Population	Types							
		A	B	C	D	E	F	G	H
Northeast	23%	20%	24	29	30	25	23	16	20
South	32	35	21	23	25	33	37	41	37
Midwest	31	28	33	33	29	29	30	29	34
West	14	17	22	15	16	13	10	13	9

Chi-square test, d.f. = 21, $p \leq .0000$.

If new fashions and trends "catch on" first in the West and then migrate eastward, such a geographic trend pattern can be discerned in the likely solar markets, over time. However, the differences are not massive; the near-term solar markets exist all over the country. They represent a potential business opportunity throughout the nation.

ANALYSIS OF MARKET SEGMENTS BY KEY VARIABLES

Solar market segments were analyzed (through analyses of variance and cross-tabulations) by key study variables. Details of these analyses are presented for definition of the energy situation, preference for energy supply sources, perceived solar feasibility, awareness and knowledge, information needs and sources, perceived social and governmental support, concern about risk, motivation, and business and policy knowledge and preferences.

Definition of the Energy Situation

In Chapter 3, we noted that concern about a national energy crisis did not appear to be an important motivation for widespread residential solar use. Most homeowners are optimistic about the future national energy situation, and most indicate they have not been seriously affected by prior energy shortages. These variables were analyzed by market segments to determine whether likely solar adopters have views different from others.

Type H consumers, the unlikeliest of solar adopters, are the most pessimistic about the future energy situation. Type G consumers, also unlikely adopters, are as optimistic as everyone else. Types A and B are also as optimistic as other groups. Clearly, pessimism about the future energy situation is not currently associated with likelihood of adoption.

However, the effects of prior energy shortages on households is associated with likelihood of adoption. As shown in Table 7.11, the majority of Type A and of Type D consumers report that they have had to make a few adjustments in response to energy problems. Type B consumers report the greatest impact, with about a third reporting that they have been less comfortable and 14 percent reporting serious impacts. Conversely, Types G and H—who have the lowest incomes and could be expected to report greatest impact—are more likely than others to report "no effect" of energy problems on their households. Since Types A and D have higher income and educational levels, it

TABLE 7.11. Perceived Impact of Energy Situation by Market Segment

Response	A	B	C	D	E	F	G	H
	Types							
No effect	10%	4	11	9	11	13	24	24
A few adjustments	53	46	50	52	47	49	40	41
Less comfortable	30	36	32	35	33	31	25	24
Serious changes	7	14	7	4	9	7	12	11

Chi-square test, d.f. = 21, $p \leq .0000$.

appears that perception of the impact on one's household is related to a degree of sophistication in energy-related matters. To speculate, these consumers have connected rising energy prices with inflation, and they judge that they are being harmed—some say seriously harmed—by inflationary pressures. To these consumers, then, solar energy may signify a potential means of obviating some of inflation's effect on their household budgets. While low-income consumers may, in fact, experience more significant impacts of energy price increases—a higher proportion of their budgets must be used for energy costs—some of them do not appear to make this connection (although 65 percent indicate moderate effects). Perceived impact is greater among more educated and affluent consumers than among older, poorer consumers.

An alternative (or perhaps complementary) interpretation is that more affluent consumers may drive more and may have been more affected by gasoline shortages. More elderly, poorer consumers are less likely to own and drive cars and, thus, might not have been as affected as others by gasoline supply problems.

In conclusion, perception of more serious impact of the energy situation on one's household is associated with likelihood of solar adoption. This appears to hold, regardless of whether transportation or residential energy supply is involved. If more consumers perceive serious effects of energy supply problems and price increases on their households, their motivation to adopt solar energy will be enhanced, with future growth of the likely solar markets. With significantly increasing utility price increases occurring almost everywhere, this is a likely development.

Preference for Energy Supply Sources

We reported in Chapter 3 that homeowners prefer to see solar energy and energy conservation developed more than any other energy

TABLE 7.12. Preferences for Energy Supply Sources by Market Segment

Energy Source Chosen in Top 3	A	B	C	D	E	F	G	H	p ≤
Solar energy	85%	81	76	76	72	57	51	42	.0000
Energy conservation	51	56	47	47	46	36	36	40	.0065
Synfuels	44	44	45	39	34	40	35	33	.0173
Coal	31	27	33	36	37	43	36	41	.0015
Nuclear energy	32	29	24	29	27	28	34	19	.0072
Oil and natural gas	16	13	31	26	31	43	45	49	.0000

Chi-square test for each source, d.f. = 7.

option. We analyzed these comparative energy preferences by market segment to determine if significant differences exist among types of consumers. Differences occur for six of the seven sources listed; no difference in preference occurs for hydropower. Results are presented in Table 7.12.

Solar energy and energy conservation are most preferred by Types A and B (as would be expected), and these sources are also strongly preferred by all others except Type H. Type H consumers tend to prefer oil and natural gas; they are also least likely to select nuclear energy as a preferred source. Coal is chosen by Types F and H most frequently; nuclear energy is selected most frequently by Types A and G. As likelihood of solar adoption increases, preference for oil and natural gas decreases. To speculate, this may be due to concern about dwindling supplies of fossil fuels, effect of their combustion on the environment, and the negative impact of oil imports on the national economy. However, these attitudes may pertain more to oil than to natural gas (the two were combined in the item).

In general, then, preference for solar energy and energy conservation is greatest among the next likely solar buyers, as would be expected. These sources are also the favorites of most other market segments. Such favorable attitudes toward the solar option constitute a fertile ground for market development. Preference for solar energy does not mean that other supply sources are not also recognized as needed by sample pluralities.

Perceived Feasibility

How feasible homeowners think solar energy currently is was a variable used to derive the market segments. Detailed cross-tabulations are reported in Table 7.13. As would be expected, Types A and B exhibit the highest degree of perceived feasibility, while Types G and H tend not to see solar energy as feasible. Type D consumers—older, solid families who are later solar adopters—tend not to see solar energy systems as technically and economically practical today for homes. These consumers need to see solar energy systems in operation and to understand their cost effectiveness before they will be ready to purchase systems. Demonstrations, model homes, exhibits, and systems on friends' or neighbors' homes would provide the kinds of "hands-on" knowledge especially needed by Type D consumers.

TABLE 7.13. Perceived Feasibility by Market Segment

Item/response	Types							
	A	B	C	D	E	F	G	H
In general, solar energy systems are technically and economically practical today for homes. (d.f. = 7, $p \leq .0000$)								
Agree	92%	68	47	18	65	27	44	15
Don't know	6	22	33	34	29	41	43	59
Disagree	2	11	21	48	6	32	13	26
If practical at all, solar energy systems are only practical for homes in warm, sunny climates. (d.f. = 7, $p \leq .0000$)								
Disagree	74	72	54	49	49	31	29	12
Don't know	11	16	19	18	22	19	28	37
Agree	14	12	27	33	29	50	43	51
Solar energy systems in general will perform satisfactorily over a long period of time. (d.f. = 7, $p \leq .0000$)								
Agree	94	74	54	54	58	41	47	19
Don't know	5	25	43	40	40	50	49	70
Disagree	1	1	3	6	2	9	4	11

Awareness and Knowledge

Awareness and knowledge variables were used to derive the market segments. An analysis of variance was performed by market segment using the Awareness Scale, with significant results. Types D, A, and B, respectively, have the highest awareness of residential solar technologies, while Types G and H have the lowest. Cross-tabulations were run for each item comprising the Awareness Scale. Type A consumers are equally aware of all the technologies listed. Types B, C, and D consumers are slightly more aware of solar domestic hot water, solar space heating, wood stoves, wind machines, bioconversion, and passive design than they are of hot tub or spa heaters, solar greenhouses, and photovoltaics.

An analysis of variance was performed by market segment on the likelihood of seeking information on solar energy, with significant results. Types B, C, and A, respectively, are highest in likelihood of information seeking, while Types G and H are lowest. Types A and B are also more likely than others to have seen one or more solar buildings in their own neighborhoods.

An analysis of variance was performed by market segment using the Knowledge Scale, also with significant results. Types A, B, and D, respectively, have highest knowledge, while Types G and H have the lowest. Cross-tabulations were run for having enough knowledge to decide on solar use by market segment. About 79 percent of Type A consumers feel that they know enough to make a solar adoption decision, compared with 55 percent of Type B, 15 percent of Type C, and 36 percent of Type D consumers. Type A consumers clearly feel more ready to decide than do other groups.

Information Needs and Sources

Consumers were asked whether they would like to receive more information on solar energy, as reported in Chapter 3. Responses were cross-tabulated by market segment, with results presented in Table 7.14. Type A, B, C, D, and E consumers are interested in receiving more solar information, while Types F, G, and H tend not to be interested.

The kinds of information needed (volunteered by homeowners) are shown in Table 7.15 for Types A through E, who more clearly define their information needs. Type C, D, and E consumers are particularly interested in receiving financial information, such as solar operating cost, initial cost, and savings. Type B consumers are particularly interested in technical information, such as feasibility

TABLE 7.14. Interest in Solar Information by Market Segment

Item/response		Types							
		A	B	C	D	E	F	G	H
Would you like to get more information regarding solar energy systems?									
	Yes	75%	86	80	75	72	32	34	24
	No	25	14	20	25	28	68	66	76

Chi-square test, d.f. = 7, $p \leq .0000$.

TABLE 7.15. General Solar Information Needs by Market Segment

What information would you like to have? (Open-ended; see Table 3.12 for full item and response codes.)

Information Need	Type A	Type B	Type C	Type D	Type E
Financial					
Operating cost	29%*	29	55	58	82
Initial cost	27	34	41	44	42
Savings	15	25	29	29	24
Payback	10	11	18	15	7
Financial incentives	5	5	1	3	3
Technical					
Feasibility of retrofit	13	29	17	19	18
How-to-do-it	19	19	13	18	18
How it works	17	9	21	13	19
Performance	6	15	11	12	12
Solar in my climate	9	9	5	9	8
Reliability	2	6	4	5	5
State-of-the-art	6	3	3	7	1
Durability	2	1	3	3	3
Weather effects	1	2	1	1	2
General					
General information	19	21	25	19	30
Types of systems available	11	11	10	8	6
Business and Policy					
Company names	6	5	2	6	2
Information sources	2	1	4	5	2
What government is doing	5	2	1	3	2
Maintenance service	1	3	3	2	2
What people think	4	2	2	1	--
Reputation of firms	1	3	2	2	--
Audits	1	1	1	1	--
Living with Solar Energy					
Advantages and disadvantages	1	2	5	3	5
How difficult to maintain	1	3	1	1	2
Safety	1	1	1	4	1
Aesthetics	--	1	1	1	3
N	(151)	(140)	(195)	(182)	(261)

*Percent based on N answering.

of retrofit, system performance, and how-to-do-it information. Type E consumers particularly need general information, such as types of systems available. Information on business, policy, and living with solar energy is mentioned by about the same proportions of each group.

As shown in Table 7.16, the solar markets want to know most about active systems, with greatest interest in water and space heating applications.

What sources are the market segments most likely to use for solar information? Results of cross-tabulations are presented in Table 7.17. The solar markets are most likely to employ solar homeowners, demonstrations, books, magazines, energy fairs, solar companies, and state energy offices. In addition to these, certain types of consumers are more likely to use other sources:

- Type B consumers are more likely than others to employ utility companies, consumer groups, federal energy agencies, public libraries, federal information organizations, and broadcast media.

TABLE 7.16. Technical and Applications Information Needs by Market Segment

What information would you like to have? (Open-ended; see Table 3.12 for full item and response codes.)

Information Need	Type A	Type B	Type C	Type D	Type E
Technologies[a]					
Active systems	10%*	9	15	10	7
Windmills	8	6	4	6	5
Passive systems	5	9	2	7	1
Greenhouses	8	6	2	3	2
Photovoltaics	3	4	3	4	2
Storage	3	2	2	1	2
Back-up systems	2	3	3	2	--
Woodstoves	1	--	--	2	2
Gasohol	1	1	1	1	--
Applications[b]					
Space heating	23	19	17	9	13
Water heating	26	17	15	7	11
Electricity	5	4	2	2	3
Cooling	3	1	3	2	3
Community systems	1	1	--	2	1
N	(151)	(140)	(195)	(182)	(261)

*Percent based on N answering.

[a]Also mentioned by a few were bioconversion systems, small-scale hydropower systems, underground homes, berming, and heat pumps.

[b]Also mentioned by a few were outbuildings, commercial buildings, pumping, grain drying, and electric cars.

TABLE 7.17. Preferred Information Sources by Market Segment

Information Source*	Types A	B	C	D	E	F	G	H
Solar homeowners	94%	95	93	90	87	75	65	60
Demonstrations/ model homes	94	95	90	89	86	62	60	50
Books/journals/reports	92	94	91	87	76	66	56	47
Magazines/newspapers	83	90	86	78	74	64	57	48
Solar companies	84	82	84	80	77	63	53	44
Energy fairs/exhibits/ home shows	86	88	87	76	74	53	47	36
Television/radio	66	75	84	66	74	57	59	56
State energy offices	78	80	80	71	69	50	39	39
Public library	67	77	73	68	58	46	37	37
Federal info organizations	70	75	70	66	63	43	41	34
Federal energy agencies	66	78	71	63	58	40	38	33
Utility companies	52	73	69	62	60	45	41	39
Friends/neighbors	60	65	63	63	61	43	42	35
Local contractors	59	66	62	56	63	48	42	36
Homebuilders/architects	63	67	57	56	59	51	34	33
Agricultural extension	58	58	63	56	57	40	37	33
Consumer groups	62	73	66	59	56	34	29	29
Seminars/workshops	63	65	53	52	50	30	31	22
Local government	46	55	75	40	50	29	32	33
Trade/professional organizations	57	57	50	46	46	31	21	20
Solar clubs	57	55	49	38	41	26	18	18
Adult education classes	49	50	51	45	46	24	21	21
Environmental organizations	36	50	42	31	34	20	15	17
Lobbying organizations	23	24	21	13	19	14	6	9

Chi-square test for each source, d.f. = 7, $p \leq .0000$.

- Type C consumers are more likely to contact local governments, state energy offices, and federal energy agencies and to use television and radio and public libraries.
- Type E consumers are more likely to use television and radio.

Information on specific magazines read most frequently by the market segments is shown in Table 7.18.

TABLE 7.18. Magazine Readership by Market Segment

	Proportion of households in which these magazines are read frequently*							
	Types		Types		Types		Types	
Magazine	A	B	C	D	E	F	G	H
Reader s Digest	54%	55	53	47	55	54	47	43
National Geographic	40	38	24	37	22	28	20	13
Better Homes and Gardens	34	42	43	43	40	37	33	31
Time	29	29	27	23	19	14	13	10
Consumer Reports	21	32	15	22	12	12	8	5
Newsweek	27	25	18	22	17	14	13	11
Family Circle	25	26	28	35	28	31	19	20
Woman.s Day	21	27	27	31	28	27	19	22
Popular Mechanics	23	17	15	18	15	14	9	8
Popular Science	19	15	11	11	9	12	5	5
U.S. News and World Report	17	14	9	16	10	11	8	5
Playboy	13	9	8	12	9	7	4	2
Redbook	11	10	15	16	13	6	8	11
People	9	11	20	9	11	7	7	6
Psychology Today	8	11	5	5	4	2	2	1
Field & Stream	9	9	15	9	15	12	8	13
Mother Earth News	11	6	3	6	2	1	1	1
Parents	5	11	9	5	6	2	5	4
Cosmopolitan	7	9	11	11	8	6	5	3
Sunset	6	6	2	5	2	2	2	2
Solar Age	2	2	—	—	—	1	—	—

*Chi-square test for each magazine, d.f. = 7, $p \leq .03$.

Social and Governmental Support

The six items on perceived position of "significant others" toward solar energy were scaled and used to derive the solar market segments.* Table 7.1 shows that several of the consumer group

*A coding error resulted in the serendipitous discovery that the significant difference between types of consumers was whether they perceived a mixed or neutral social milieu or did not know what others thought about solar energy, or whether they perceived definite opinions around them. The responses were originally recoded so that a "favorable" response received a 5, a "mixed/neutral" response a 3, and an "opposed" response a 1. In the scale program employed, "don't know"

groups score "average" in perceived social/governmental support. Detailed cross-tabulations, shown in Table 7.19, refine the analysis and reveal distinct differences among the market segments.

The likely solar markets—Types A through D—perceive a sociopolitical environment generally favorable toward residential solar use. Little difference occurs among market segments on perceived opposition to solar use. Type F consumers are most likely to perceive a mixed or neutral environment, while Type E consumers are least likely to perceive this. Type G and H consumers are least likely to know what others think. Since Type F consumers may or may not adopt solar energy and are average in opinion leadership and low in innovativeness, they may be waiting until others try solar energy before becoming involved themselves. This would be characteristic of the "later adopters" of innovations. Type E consumers, however, perceive as favorable a milieu as Types A and B; therefore, their social world is encouraging them to become solar adopters. Other barriers are keeping them from early adoption.

Perceived Barriers to Solar Adoption

The Concern about Risk Scale was used to derive the market segments. Each of the 18 concern-about-risk items was cross-tabulated by market segment, with significant results shown in Table 7.20. The likeliest solar adopters, while they have some concerns, are not the most concerned groups. Type A consumers are most worried about warranty coverage, initial cost, reputability of solar firms, and operating reliability. Type B consumers are also concerned about potential damage to systems through storms, vandalism, and so on, and about system safety. Type Ds have about the same concerns as Type As, including insuring the system; their

---

responses were treated as missing data; the mean score for the sample was automatically inserted for respondents with missing data. The coding error resulted in scores of 3 for "mixed/neutral," 1 for "opposed," and 0 for "favorable." When the error was corrected and the analyses redone, the results were no longer interpretable. The detailed cross-tabulations showed that the large proportions of "don't know" responses skewed the results; analyses performed with the data coded with "mixed/neutral" as the high end of the scale proved to be the better solution. The predominant "real" difference is apparently between those who perceive a position (whether negative or positive, although most are positive) and those who do not.

TABLE 7.19. Perceived Sociopolitical Environment by Market Segment

Item/Response*	Types A	B	C	D	E	F	G	H
Is each of the following favorable to or opposed to using solar energy for homes?								
The federal government:								
Favorable	71%	74	83	76	74	51	67	57
Mixed/neutral	13	12	3	6	3	29	3	2
Opposed	11	8	8	8	11	11	5	4
Don't know	4	6	6	10	12	10	25	37
Your state government:								
Favorable	65	69	72	64	66	38	56	45
Mixed/neutral	14	13	5	14	5	38	6	5
Opposed	11	8	10	8	10	8	6	6
Don't know	9	11	13	14	19	16	31	44
Your local government:								
Favorable	58	57	60	51	61	22	45	36
Mixed/neutral	21	22	12	21	9	56	11	6
Opposed	11	7	9	7	9	8	7	7
Don't know	10	15	19	22	20	14	37	52
Your neighbors and friends:								
Favorable	67	59	53	52	58	14	28	17
Mixed/neutral	17	21	21	20	10	58	17	17
Opposed	2	6	4	7	5	5	7	13
Don't know	13	14	22	22	27	23	48	53
People you work with:								
Favorable	70	67	56	53	57	13	27	13
Mixed/neutral	11	14	17	19	9	55	14	13
Opposed	4	4	3	6	4	8	4	9
Don't know	16	15	25	22	31	25	55	65
Your spouse/housemate:								
Favorable	87	82	68	69	72	21	40	29
Mixed/neutral	6	9	16	9	6	57	17	14
Opposed	1	2	3	7	5	13	12	18
Don't know	6	8	13	16	17	9	31	39

*Chi-square test for each item, d.f. = 21, $p \leq .0000$.

concerns are clearly financial. Types C and E are the most concerned groups among potential adopters. Type Cs have the same concerns as Bs, as well as concern about system obsolescence and prohibitive local codes. Type Es are additionally concerned about climate feasibility, lower resale value of the home, and purchase problems; they are the group most concerned about possible problems with the utility company. Type Fs have the same worries about solar energy as Type Bs, although with slightly less frequency. Types G and H tend not to know what potential barriers are important; Type G is relatively low in concern, while Type H exhibits relatively high levels of concern across the board.

TABLE 7.20. Concern about Risk by Market Segment

Defined as important/ very important*	A	B	C	D	E	F	G	H
	Types							
Warranty coverage	82%	93	97	95	96	83	60	83
Initial cost	80	90	95	92	93	84	68	83
Operating reliability	78	91	97	94	97	85	57	81
Reputability of solar firms	80	95	94	91	91	84	58	77
Damage to system	66	81	92	82	92	78	54	16
Safety concerns	62	80	94	77	94	80	51	80
Insurance coverage	59	79	91	85	91	74	45	77
Maintenance costs	58	72	84	82	87	70	51	74
System obsolescence	52	67	78	75	78	67	37	63
Difficulty in obtaining loan	57	73	79	59	77	62	40	69
Climate concerns	33	54	71	63	76	58	40	67
Prohibitive local codes	51	61	79	60	76	54	28	56
Lower home resale value	23	43	68	50	74	51	22	66
Purchase problems	24	35	51	39	58	46	24	53
Aesthetics	24	40	52	37	54	39	24	51
Problems with utility company	20	27	38	29	47	34	16	47
Lifestyle effects	7	15	23	15	26	10	20	30
Social effects	3	9	9	5	12	5	4	11

*Chi-square test for each item, d.f. = 7, $p \leq .0000$.

As perceived difficulty of retrofit increases, likelihood of solar adoption decreases. Table 7.21 shows the cross-tabulation by market segment. Most likely to assess a retrofit project as easy are Type As; Type Bs are less sanguine about solar retrofits. Perceived difficulty is higher for Types C and D. Types F through H tend to perceive retrofit as difficult or impossible, or not to know.

TABLE 7.21. Perceived Difficulty of Retrofit by Market Segment

Response*	A	B	C	D	E	F	G	H
	Types							
Could not use	5%	2	5	5	9	12	18	19
Major renovation	29	37	48	48	39	38	40	32
Don't know	6	4	12	9	20	19	23	39
Some difficulty	19	30	24	22	17	18	11	7
Easily	40	27	11	16	16	13	8	4

*Chi-square test, d.f. = 28, $p \leq .0000$.

Motivation

The Relative Advantage Scale was used to derive the market segments. Each of the 15 relative advantage items was cross-tabulated by market segment, with significant results, shown in Table 7.22. Most important to Type A likely solar adopters are saving money over the long term, protecting against rising costs, conserving natural resources, and protecting the environment. To Type As, the long-term benefits of solar adoption seem most important. Type B consumers have the same motivations as Type A and, in addition, want to reduce utility bills now and have a more reliable energy supply. Type Bs are looking for both immediate and long-term benefits. Types C and E, mainstream families, have the same pattern of motivation as Type B consumers, yet they feel that additional benefits of solar adoption, such as increasing the home's comfort and overall self-reliance, are important. These two types rank highest in motivation to adopt solar energy, especially Type C. Type E is the most interested group in innovation and experimentation and in increasing status and prestige through solar adoption. Types F and H are interested mostly in short- and long-term economic benefit, while Type G is not strongly motivated by any of the potential advantages of solar adoption.

Current utility costs, if high, might be an important motive for consumers to consider solar adoption. Average winter and summer

TABLE 7.22. Motivation by Market Segment

Define as important/ very important *	Types							
	A	B	C	D	E	F	G	H
Saving money, long term	86	92	97	89	94	83	52	81
Reducing utility bills now	78	90	96	83	92	83	56	83
Protecting against rising costs	86	95	96	87	93	79	50	76
More reliable energy supply	84	90	94	87	93	71	43	72
Conserving natural resources	85	91	93	81	91	69	43	67
Increasing home's comfort	68	80	89	68	89	63	34	75
Easing energy shortage	73	88	84	74	63	68	37	61
Increasing self-reliance	80	89	87	73	84	55	26	59
Independence from utility	68	83	87	65	81	56	29	60
Income tax credits	64	74	87	67	83	54	26	58
Fewer large power plants	64	78	82	64	82	58	29	57
Independence from federal government	66	69	81	66	77	60	32	56
Increasing resale value of home	56	60	81	58	81	56	26	65
Chance to innovate	34	32	40	19	44	16	10	19
Social attention	12	11	19	6	24	11	7	30

*Chi-square test for each item, d.f. = 7, $p \leq .0000$.

monthly utility costs were analyzed by market segment, with no significant results. Utility costs are apparently not currently high enough to motivate strong interest in solar energy as an alternative.

Business and Policy Knowledge and Preferences

Analyzed by market segment were attitudes toward local and individual control of energy, assessment of solar energy as an investment, perceived system cost, preferred financing and financial incentives, knowledge of existing tax credits, and perceived local availability of solar products and services.

Preferred Locus of Energy Decision Control

Four survey items measured attitudes toward individual involvement in energy. These were cross-tabulated by market segment, with significant results. In each case, Type A consumers score high in preference for individual and local control of energy supply. Type B consumers also tend to score high, although not as high as Type A. Types G and H tend to score lower than the other groups.

Agreeing with the statement favoring individual energy conservation are 92 percent of Type A and 97 percent of Type B consumers, compared with 79 percent of Types G and H. Type H (50 percent) is most likely to agree that we should trust the federal government to solve our energy problems, while Types A (58 percent) and D (54 percent) tend to disagree. Similarly, Types G and H are most likely to approve giving energy companies a free hand (48 and 42 percent, respectively), while Types A and B tend to disagree (46 percent each). Finally, Types A through E tend to agree that energy needs should be met through local resources, while Types F, G, and H are lower in agreement (although three out of four still agree) and are more unsure.

In short, the hypothesis that preference for individual and local control of energy supply is associated with the likelihood of solar adoption is supported by the analysis.

Perceived Investment Potential

Not surprisingly, Type A and B consumers are significantly more likely than others (57 and 52 percent, respectively) to assess solar energy as a good investment. The plurality of Type C think that it is, with about a third unsure. Type D consumers, whom we have already seen to be concerned about feasibility, are most likely not to consider solar energy as a good investment (45 percent, with 30 percent unsure); Types F, G, and H also tend not to see solar energy as a good investment.

TABLE 7.23. Perceived System Cost by Market Segment

Item/response*	Types A	B	C	D	E	F	G	H
Solar domestic hot water system cost estimate								
Don't know	20%	22	37	27	38	50	59	69
Low estimate	37	35	36	34	35	25	27	17
About right	38	30	20	29	18	15	11	7
High estimate	6	13	7	10	9	9	3	6
Solar space heating system cost estimate								
Don't know	21	22	36	29	40	51	61	67
Low estimate	51	45	42	38	37	28	26	18
About right	12	9	9	13	9	6	5	6
High estimate	17	24	12	20	13	14	7	10

*Chi-square test for each item, d.f. = 21, $p \leq .0000$.

Type A consumers, at 38 percent, can best estimate solar domestic hot water system costs, as shown in Table 7.23, but 37 percent estimate low. Type B consumers are also not well informed about system costs, although 30 percent estimate in the correct range, as do Type D consumers. Type Ds are the most well informed about solar space heating costs, with 13 percent giving estimates in the correct range. Likely solar adopters tend to underestimate space heating system costs. Types F, G, and H simply don't know what space heating systems cost. Thus, while likely adopters are more informed than other consumers, they still tend to underestimate sys-costs. When they learn what systems actually do cost, some potential buyers may defer purchase or abandon their plans altogether.

Paying for Systems

The bulk of consumers would need to finance at least part of a system purchase. About one in five Type A consumers would pay cash for a system, 46 percent would finance the entire system cost, and one in three would use both a down payment and a loan. Of Type B consumers, 56 percent would want to finance the entire cost, 39 percent would use both a down payment and a loan, and 6 percent would pay cash. Almost four out of ten Type G and H consumers state that they are not interested in a system purchase, compared with 4 percent of Type A and none of Type B consumers.

The possible financing mechanisms listed included home improvement loan, mortgage extension, increased mortgage payment, paycheck deduction, increased monthly utility payments, utility leasing, rent from solar contractor, and a utility lien mechanism. All market segments prefer the home improvement loan over every other financing alternative listed, with 40 percent of Type A, 36 percent of Type B, 40 percent of Type C, and 33 percent of Type D consumers choosing this option. Other mechanisms are selected by fewer than 15 percent of each consumer group.

Types A and B are most likely to be willing to pay something more than their present monthly utility costs to purchase a system. Half of Type As and 46 percent of Type Bs indicate a willingness to to do so, as shown in Table 7.24. Conversely, Types G and H are least likely to consider paying more. The willingness to pay more probably relates to consumers' belief that solar energy is a good investment and also to perceived noneconomic benefits that solar use offers.

Preferences for financial incentives are cross-tabulated by market segment. Low-interest loans are most preferred by Types A, E, F, G, and H. Types B and D slightly prefer tax credits; Type Cs prefer the two forms of incentive about equally. Less popular among all groups are grants and rebates.

Knowledge of Tax Credits

Not surprisingly, Types A and B are most aware of the existing 40 percent federal tax credit for most solar purchases (76 and 73 percent, respectively). Of Type C consumers, 47 percent are aware, and of Type D consumers, 62 percent know of the credits. Least aware of the credits are Type H consumers, at 22 percent.

Nearly the same proportion of each consumer type lives in states with solar tax credits, ranging from 28 percent of Type F to

TABLE 7.24. Willingness to Pay More for Solar Energy by Market Segment

	Types							
Response*	A	B	C	D	E	F	G	H
Pay from $8 to over $85/month more	50%	46	40	31	36	16	16	8
Pay nothing more per month	42	46	51	56	48	51	48	52
Don't know	7	5	5	5	9	11	10	11
Don't want a system	2	3	4	8	7	22	27	30

*Chi-square test, d.f. = 21, $p \leq .0000$.

TABLE 7.25. Knowledge of Solar Tax Credits by Market Segment

Federal Tax Credit *	A	B	C	D	E	F	G	H
Aware of credit	76%	73	47	62	40	39	36	22
Unsure	17	21	36	27	40	41	42	55
Unaware of credit	7	6	17	11	20	20	22	24
State Tax Credits **								
Unaware, state has no credit	15	16	13	16	16	16	18	11
Unaware, state has credit	5	6	3	2	4	6	5	7
State has no income tax	13	16	10	12	8	11	7	7
Unsure, state has no credit	28	26	40	35	40	40	37	47
Unsure, state has credit	15	17	20	18	19	16	24	20
Aware of credit, state has no credit	9	6	7	7	6	6	5	4
Aware of credit, state has credit	13	14	8	11	7	6	5	4

*Chi-square test, d.f. = 14, $p \leq .0000$.
**Chi-square test, d.f. = 42, $p \leq .0000$.

37 percent of Type B. Most aware of existing state tax credits are Types A, B, and D, but only by small minorities. Data on knowledge of tax credits by market segment are shown in Table 7.25. While knowledge of federal tax credits is clearly associated with likelihood of adoption, the same cannot be said for knowledge of state tax credits. Probably too few consumers are yet aware of state solar tax credits for them to be strong incentives for solar adoption; however, once potential buyers are at the Decision-Intention stage, state tax credits could help "seal" the purchase.

Most aware of the local availability of solar products and services are Type B, then Type A, consumers; least aware are Type H consumers, shown by analysis of variance ($F = 70.9493$; $p \leq .0000$). Even so, fewer than half of likely solar adopters are aware of new solar homes, reliable solar distributors and installers, and small wind machines available to them locally. About half are aware of solar energy systems for purchase, architects who can do solar design, contractors to build or convert homes to solar energy, and materials and products for solar energy systems available in their own areas. The majority of all consumer types think wood stoves are locally available, although 89 percent of Type Bs think this, compared with 66 percent of Type Hs. Type D consumers tend to have less knowledge of local availability than do Types A and B but more than other groups. Since the geographic distribution of the types is general, it is fair to conclude that knowledge of the local availability of solar products and services, while limited, is significantly higher among likely solar adopters.

RELATIONSHIP OF VALUES AND LIFE-STYLE PREFERENCES TO LIKELIHOOD OF SOLAR ADOPTION

In Chapter 4, we posed hypotheses about the relationship of values and life-style preferences to solar adoption. These hypotheses are supported by analyses described here. One-way analyses of variance by market segment were performed on innovativeness, opinion leadership, materialist and postmaterialist values, energy conservation actions taken, and preference for voluntary simplicity as a life-style.

Table 7.26 presents the results for innovativeness and opinion leadership. In both cases, the next likely solar adopters rate high, while consumers unlikely to adopt rate low. The second and third tiers are distributed between the two extremes. Both innovativeness and opinion leadership are associated with likely solar adoption.

The values scales analyses are shown in Table 7.27. The consumers to whom materialist values are most important are those least likely to adopt solar energy. Conversely, those to whom materialist values are least important are most likely to adopt. Similarly, likely solar adopters rate postmaterialist values highly, while consumers unlikely to adopt solar energy tend not to hold them as important values. These results lend support to the idea that solar adoption is associated more strongly with postmaterialist than with materialist values.

We predicted that the more energy conservation actions taken, the more likely the homeowner would be to adopt solar energy. This idea is supported by analysis of the energy conservation index presented in Table 7.28. The greatest number of conservation actions have been taken by consumers in the "next tier," while the least conservation action has been taken by consumers unlikely to adopt solar energy. Conserving energy is associated with likelihood of solar adoption.

TABLE 7.26. Analyses of Variance: Innovativeness and Opinion Leadership

Innovativeness

Type	Rating*	$\bar{X}$
B	High	.9814
A	High	.9653
C	High	.9536
D	Avg.	.8490
E	Avg.	.6915
F	Avg.	.6547
H	Low	.4207
G	Low	.4205

Range: 0-5 $\bar{X}$: .7084 SD: .9880
F: 14.4257 $p \leq .0000$

Opinion Leadership

Type	Rating	$\bar{X}$
A	High	1.9604
B	High	1.9068
D	High	1.7143
C	High avg.	1.3418
E	Avg.	.9890
F	Avg.	.9865
G	Low avg.	.8826
H	Low	.5518

Range: 0-9 $\bar{X}$: 1.2032 SD: 1.5668
F: 29.2267 $p \leq .0000$

*For all analysis of variance tables a qualitative rating is shown. The rating was derived from graphically plotting the mean scores.

TABLE 7.27. Values Scales Analysis of Variance

Materialist Values

Type	Rating	$\bar{X}$
F	High	6.4081
G	High	6.4053
E	Avg.	6.3113
H	Avg.	6.3079
C	Avg.	6.2447
D	Avg.	6.0898
A	Low	5.6089
B	Low	5.2795

Range: 0-12 $\bar{X}$: 6.1468
SD: 2.0386
F: 8.2387 $p \leq .0000$

Post-Materialist Values

Type	Rating	$\bar{X}$
B	Very high	3.6335
A	High	3.2475
D	Avg.	2.8041
C	Avg.	2.7173
E	Avg.	2.5592
H	Avg.	2.5091
F	Low	2.3946
G	Low	2.3788

Range: 0-12 $\bar{X}$: 2.7118
SD: 1.9396
F: 10.6129 $p \leq .0000$

TABLE 7.28. Analysis of Variance: Conservation Actions Taken

Type	Rating	$\bar{X}$
B	High	3.5963
A	Avg.	3.4554
D	Avg.	3.3020
C	Avg.	3.2785
E	Avg.	2.9835
F	Avg.	2.9596
G	Low	2.6364
H	Low	2.6189

Range: 0-7 $\bar{X}$: 3.0455 SD: 1.2513
F: 21.3612 $p \leq .0000$

TABLE 7.29. Voluntary Simplicity Analysis of Variance

Type	Rating	$\bar{X}$
B	High	12.1118
A	High	12.0653
C	High avg.	10.7076
D	High avg.	10.4733
E	Avg.	9.5994
F	Avg.	8.9593
G	Low	8.0837
H	Low	7.2407

Range: 0-28 $\bar{X}$: 9.6317 SD: 4.9686
F: 33.4877 $p \leq .0000$

We also anticipated that likelihood of adoption would be higher among consumers preferring a "voluntary simplicity" life-style. The voluntary simplicity items were recorded to form an index, and an analysis of variance by market segment was performed (Table 7.29). As predicted, Types B and A—consumers nearest solar adoption—have the highest preference for voluntary simplicity, while Types G and H—those unlikely to adopt—have least preference for it. Clearly, preference for a voluntarily simple life-style is associated with likelihood of adopting solar energy.

In sum, the relationships with likelihood of adoption found in these analyses for innovativeness and opinion leadership bear out results in the existing literature on diffusion of innovations: that innovativeness and opinion leadership characterize earlier adopters of innovations. The results showing that preferences for postmaterialist values and voluntary simplicity life-styles are associated with earlier adoption are a new contribution to the diffusion literature.

ATTITUDES TOWARD UTILITY COMPANIES BY MARKET SEGMENT

Some of the consumers most likely to purchase residential solar energy systems are also highly favorable toward utilities. The Type B consumer is the most favorable toward utility involvement in solar energy. Type Bs consistently opt for utility involvement more frequently than any other group in all roles promoting solar use. For example, 79 percent choose utilities to educate consumers on solar energy, 61 percent to provide financial assistance, 76 percent to conduct audits, and 58 percent to inspect systems. The majority of this group does not view possible problems with the utility company as an important barrier to solar adoption.

Type B consumers are very likely to be in touch with utility companies for solar information (73 percent), and they are also the group most likely to use an audit for solar information. In fact, 91 percent are interested in receiving a home energy audit. Type B consumers are one of the most favorable groups toward utility financing mechanisms. Nevertheless, while Type B consumers want utilities involved, their ultimate goal may be to become more independent of them. The largest proportion of any group, 65 percent of Type B, identify increasing such independence as an important advantage of solar adoption.

Type B consumers are probably already contacting utility companies with requests for audits, energy conservation, and solar information. They are highly favorable to utility involvement in solar programs.

TABLE 7.30. Attitudes of Residential Solar Market Segments toward Utility Involvement*

Proportions selecting utilities to . . .	Types A	B	C	D	E	F	G	H
Provide financial incentives	14%	61	33	33	16	17	12	10
Inform consumers	40	79	50	54	30	30	25	19
Conduct home energy audits	41	76	50	61	39	34	28	18
Cover solar warranties	3	22	6	12	9	6	6	4
Mandate passive design or solar DHW in new homes	4	15	7	8	9	7	5	6
List solar contractors	18	51	33	38	20	20	14	13
Inspect systems	22	58	41	42	32	31	20	19

*High and low percentages are underlined. Chi-square tests for each item, d.f. = 7, $p \leq .0000$.

Type A consumers are different: they display some antiutility attitudes. They consistently choose utility companies for roles in promoting solar use less frequently than do several other groups, but not as infrequently as consumers who will not adopt solar energy. About 40 percent of Type A consumers would favor utility involvement in educating consumers and conducting home energy audits, while about one in five would approve of utilities inspecting systems. They are otherwise not favorable toward utility involvement. However, most of them do not define possible problems with the utility company as a significant barrier to solar adoption. About a third of them—far fewer than Type B—see increasing independence from their utility company as an important advantage of solar adoption.

Nevertheless, over half of Type A consumers indicate that they would probably contact a utility company for information on solar energy, while 80 percent are interested in receiving a home energy audit. However, they tend not to prefer utility financing arrangements.

Data on attitudes of the market segments toward utility involvement in solar promotion are presented in Table 7.30. To summarize the findings:

- Type B (likely solar adopters) are consistently at the top in choosing utility involvement—far above the other market segments.
- Type A (likely solar adopters) are low average across the board.
- Types C and D (potential solar adopters) are favorable, although not as much as Type B.
- Types E and F (may or may not adopt) are average in favorability and less favorable than Types C and D.
- Types G and H (will not adopt solar energy) are consistently low in favorability toward utility involvement—they are at the bottom across the board.

TABLE 7.31. Utility-Related Attitudes by Solar Market Segment*

Item	Types							
	A	B	C	D	E	F	G	H
Defining as important possible problems with the utility company (as a barrier to solar adoption).	<u>20%</u>	28	38	30	49	37	<u>17</u>	<u>59</u>
Defining as important increasing independence from the utility company (as an advantage of solar adoption).	<u>35</u>	<u>65</u>	57	48	61	55	<u>29</u>	58
Likelihood of contacting the utility company for information about solar energy.	52	<u>73</u>	69	62	60	45	41	<u>39</u>
Likelihood of using a home energy audit to get information about solar energy.	<u>59</u>	<u>74</u>	63	58	51	32	27	<u>23</u>
Interested in getting a home energy audit.	<u>80</u>	<u>91</u>	71	70	59	30	16	<u>14</u>

*High and low percentages are underlined. Chi-square test for each item, d.f. = 7, $p \leq .0000$.

Data on other utility-related attitudes are presented by solar market segments in Table 7.31. To summarize these findings:

- Consumers most likely to adopt solar energy (Types A and B) are least likely to define possible problems with the utility as an important barrier to solar adoption. Consumers least likely to adopt solar energy (Type H—the frightened group) tend to see this as an important problem.
- The Type B market segment is most likely to define increasing independence from the utility company as an important advantage of solar adoption; the Type A segment is about half as likely to do so.
- The Type B market segment is most likely to contact the utility company for information about solar energy. Consumers least likely to adopt solar energy are least likely to be in touch.
- Both Type A and Type B segments are more likely to request a home energy audit than are other groups. Those least likely to adopt solar energy are also least likely to request an audit.

In general, interest in utility involvement in solar energy tends to be highest among consumers in the immediate and intermediate solar markets. The social acceptability of utility solar programs is especially high among those most likely to use them.

VERIFICATION OF THE "NEXT TIERS" ANALYSIS

An attempt was made to verify the results of the "next tiers" analysis, using the solar homeowners data in combination with the homeowners data. The basic idea for the verification was that if consumers who already own solar energy systems formed clusters with the next likely solar adopters among homeowners, this would provide some empirical verification for the analytical scheme used to predict the solar markets. The approach used has certain limitations, and the results must be considered preliminary.

The most powerful predictors of the solar markets were likelihood of seeking information about solar energy and knowledge levels. Items measuring these variables were judged not relevant for solar homeowners and were not replicated in the solar owners survey. Therefore, in attempting an analysis using both data sets, we were limited to six variables with relatively weaker predictive power. These six were:

- Preference for industry and government involvement (combined)
- "Intellectual" image of solar users
- Perceived feasibility of solar technologies
- Perceived governmental and social support of solar use
- Concern about risk
- Perceived relative advantage

The centroid measurements characterizing the homeowner clusters were defined and used to search the combined data base, to capture respondents with scores similar to each centroid. The resulting eight groupings had characteristics similar to the original homeowner clusters, with some exceptions to be discussed below.

When "old" clusters were cross-tabulated by "new" clusters, homeowner types tended to cluster with the same types derived from the combined data. The explained variance in new cluster membership is 0.45, with a high level of statistical significance. About two-thirds of the solar homeowners cluster with Type A consumers, with characteristics basically similar to those describing Type A consumers. The combined Type As are more likely to be located in western states, have higher incomes, and have taken more energy conservation action than the Type A consumers among homeowners. These differences would be expected among solar homeowners.

About one in five solar homeowners clusters with Type E consumers, with characteristics quite different from Types Es. Demographically, the new Type E cluster moved in the direction that would be expected of solar homeowners in income (higher), education (higher), region (western), and primary heating fuel (electricity). However, the new Type E cluster is older and has taken relatively less energy conserving action than has the old Type E cluster; these changes are not in the predicted direction. Further, the new Type Es have scores on some predictor variables that are not generally associated with solar adoption. They have lower "intellectual" image scores, lower perceived feasibility scores, and lower perceived relative advantage scores than the homeowner Type E consumers. They also exhibit lower concern about the risk associated with solar adoption.

Possible explanations for this finding are:

- Some solar owners in Type E may have been disaffected by solar ownership.
- They may have adopted solar energy for a set of reasons not covered adequately in the present survey, although this is unlikely.
- They may have been solar owners longer than others, thus exhibiting different attitudes from currently relevant ones; for example, some may be farmers whose windmills are several decades old.
- Some may be do-it-yourself solar owners.
- Some other effect of the sampling procedure could account for the difference.

Type B consumers are not empirically validated by the analysis. One of the distinguishing characteristics of Type B consumers is their desire for strong institutional backing of solar promotion, both by industry and government. This institutional backing has been relatively weak to date, perhaps preventing most of this consumer type from already adopting solar energy. Also, Type B consumers are somewhat fashionable and trendy, while Type As seem more settled and established. It appears that solar adoption has so far been associated with a sense of practicality and levelheadedness.

Resources available for data analysis were exhausted by the completion of this preliminary analysis. Further verification of the solar adopter types could be accomplished with the existing data base but must await future research.

SUMMARY

The acceptance of new technologies, and their ultimate spread throughout the market, has been a topic of study for many years. This study, carried out in the diffusion of innovations tradition, has found that the diffusion of residential solar technologies is likely to follow a pattern similar to that of other technologies. We have successfully segmented the likely solar markets and described them in relevant detail. The next solar markets will be "early adopters" of solar innovations rather than "innovators"; they have the characteristics associated with early adopters of innovations.

Likelihood of residential solar adoption is associated with the following personal characteristics:

- Younger age
- Male gender
- Higher educational attainment
- Professionals, managers, businessmen in occupation
- Higher income
- Opinion leadership
- Innovativeness
- Prior energy conservation activity
- Younger families
- Electricity and fuel oil as a primary heating fuel
- Preference for postmaterialist values
- Preference for voluntary simplicity as a life-style

Likelihood of residential solar adoption is associated with the following attitudes, knowledge, and preferences:

- Perception that one's household has been somewhat seriously affected by the national energy situation
- Perceived feasibility of systems
- A largely favorable and noncontroversial social milieu, particularly among co-workers, friends, and spouses
- Seeking information about solar energy
- Relatively informed already
- Knowledge of federal solar tax credits
- Interest in long-term economic benefit and in conserving natural resources
- Preferences for individual and local control of energy decision making
- Average to very high preference for government and industry involvement in the promotion of solar use

The next likely solar adopters are younger, highly educated, with higher incomes than other consumers. They tend to be professionals, managers, and businessmen. They are more likely to be male than female. Political orientation does not appear to be important in distinguishing them from others. They are high in both opinion leadership and innovativeness. Since innovators are not particularly high in opinion leadership, this finding suggests that nationally the "next tiers" are "early adopters" whose use of residential solar technologies will influence the "early majority" to adopt solar energy. In certain locales where little or no solar activity has occurred, the next solar users are likely still to be the "innovators," who are not opinion leaders but will initiate the solar diffusion process in their own areas.

A new contribution to the diffusion literature is the findings on values and preferred life-style: likely solar adopters are highest in postmaterialist and lowest in materialist values of all homeowners. They are also conserving more energy and living a more voluntarily simple life-style. Life-style and values preferences distinguish these consumers from others.

With such widespread feasibility toward solar energy, why haven't more homeowners installed systems on their homes? The process of diffusion follows a well-known pattern, described in Chapter 1, and the "next tiers" analysis shows which homeowners are likely to adopt solar energy the soonest, which will adopt later, and which are unlikely to adopt at all. Since the process is "natural," we cannot expect to intervene in it to change, say, a Type D consumer into a Type B consumer. All we <u>can</u> do, with business and policy decisions, is to speed up the rate with which technologies are adopted. To attempt this, we need to know what is keeping consumers from adopting, especially the next likely markets. We can also ask what we can do to bring along later solar adopters somewhat more rapidly. The "next tiers" analysis pinpoints the areas where policy can make a difference.

For Type A consumers, perceived relative advantage is not yet sufficiently strong to prompt immediate action. These consumers are motivated primarily by long-term economic benefit and by such noneconomic interests as protecting the environment and conserving natural resources. They do not perceive serious impacts of energy problems on their households. In every other way we have measured, Type A consumers are ready to adopt solar energy. Their concerns revolve primarily around purchase problems and could be solved by utility marketing of systems, government-backed solar warranties, or solid industry backing of solar products and installation. Rising utility prices will increase the perceived relative advantage for this group. The availability of home improvement loans, especially low-interest loans, would also increase the relative advantage of solar adoption for Type A consumers. This would speed the rate at which they are likely to purchase systems.

Type B consumers still need more information about solar energy. They particularly want some immediate economic benefit, through reduced utility bills, from adopting solar energy. They have somewhat stronger motivation to adopt than Type A consumers, since they perceive more serious impacts of the energy situation on their households. However, they also have stronger concern about the risks of adoption than do Type As. Type B consumers, particularly, want involvement by government and industry in promotion of solar use. Consumer information programs, especially on costs, savings, operating reliability, and safety of solar energy systems, whether from industry or government, are needed to encourage Type B consumers. Type Bs would also be encouraged to act by low-interest loans. They would be particularly responsive to utility solar programs.

Both Type A and Type B consumers need to be "sold" on solar energy. The detailed presentation of specific products and their advantages would provide much of the information they need to make a final purchase decision.

Type C and D consumers need more information, especially on the readiness of solar products for application, their performance, and cost-effectiveness. Type C consumers also have high motivation but higher concern about risk. They need more information about tax credits. Type C consumers would be very interested in government programs to promote solar use. Both Types C and D would also be motivated by low-interest loans.

Type D consumers, besides perceiving solar energy not to be cost-effective, also have some concern about risks of adoption. They are not as likely to seek information as Type C consumers, but they are open to information. Type D consumers would be very interested in industry programs to promote solar use. Carefully designed consumer information programs would prepare Types C and D to adopt solar energy more rapidly than they otherwise will.

8

COMMENTARY ON IMPLICATIONS

INTRODUCTION

What one gains from the National Study depends on what one's interest is. This book is addressed to the many different audiences concerned about how solar energy can be transformed into a significant supplier of energy in the United States. The residential use of solar energy is one key to accomplishing this goal, and solar technologies are well suited to dispersed application. Business, industry, government, media, and analysts are interested in how we can speed up the use of solar energy in the nation's homes. This study has some of the answers to guide us toward this goal.

Over the last few years, a rather naive hope seems to have informed government policy on solar energy. The "Manhattan Project" mentality—the quick tech-fix—appears to have dominated solar policy. Little appreciation has existed for the behavioral elements of a major shift in energy supply sources from nonrenewable to renewable resources and from central to dispersed power sources. The federal government's programs are organized and funded technology by technology (wind, active solar heating and cooling, passive solar heating and cooling, photovoltaics, and so on). Consumers, however are concerned about heating, cooling, hot water, and electricity. The interface of technology with users has been largely ignored: government tends to view this as private industry's domain, and it is difficult for industry to afford the costs involved.

Nevertheless, millions of dollars in federal funding have been spent on efforts to "commercialize" solar energy, with the expectation that if enough money were thrown at the problem, things would begin to happen "overnight," that is, in one to two years' time. This was done in the absence of theory and any systematic empirical understanding of the potential solar markets. The "energy crisis" perspective demanded immediate response in the marketplace, a demand that was misguided from the beginning. The few researchers looking at the question determined whether or not consumers "believed in the energy crisis" instead of determining what they assessed the energy situation to be, so that policy decisions and business thrusts could be accurately formulated and targeted.

Very few in either business or government had the perspective that conceptual and behavioral change (as any behavioral scientist or clinician will attest) requires time and energy to achieve. To accomplish society-wide change in energy supply is a massive goal, one which could easily be expected to take a relatively long period—say, 10 to 15 years with a carefully planned, well-coordinated effort involving both industry and government.

In addition, the federally funded solar establishment was structured to be at odds with itself rather than to work through intramural cooperation. The national center for research in solar energy, SERI, was administratively separate from the regional solar energy centers; the national and regional centers competed with each other for funding, as well as with the national laboratories. Funds for solar consumer information went both to the Franklin Institute, where the function had been housed, and to SERI, which had a congressional mandate to provide public information on solar energy. In effect, a great deal of taxpayer resources went into domain struggles rather than into accomplishing the work at hand.

While this was going on, business was not pleased with federal solar decision making, which was not perceived as helping the solar industry. Even the solar tax credits were maligned, since they had been promised and then delayed a year: solar businessmen said that they had lost sales or had gone out of business because customers were waiting for the tax credits to be passed by Congress. Business wanted results immediately and judged long-range planning and strategies virtually useless. Business has generally not been supportive of the federal programs that were supposed to be designed to increase marketplace acceptance of the products they were trying to move.

In addition, utility companies were, in many cases, reluctantly forced into dealing with energy conservation and renewables by federal laws requiring them to provide home energy audits. Home builders have successfully fought a federal law requiring them to build more energy-efficient homes.

Sooner or later, most thoughtful people will agree, we must shift to renewable energy sources; this isn't arguable any longer. An intelligent approach to how that shift will be accomplished is needed. Any successful approach has to take into account the fact that millions of consumers will make decisions and take action that furthers or delays the transition. Such a process is naturally going to take not only time but also careful business and policy decision making.

To plan intelligently, we must know the audiences to whom our plans are addressed. How do they define the situation? Do they even perceive a problem? What interests or motivates them about solar technologies? What, other than habit, repels them about solar use? Until we can accurately come to grips with these "social facts," we will probably continue to spend on misdirected programs and domain struggles the funding needed for more constructive aims.

In this commentary, we discuss some implications of the National Study's findings. The commentary is speculative, interpreting findings in the broader context of the marketplace, and does not represent survey findings directly. Some ideas pertinent to each of the potentially interested audiences—solar companies, architects and homebuilders; utility companies; federal, state, and local governments—are presented. Now that we know who is likely to adopt solar energy, the questions become who is going to reach them, and how?

WHAT IS RELATED TO LIKELIHOOD OF ADOPTION?

The guiding ideas of the study, by and large, stood the empirical tests, with few exceptions. Implications from the study's findings on likelihood of solar adoption are discussed in this section.

Definition of the Energy Situation

We found that the anticipated future energy situation did not increase likelihood of adoption. Solar owners are not only optimistic about the nation's future energy situation, but also tend to report no serious effects of the prior energy situation on their households. To speculate, they may feel somewhat protected from vagaries in energy supply by virtue of their solar ownership. Most homeowners are also optimistic about the nation's energy outlook, and those who are pessimistic are least likely to adopt solar energy. Further, most report no significant impacts of the national energy situation on their households; however, perceiving these impacts is associated with the likelihood of solar adoption, at least for some consumers.

Other studies have shown that residents of the United States view inflation as the highest priority national problem. National Study findings suggest that most consumers do not attribute inflation's impact on their household budgets to energy price increases. Economists may disagree over the role of energy prices in driving inflation, but it is clear that, if consumers are convinced such a connection exists, the impetus for solar adoption is increased. Consumer education on the relationship between energy prices and inflation, and on how the average family's budget is being affected, would be helpful in developing a frame of mind receptive to solar energy as a solution to household energy-related problems.

These findings suggest that fear is not a strong motivation for solar adoption. Perceived impact of the energy situation on the household seems to become defined as a problem to be solved, which serves to increase the salience of energy conservation and solar energy. What seems to motivate consumers are positive, rather than negative, factors. To solar owners, long-term economic benefit, advantages to society, and psychological satisfaction were important considerations in their adoption-decision process.

Social and Governmental Support

Most consumers perceive a favorable sociopolitical environment for solar energy. Likelihood of solar adoption is associated with a favorable milieu, particularly among co-workers, friends, neighbors, and spouses—those closest to the family. Those unlikely to adopt are unsure of governmental and social positions toward solar energy, or perceive a mixed or neutral environment. The relatively noncontroversial nature of solar energy, to date, has probably contributed to perception of a favorable sociopolitical environment. If solar energy were to become more controversial politically, then some consumers might shy away from it. Solar owners help to create the favorable social milieu for more residential solar use through their satisfaction with the experience, communicated enthusiastically to others.

Knowledge

We reported that a knowledge "threshold" has been posited, below which little adoption of innovations is likely to occur. That threshold has been defined as about 20 percent of the population.* Study findings show that, while 7 percent of homeowners term themselves "very well informed" on solar energy, 28 percent say that they presently know enough to make a decision about using solar energy. Which measurement might we use to assess how close knowledge of solar technologies is to the "threshold value"? The most likely answer is that we are somewhere between the two measurements, at approximately 14 to 17 percent. To increase their understanding sufficiently

---

*See discussion in Chapter 1.

to permit evaluative processes to occur, consumers will need more exposure to solar energy. Most have not noticed a solar home or building in their own neighborhoods. The increasing presence of solar installations will have a "snowball" effect, as those nearby begin to observe solar buildings and learn about them from their owners.

While consumers in general are not yet well informed, the study's findings suggest that we are fairly near an "inflection point" on the diffusion curve—a point at which the opinion leaders among early solar adopters begin to make their influence felt, and the early majority start to think more seriously about solar technologies. At this point, the adoption curve begins to slope upward more steeply. How near in time that actually is depends upon the decisions made by business and government in the years just ahead. The consumer market, however, is there.

Information Needs

The study defines specific kinds of information that consumers need; information programs and products should be designed to deal with these consumer information needs. If only a few consumers mentioned a particular information requirement, this does not mean it is unimportant. Those more informed are more likely to bring up questions that others haven't yet developed.

Realistic solar cost information is needed by consumers. Even those who feel they are informed, while they do know more than others, tend to underestimate system cost. Information on where to purchase systems and system components and solar homes is not widely available; solar companies and builders need to advertise more than they have. Utility company or state-maintained lists of solar contractors meeting standard business practices would also help in this regard.

Consumers need to know that solar energy works—that it actually produces heat and hot water, for example. "Hands-on" demonstrations of systems where consumers can see them and feel the heat and hot water they produce are important in conveying solar energy's practicality. Model solar homes usually attract thousands of visitors, and solar tours are popular. Exhibits at home shows, fairs, and shopping centers attract crowds. Workable, attractive systems and homes that consumers can visit are an important part of the education process and contribute to credibility, interest, and persuasion. These demonstrations should not be "freak shows" but should convey a sense of competence, realism, and practicality.

Information Sources

Present solar owners found books, journals and reports, and solar homeowners to be the best information sources when they were deciding on solar use. Other valued sources were energy fairs, exhibits and home shows, and newspapers and magazines. The importance of printed material, "hands-on" knowledge, and persuasion by those familiar with systems is made clear through these choices. Among organizations, solar owners relied heavily on solar companies, state energy offices, and federal agencies. Both business and government sources are apparently needed by consumers considering residential solar use, perhaps to achieve balance of perspective. Solar owners relied heavily on solar companies for their information, but this represents a business burden for companies who not only have to sell, install, and service their products but also have to educate consumers, as part of the process.

The most important information source defined by homeowners—federal organizations—are not now likely to fulfill consumer information needs, owing to the administration policy of federal budget cuts targeted specifically at consumer education and information programs. If consumers are to have the information they need, other organizations will have to perform this role. Media are important, especially print media in the near term, with television and radio becoming more significant as the market matures. Some public service broadcasting on energy conservation and solar energy is serving to increase public awareness on where to go for solar information. Utility companies can also be a crucial information source, as important in dissuading as in persuading consumers that solar technologies are worth considering.

Local expertise is crucial in persuasion. Home builders; architects; and heating, plumbing, and air conditioning contractors can be informed persuaders (and potent dissuaders if they are unfamiliar with solar technologies). Solar homeowners themselves, largely satisfied with their solar experiences, are important word-of-mouth persuaders. Local solar voluntary organizations have filled the role of citizen educators and represent an important stock of solar expertise and experience. Home improvement centers, lumber yards, and the new energy conservation and energy auditing firms can also be key channels for solar information and persuasion when their personnel are solar informed.

Educational settings are important. Local community colleges, agricultural extension services, and utility companies are presenting workshops, seminars, and courses on energy conservation and solar energy, especially with how-to emphasis. These will continue to be useful to a certain proportion of the market.

Advocacy organizations, such as the Solar Lobby, receive many requests for solar information. They may best provide information on the national solar budgets, the politics of solar energy, and the national, community, business, and individual advantages of solar use. More information is needed from credible sources that test and rate consumer products on the market.

Perceived Risk

Consumers will tolerate higher risks when they are voluntary than when they are imposed involuntarily. Adopting solar energy is still widely viewed as risky. Therefore, mandating solar use or programs, especially without persuading users of the absence of risk, would not be an effective method of encouraging adoption in its fullest sense, that is, satisfaction with the purchase decision. Although a majority of homeowners surveyed implied endorsement of mandating solar use in new home construction, more were opposed to mandating than to any other kind of policy to promote solar use. At the present time, mandating runs the risk of creating a negative backlash against solar use by forcing organizational and individual acceptance of risks they may not view as tolerable. The response to mandating, in this context, tends to be a demand for a "sure thing."

Perceived risk can better be dealt with by sound information programs and strong organizational backing for marketed solar energy systems. Solid warranties and reliable maintenance and repair services will alleviate some of the most serious consumer concerns. Suitable financing mechanisms will also reduce concern about initial cost.

Motivation

Important motivations for solar adoption are not strictly economic. Clearly, financial motives are important in consumer purchases; yet, a wide array of products that have little or nothing to do with financial motives are successfully marketed, for example, cosmetics, pet food, and jewelry. For consumers with large discretionary incomes, economics may not be a prime motivator at all. Among solar owners, initial cost was not the important barrier that it is for others. Three major types of motives to adopt solar energy were found in this study, and there are probably four. In addition to seeing solar adoption as an economic investment for the future, solar owners and likely solar adopters are interested in protecting the natural environment and conserving resources through solar use. Consumers are also interested in the personal satisfaction they derive from solar adoption. Solar owners report feelings of happiness, self-reliance and independence, being more related to nature, educating their children in natural processes and how man interacts with them, satisfaction that they are doing the "right thing," and patriotism. Some enjoy the attention they receive from others, even showing strangers through their homes. It is often forgotten that people are motivated by altruistic as well as more "selfish" desires. Recognizing the existence of these altruistic motives is sometimes viewed as idealistic, yet to deny their existence is to be out of touch with reality. The reasons for personal satisfaction vary, but it is a fact that they are not all economic. The near-term markets want to know about costs and savings, and their other motives should not be ignored in any successful marketing program.

For most consumers, it is important to show that a solar purchase is a good investment. Tax credits are simply not understood; most consumers are unaware they exist. They should be much more widely advertised and explained. Increasing the number of deductions on one's paycheck to increase take-home pay is legitimate when a solar tax credit is expected at the end of the year. The additional monthly take-home pay this yields can be used to retire a loan that was used to finance a solar energy system. More consumers could be motivated to adopt solar energy if they knew that the tax credits exist and how they work.

Low-interest loans are the preferred form of financial incentives, and home improvement loans the preferred financing mechanism to pay for systems. Since tax credits aid only those paying fairly substantial income taxes to obtain systems, it is easy to understand why low-interest loans would be favored more widely than the credits. The Solar Bank, earlier proposed, delayed, and likely to be defeated by the Reagon administration, would provide the kind of financing assistance to residential and commercial customers needed to trigger the immediate market into action. This is interesting because, while the tax credit amounts to a direct grant that offsets system cost, the loan does not offset anything but part of the interest cost. The implication is that customers might be willing to pay full system cost if there were a way to do so. Banks and savings and loan associations in areas of high solar activity have developed solar loan programs; however, these are not yet widely available nationally. The Solar Bank would speed up the rate at which loans to purchase systems became available to interested consumers nationwide.

THE BUSINESS OPPORTUNITY

The next likely solar market is about 18 percent of owner-occupied households. Many of them could adopt solar energy over the next two to three years. They are primarily interested in solar domestic hot water systems. Not all of these potential adopters will actually adopt solar energy, yet they represent a sizable market for any solar business venture.

The likely solar market needs to be "sold" on solar energy. Many of these consumers, although having some knowledge and all the "right" attitudes, have still not seriously considered a solar purchase. They are expected to respond most favorably to a soft-sell, low-pressure approach. They are readers, so printed material is important; references for further reading will enhance the literature provided to them. Some magazine or newspaper advertising would be helpful. Once the market has matured, television advertising could be more appropriate.

Fuel oil and electricity customers are more likely to adopt solar energy, and, as natural gas prices increase markedly owing to deregulation, natural gas users will become more likely solar adopters in the future.

Consumers conserving energy are more likely to adopt solar energy. Solar adoption itself does not imply a cessation of energy-conserving activities. In fact, energy conservation appears to mesh well with preference for voluntary simplicit—doing more with less—as an interesting challenge to many homeowners. Successful solar marketing programs will appeal not only to the near- and long-term economic benefits of solar use but also to the postmaterialist values held by the likely solar markets. Old-fashioned ideas, expressed in maxims like "a stitch in time saves nine" and "a penny saved is a penny earned," partially reflect voluntary simplicity as a preferred life-style. While these sayings have become clichés, new language expressing their basic ideas is needed. The notion of individual control over one's own energy situation expresses this preference for increased self-reliance. The wise use of resources is valued, whether these are national or household resources, natural or financial resources.

While there is greater favorability to solar energy in the western United States, the solar markets are located throughout the country. Since the South is doing the most to conserve cooling energy and exhibits relatively less interest in solar energy generally, it appears that the South equates solar technologies with heating. Active solar cooling technologies are still being developed for commercial readiness. The logical remaining immediate solar application in the South would be solar domestic hot water. Certain passive solar cooling retrofits, such as reflective glazing and use of foliage for shading, might also be of interest in southern climates.

Elsewhere in the nation, where heating is of importance, solar applications likely to be adopted first are wood stoves, active and passive solar domestic hot water systems, and certain passive retrofits, such as increased glazing on southern exposures and thermosyphon air panels. In new home construction and in home additions, effective passive features can be incorporated immediately. These include orienting the home or addition properly, use of glazing for solar gain and of thermal mass to store and release heat and cool, overhangs and foliage for shading, and a variety of window insulation treatments.

Where wind regimes are suitable, likely adopters of small wind energy conversion systems are consumers who already own one or more solar energy systems. These owners are likely to heat their household's water using the sun and may have some form of solar space heating. The next solar application of interest to them would provide electricity. Photovoltaic manufacturing processes are being refined and cell efficiency is improving, with the result that competitively priced photovoltaic systems are not many years away. Residential photovoltaics will likely follow the present solar markets. The same kinds of consumers will form the early markets for residential photovoltaic systems, with a greater likelihood of adoption among consumers who already own a solar feature or home.

BUSINESS POLICY IMPLICATIONS

The business problem appears to be that the smaller innovative firms producing and installing solar energy systems frequently have neither the capitalization nor the marketing capability to provide products they can guarantee and service to the satisfaction of the "early majority." Further, while innovative systems may be quite effective from a technical standpoint, without testing and certification, most consumers have few ways of knowing which are the best systems or what the trade-offs are between more and less expensive systems. Appropriate institutional roles are still being developed in the solar field.

Solar Companies, Architects, and Home Builders

Findings support the notion that, by and large, solar companies have been doing a reputable job in producing, installing, and marketing systems. The industry is small but has been active in working on standards, testing, and certification of its equipment. While there have been some problems, the industry is clearly making an effort to improve its products and services. Some suggestions to solar companies, architects, and home builders on marketing solar products and homes are:

- Know your market. Findings from this study can help you define what the market is likely to be in your local area. Realize that there are different types of adopters and devise different marketing strategies for each of them.
- Seek present solar owners as potential purchasers of additional systems.
- Be certain that your company is listed with the state energy office, the local government, the Better Business Bureau, and utility companies maintaining lists of solar contractors.
- Advertise in the yellow pages, magazines, and newspapers in your area. Remember that word-of-mouth advertising by satisfied customers is important during the early stages of solar adoption.
- Demonstrate your equipment whenever possible at shopping centers, trade fairs, exhibits, local community events, and home shows. Have flyers that describe your products and services and distribute them widely.
- Emphasize the quality of your products.
- Understand the risks that consumers perceive and be prepared to answer these concerns.
- Use satisfied customers as references; opinion leaders would be particularly important customers in this regard.
- Provide adequate guarantees and warranties as well as good repair and maintenance services. Stand behind your products.
- Be certain that claims made for system performance are realistic and accurate.
- Wind machine manufacturers and distributors need to examine problems that consumers might be having with their products and correct them if necessary; further, they should obtain reliable data on wind regimes in the nation to be certain that wind systems are marketed where they are most effective. Wind companies could loan anemometers to dealers and potential or actual customers to measure the wind resource in local areas.

Solar companies have been doing a good job, as evidenced by satisfied customers; there is still room for improvement of system design, quality, and installation, in many cases.

Recent studies have shown that energy efficiency is becoming a much more significant selling point for new home buyers. A new house, properly oriented on the lot, incorporating some passive solar design features, and including an active or passive solar domestic hot water system is a highly energy-efficient building. Home builders may be interested in participating in workshops and other programs that show them how to incorporate solar energy into their standard building designs. The Denver Metro Homebuilders Program was highly successful in working with home builders in the Denver area to build solar homes during 1981. Nearly 100,000 consumers visited these homes over a 16-day period, resulting in $2.5 million in sales. Another $4.5 million in serious business interest was generated.

Home improvement centers can also benefit by becoming more solar informed. They can market many of the supplies that do-it-yourself consumers need to build solar projects in their homes. Customer materials on the kinds of projects that can be built—such as sunspaces—can help increase consumer interest and sales.

Utility Companies

Many consumers would welcome utility involvement in the solar field. Utilities have a history of service to customers. Older consumers will remember when utility companies marketed and serviced home appliances. Utilities are generally regarded as having technical expertise on energy use in buildings, and they have data bases useful for solar programs. They could provide the larger organizational backing consumers apparently desire.

Alternatives for utility involvement in solar energy range from minimum to maximum. A minimum level, with primarily public relations advantages, would involve solar information and education programs, in addition to home energy audits. A maximum level, with both business and public relations payoffs, would involve marketing of equipment, inspection of systems, financing arrangements with repayment on utility bills, and maintenance service. A few utility companies around the nation are already involved at these various levels. A utility company may wish to have a contingency plan for a solar program to be implemented in the future when business conditions warrant it. A utility company considering a solar program should:

- Assess the market within and near its local service area in light of this study's findings.
- Assess local solar businesses and manufactured solar products.
- Assess the financial feasibility of the program.
- Plan the program over the long term, keep records, and evaluate its results. This makes for informed management decision making and answers questions about participation rates, solar market

penetration, customer acceptance and satisfaction, and amounts of energy conserved. The responsive provision of this information can be impressive to a regulatory body.
- Design information and advertising campaigns for the program.

What options can a utility consider? Six distinct possibilities are discussed here: information programs, financing, leasing, selling, system inspections, and maintenance.

Information Programs

Helping consumers become more informed about solar energy prepares them for a future purchase decision; an information program can be viewed as an investment in a future market. Packaging information attractively is important. Some ideas that have proved successful are:

- Bill stuffers to let consumers know what information is available
- Various brochures and booklets on residential solar applications
- Seminars on energy conservation and solar energy
- How-to workshops for do-it-yourselfers
- Newspaper articles covering utility company activities in solar energy
- Items on the company's solar activities in the monthly newsletter
- Information in the annual report

Financing

By providing financing for purchased systems, a utility company can reduce the first-cost barrier and make sure customers understand how to use the existing tax credits. Repayment through the utility bill is advantageous to the customer since the cost of the energy saved helps to pay for the system in a direct way.

Leasing Systems

In general, consumers prefer to own their own systems. Yet, owing to perceived risks involved or a desire for convenience, some customers might be more interested in leasing a system for a set fee than in purchasing one outright. If a utility company were to offer a solar leasing service, the commercial market should also be assessed. For several commercial low-temperature applications—such as laundries, beauty shops, and car washes—a solar hot water system similar to those used for residences, only larger, could be advantageous.

One problem with leasing is that if a house sells and the new owner does not want the system, the utility company would have to remove it. The question of tax credits would have to be worked out for both individual homeowner income taxes and company investment tax credits.

Selling Systems

Systems must be designed and sized for specific households. Fortunately, home construction practices tend to be similar within local service areas, although they vary widely in different parts of the country. Life-style of the occupants can mean a difference of a factor of two in energy use. Part of a successful selling program, then, is in proper design and sizing of systems.

The solar equipment itself must be reliable. The elements of the system—collectors, controllers, and storage—must be designed and produced to meet the climate conditions where they will be installed. Solar manufacturers can provide warranties on their products. A utility company selling systems would want reliable equipment that it had evaluated before putting it on the market.

System installation has been one of the sources of trouble with early systems. To make the most of available solar radiation and thus function at peak efficiency, collectors must be installed at the proper angle, facing the right direction, in a place where sunshine isn't blocked by trees, other buildings, or the like. Technical competence is required to achieve successful system installation.

Creative working arrangements with solar firms are very important. The solar industry has been understandably reluctant, or even hostile, about the utility "giants" entering the solar field. Solar firms may fear that utility companies will tend to market the most expensive and least cost-effective systems out of concern for problems that could arise from systems in anything but the "Cadillac" class. They may feel that utilities run the risk of diminishing the market by making solar energy too expensive for most people. Having a range of equipment options would help solve this problem.

The many successful utility home insulation programs could serve as a model for working with solar companies. If a utility company works with its local solar infrastructure in creative ways, permitting them to benefit from the services the utilities can do best (such as marketing and financing), while the utility benefits from their technical solar knowledge, everyone, including the customers, can come out ahead. The experience and knowledge of local solar firms can be genuinely important to utility program success. Their interests and the utility company's should be the same—to get as many systems installed as possible, as a profitable enterprise. Creative, open-minded approaches are essential to form the workable business liaisons that would make such a program function successfully.

System Inspection

Once a system has been installed, the utility company might find it prudent to inspect the installation and check out system functioning before the system is "turned over" to the customer. Some solar owners thought they were saving conventional energy with solar energy systems that weren't even turned on. If any problems are discovered, they can be corrected before the customers accept the system. Program credibility, customer satisfaction, and system performance are all enhanced by system inspection. It saves future headaches for everyone. Inspectors could ascertain, as well, that customers know how to operate their systems properly.

Maintenance

Utility staff or subcontractors could provide regular maintenance programs for customers desiring them for an annual fee, or, service could be provided on call. Utility maintenance programs would serve to reduce customer anxiety should problems arise with the system. It is in the utility company's, as well as the customer's, best interest to have reliable, properly designed and installed systems to save everyone costs and difficulty. In addition, customers should be accurately informed about what to expect from their systems: <u>not</u> 100 percent of their hot water, for example. Degree day information could be routinely printed on utility bills along with CCF of natural gas or kWh of electricity used so that customers could calculate their own system performance if they were interested. This kind of feedback reinforces the purchase decision and contributes to customer satisfaction.

Utilities could automate a solar maintenance program. Once a system has been installed in a home, the natural gas or electricity use should decline. Computers could be programmed to alert maintenance staff if usage at that address reaches or surpasses normal—an indication that the system may not be functioning properly.

PUBLIC POLICY IMPLICATIONS

The theory of government intervention suggests that, when a policy is judged for the benefit of society as a whole, it is appropriate for government to enact it. Development and application of solar energy to supply part of the nation's energy needs has several advantages for the public good:

- It can reduce the balance of payments and national security problems arising from the importation of oil.

- It can help conserve reserves of fossil fuels so that they will last longer and be available for important uses.
- It can result in more efficient use of energy; when energy is produced and used on-site, long-distance transmission of power or fuels with attendant losses of efficiency is avoided.
- It can delay the necessity to construct large, capital-intensive power plants, the expense of which is passed on to consumers.
- It can create jobs, especially for blue-collar workers.
- It can heighten individual, community, and national self-sufficiency.
- It can contribute to environmental protection.

These advantages for the common good form the rationale for public policy in support of solar energy. Further, government aid to other power sources, such as oil depletion allowances and nuclear subsidies, reveals the recognition that abundant energy supplies contribute to the nation's economic well-being. Taxpayers and utility customers regularly support the development of energy supply in the United States—such development has not been strictly a free market enterprise.

Financial Incentives

- Government should advertise tax credits just as they advertised the No-Cost, Low-Cost Energy Conservation Program in New England. Evaluation of the advertising effort showed this program to be successful in increasing energy conservation behaviors.
- The federal solar tax credits should be kept in effect beyond 1985. Some solar technologies, like active solar cooling and photovoltaics, will not be widely available commercially before 1985. The entry of these new technologies to the market will be benefited by the tax credits. A way should also be found to extend tax credits for passive systems. Business would be wise to lobby for the maintenance and extension of the tax credits.
- The Solar Bank plan to provide low-interest loans to residential and commercial consumers should be implemented.
- Types G and H consumers will not adopt solar energy without a government program to install systems on low-income homes. Such a program would be a logical extension of the Low Income Weatherization Assistance Program currently in effect. Installing measures that reduce energy use are clearly a more effective use of tax money than helping low-income citizens pay for escalating utility costs.

Consumer Information

A widely avertised and well-run government solar information center for consumers would be useful in meeting their needs for sound information. State energy offices would be a favorable location for this function. State offices, however, should be backed by high quality information from federal and industry sources.

Solar products and services are inadequately advertised. Small solar businesses can't afford expensive advertising, so government involvement would probably help. Magazine and newspaper ads and articles, especially local feature stories, would help. Solar tours, exhibits, and demonstrations should be advertised. Brochures on solar energy systems and on where to get more information about them should be produced. Lists of solar contractors, architects, builders and suppliers, and lending institutions willing to make solar loans should be maintained and distributed. Government's role in consumer education is important since consumers need relatively unbiased information sources.

Other Roles

- Government should support testing, certification, and standards programs for solar equipment to reduce consumer concern, assist industry, and improve quality and performance of systems.
- Public statements by responsible political leaders and government officials on the importance of energy conservation and solar use to national well-being will provide needed leadership.
- In dealing with utility companies, home builders, architects, local building inspectors, and others interested in solar energy, government can support programs to demonstrate how solar technologies can be used properly and in a cost-effective manner. Information programs and workshops should be designed for these special audiences to aid them in their own design and marketing efforts.

State Government

States that have solar tax credits and other forms of financial incentives should advertise them more widely. State solar incentives are the least widely known of the solar incentives.

State energy offices are a very important consumer information source. Consumers and solar owners alike selected them as a solar information source they would be very likely to use. State energy office staffs should heighten their knowledge of solar energy applications for residential use, produce information products, provide hot lines for consumers, maintain lists of reputable solar contractors, and work with local solar energy voluntary associations in consumer education efforts.

States could have solar demonstrations through retrofit programs for state buildings. This would provide the kind of hands-on knowledge needed by consumers, while impressing taxpayers with wise use of state resources through reduced energy bills.

States also frequently have a role in consumer protection. Consumer protection office staffs need more training on solar-related business problems, warranty issues, system installation and maintenance, and solar standards and certification processes. A sound consumer protection effort will increase consumer confidence in making their solar purchases.

Local Governments

Consumers show strong preferences for local government involvement in promotion of solar use. Further, consumers are least likely to know the position of their local government, of all levels of government, toward solar energy. More local government officials are becoming concerned about losing from the local economy the money spent for energy. Clearly, local governments have reason to do more to promote residential solar use. Some suggestions for local governments are:

- Set up a local energy office. A citizen advisory panel could be appointed to serve with the office director. The first task of this entity would be to identify energy sources for the community and to define energy use characteristics. The second task would be a set of recommendations comprising a community energy plan for the next 20 to 25 years.
- Plan and implement a community information program that conveys the community's energy plan and informs consumers about their role in it. Maintain lists of local solar contractors, architects, builders, and suppliers, as well as other local information needed by consumers.
- A municipal utility can engage in solar programs, just as investor-owned utilities can. (The potential utility role was discussed in the section on utilities.)
- Local governments can demonstrate solar feasibility by installing solar features on municipal and county buildings.
- Local governments can offer incentives. Sales tax exemptions are one; another is solar design contests for architects, students,

home builders, and interested citizens. Public recognition of outstanding citizen contributions to energy conservation through awards is also an incentive to put local creativity to work.

- Local building inspectors must be trained in solar inspections to conduct adequate inspections of the many different types of solar installations.
- Solar access laws, guaranteeing the right of installed system owners to sunlight, can be passed as local ordinances.
- Energy efficiency can be used as a criterion in the granting of building permits for new home construction. Similarly, a municipal utility could lower tap fees for energy-efficient building.

Local governments, by encouraging solar use, can keep money in the local economy, provide jobs locally, and improve the local environment. Both state and local governments should become more active in promoting solar use, with a diminishing federal role in the near term.

Analysts

The study provides evidence that diffusion of solar energy is following a more horizontal than vertical diffusion pattern. The vertical pattern of innovation diffusion is exemplified by the agricultural extension service. Findings from research at land-grant universities have been promulgated by the agricultural extension system to farmers, through their county agents. Accuracy and quality of information following this communication channel has been controlled.

Horizontal diffusion is decentralized and follows a network pattern. Innovations developed in one community are transferred within that community or to another through informal interpersonal connections. Recipients of the new idea are likely to modify it to suit local conditions as they implement it in their own community. The modified idea is, in turn, borrowed (and changed) by others. While the horizontal diffusion pattern can be quite effective in rapidly spreading adoption of innovations, the quality of information transmitted through horizontal communication channels is not controlled.

Analysts puzzling over the problem of how solar energy's infrastructure should be structured should be aware of this important difference in the residential solar diffusion process.

Data from the National Study are not directly transferrable to local markets. Demographic characteristics of local market areas should be compared with those describing the likely solar markets to determine the size of the local market.

A more systematic study would involve an appropriate sampling of households, replicating key items from the homeowners interview schedule. The items to include can be determined by examining findings in Chapters, 3, 4, and 7 on the important correlates of solar adoption. Researchers wishing to perform more sophisticated analyses may find the methods used in this study of value in designing their own projects.

Future research needs include:

- Further analyses of the existing data bases to permit even more precise characterization of the solar markets
- Further analyses of the existing data base to address questions not raised in the work completed
- Longitudinal replication of key parts of the national homeowners survey to determine the rate of diffusion
- Probability sampling of solar owners when a sampling frame, such as U.S. Census data, permits
- Schematic sampling of knowledge, attitudes, and practices regarding solar energy among utility companies, home builders, architects, lending institutions, and other stakeholder groups in the solar diffusion process.

Clearly, the diffusion of solar energy has not yet been institutionalized. Policy and business analysts should include findings from this study in their work on how the use of solar energy can optimally be spread in the United States.

BIBLIOGRAPHY

Anderson, J. 1980. "Tax Incentives, 1979 and 1980 Sessions." Golden, Colo.: Solar Energy Research Institute. Draft Manuscript.

Arndt, J. 1967. "Role of Product Related Conversations in the Diffusion of a New Product." Journal of Marketing Research, 4, pp. 291-95.

Associated Press-NBC News Poll. 1979. Denver Post, May 6, 1978.

Berning, C. K., and J. Jacoby. 1974. "Patterns of Information Acquisition in New Product Purchases." Journal of Consumer Research, 1, pp. 18-22.

Brewer, Garry D., and Ronald D. Brunner. 1978. "Citizen Viewpoints on Energy: Nineteen Citizens Sample." Ann Arbor: University of Michigan, August, 1978.

Brunner, R. 1978. "Citizen Viewpoints on Energy: Richmond, Indiana." Institute of Public Policy Studies. Ann Arbor: University of Michigan, March, 1978.

Burns, Barbara. 1980. The Relevance of Behavioral and Social Models to the Study of Consumer Energy Decision Making and Behaviors. SERI/RR-722-341. Golden, Colo.: Solar Energy Research Institute, August, 1980.

Dunlap, Riley, and Kent D. Van Liere. 1978. "The New Environmental Paradigm: A Proposed Measuring Instrument, and Preliminary Results." Journal of Environmental Education, 9, pp. 10-19.

Elgin, D. 1981. Voluntary Simplicity. New York: William Morrow.

Farhar, B. C., and J. Mewes. 1976. Social Acceptance of Weather Modification: The Emergent South Dakota Controversy. Monograph 23. Boulder, Colo.: Institute of Behavioral Science, University of Colorado.

Farhar, B. C., G. Johnson, J. A. Clark, J. Mewes, R. Rinkle, and S. Shoulders. 1978. Technology and Society: Weather Modification in South Dakota. Boulder: Institute of Behavioral Science, University of Colorado.

Farhar, B. C., C. T. Unseld, R. Caputo, and J. Easterling. 1979a. Citizen Participation in the Domestic Policy Review of Solar Energy. SERI/TR-53-126D. Golden, Colo.: Solar Energy Research Institute, January, 1979.

Farhar, B. C., P. Weis, C. T. Unseld, and B. Burns. 1979b. Public Opinion About Energy: A Literature Review. SERI/TR-53-155. Golden, Colo.: Solar Energy Research Institute, June, 1979.

Farhar, B., R. Vories, C. T. Unseld, R. Crews, and F. F. Shoemaker. 1979c. National Study of the Residential Solar Consumer, Highlights Reports. Nos. 2 and 3. Golden, Colo.: Solar Energy Research Institute.

Farhar, B. C., C. T. Unseld, R. Vories, and R. Crews. 1980. "Public Opinion about Energy." Annual Review of Energy, 5, pp. 141-720.

Hamrin, J. G. 1978. Low Energy Consuming Communities: Implications for Public Policy. Davis: University of California.

Henderson, C. 1978. "The Frugality Phenomenon." Bulletin of the Atomic Scientists, 34, pp. 24-27.

Inglehart, R. 1977. The Silent Revolution. Princeton: Princeton University Press.

The Jordan Energy Line. 1979. 2(4), p. 1.

Kaplan, L. B., G. J. Szybillo, and J. Jacoby. 1974. "Components of Perceived Risk in Product Purchase: A Cross-Validation." Journal of Applied Psychology, 59, pp. 287-91.

Krane, S. 1977. "Social Implications of the National Hail Research Experiment." Ph.D. dissertation. Boulder: University of Colorado.

Leonard-Barton, D. 1978. The Diffusion and Adoption of Solar Equipment Among California Homeowners: Report on a Pretest Study. Stanford, Calif.: Institute for Communications Research, Stanford University.

Leonard Barton, D., and E. H. Rogers. 1979. "Adoption of Energy Conservation among California Homeowners." Stanford: Stanford University, May, 1979.

Levine, M. 1979. Memorandum. Berkeley, Calif.: Lawrence-Berkeley Laboratories.

Lovins, A. B. 1977. *Soft Energy Paths: Towards a Durable Peace*. Cambridge, Mass.: Ballinger.

Lutz, R. J., and P. J. Reilly. 1973. "An Exploration of Two Effects of Perceived Social and Performance Risk of Consumer Information Acquisitions." *Advances in Consumer Research*, 1.

Margolin, J. B., and M. R. Misch. 1979. "Decision Analysis Panel Meetings Report." Washington, D.C.: George Washington University, June, 1979.

Marshall, Eliot. 1979. "Public Attitudes to Technological Progress." *Science*, 20, July, pp. 281-85.

Maslow, Abraham H. 1970. *Toward a Psychology of Being*. 2nd ed. New York: D. Van Nostrand.

Mitchell, Arnold, and D. Elgin. 1976. *Voluntary Simplicity*. Business Intelligence Program Guidelines No. 1004. Stanford: Stanford Research Institute.

Perry, M., and B. C. Hamm. 1969. "Canonical Analysis of Relationships Between Socioeconomic Risk and Personal Influence in Purchase Decisions." *Journal of Marketing Research*, 33, pp. 351-54.

Ramsey, K., and P. Niland. 1979. *California's Experience with Solar Consumer Inquiries and Complaints*. Sacramento: California Energy Commission.

Reuyl, J. S., W. W. Herman, R. C. Carlson, M. D. Levine, and J. G. Witwer. 1977. *Solar Energy in America's Future*. 2nd ed. Menlo Park: Stanford Research Institute, March, 1977.

Roessner, D. J., D. Posner, F. Shoemaker, and A. Shama. 1979. *Application of Diffusion Research to Solar Energy Policy Issues*. Golden, Colo.: Solar Energy Research Institute.

Roessner, J. D., P. Weiss, R. Saltonstall, B. Green, J. Ashworth, and B. deKieffer. 1979. Turning Laws Into Incentives: The Implementation of State Solar Energy Initiatives (Research Plan). SERI/MA-51-106. Golden, Colo.: Solar Energy Research Institute, February, 1979.

Rogers, E. M., D. Leonard-Barton, E. Rosa, T. Avi-Itzhak, and R. Adhikarya. 1979. Solar Diffusion in California: A Pilot Study. Stanford: Institute for Communications Research, Stanford University, August 14, 1979.

Rogers, E. M., and F. F. Shoemaker. 1971. Communication of Innovations. 2nd ed. New York: The Free Press.

Rokeach, M. 1973. The Nature of Human Values. New York: The Free Press.

Roper Organization, Inc. 1979a. Roper Rep. No. 79-2, New York, January.

Roper Organization, Inc. 1979b. Roper Rep. No. 79-6, New York, June.

Roselius, T. 1971. "Consumer Rankings of Risk Reduction Methods." Journal of Marketing, 35, pp. 56-61.

Sawyer, Stephen W. 1979. "A Survey of Solar Consumers in Northeastern and Southwestern United States." College Park, Md.: Department of Geography, University of Maryland.

Scott, J. E. 1976. Consumer Demand Analysis: Solar Heating and Cooling of Buildings. Newark: University of Delaware, September, 1976.

Shama, A., and K. Jacobs. 1979. Social Values and Solar Energy Policy: The Policy Maker and the Advocate. SERI/RR-51-329. Golden, Colo.: Solar Energy Research Institute, October, 1979.

Slovic, Paul. 1977. "Perception and Acceptability of Risk from Nuclear and Alternative Energy Sources." Draft report to the Sociopolitical Resource Group of the Risk Impact Panel, March, 1977. Washington, D.C.: National Academy of Sciences/National Research Council.

Solar Energy Institute of North America. 1979. State of the Union Report. Washington, D.C.

State Solar Energy Incentives Primer: A Guide to Selection and Design. 1979. SERI/SP-434-470. Golden, Colo.: Solar Energy Research Institute, December, 1979.

U.S., Department of Energy. 1978. Status Report on the Solar Energy Domestic Policy Review. Washington, D.C., August 25, 1978.

U.S., Departments of Housing and Urban Development and Commerce. 1978. Annual Housing Survey. Part A, General Housing Characteristics. Washington, D.C.

Unseld, C. T., and R. Crews. 1979. Residential Solar Energy Users: A Review of Empirical Studies and Related Literature. SERI/TR-54-245. Golden, Colo.: Solar Energy Research Institute, December.

Van Liere, K. D., and R. E. Dunlap. 1979. "The Social Bases of Environmental Concern: A Review of Hypotheses, Explanations and Empirical Evidence." Southern Sociological Society Annual Meeting, Atlanta.

Ward, D. S. 1979. Executive Summary: Solar Heating and Cooling Systems Operational Results Conference. Colorado Springs, November 17-December 1.

Yarosh, M., et al. 1978. Solar Commercialization: The Consumer Experience. Cape Canaveral: Florida Solar Energy Center.

Zimmerman, J. D. 1980. The Residential Solar Consumer: Value, Lifestyles and Experiences. Draft report submitted to Solar Energy Research Institute. Tucson: University of Arizona.

APPENDIX A: LITERATURE REVIEWS

Four reviews of the literature were conducted. The first of these resulted in a major report, Public Opinion About Energy: A Literature Review (Farhar et al. 1979b). This analysis and its update (Farhar et al. 1980) revealed the lack of national data on solar energy, as compared to the wealth of data on other energy alternatives, such as nuclear energy and energy conservation. A summary of this review is presented below.

The second literature review underscored the limitations of existing data on solar owners. This study, Residential Solar Energy Users: A Review of Empirical Studies and Related Literature (Unseld and Crews 1979), illustrated some of the factors that appeared to affect the solar adoption decision and some problems in living with solar energy, but samples were so limited in scope that no generalizations could be made about solar homeowners. This review is also summarized below.

The third and fourth literature reviews were directed at existing theory relevant to the study. Theory is supposed to describe and explain the "real world." Using theory aids in knowing the important questions to ask. One of the problems identified in the first two reviews of the literature was that most studies were conducted without regard to theory. This lack prevented them from being highly useful for policy and marketing purposes, because data on theoretically relevant variables were frequently not collected and analyzed. The third literature review, published as The Relevance of Behavioral and Social Models to the Study of Consumer Energy Decision Making and Behaviors (Burns 1980), examined existing models of consumer decision making. The fourth literature review focused primarily on diffusion theory and related literature. Its results are contained in this report.

PUBLIC OPINION ABOUT SOLAR ENERGY

Solar energy has received little attention in surveys of the U.S. public; one or two items about it have been included in about 20 national surveys. Most of the information available about public opinion on solar energy is derived from marketing studies of special, localized samples and is not generalizable to the public-at-large.

Public attitudes toward solar energy can only be described as positive. Government incentives for solar energy were favored in several studies, indicating strong citizen support for tax incentives to promote solar use.

Existing data display considerable variability on how effective (that is, feasible and reliable) solar energy is considered to be. A 1979 national survey asked to what proportion of "our energy problems in the future" solar energy would be the answer (Roper 1979b); 3 percent said "all," 18 percent said "most," 57 percent said "some," and 16 percent said "very little." The balance of 7 percent did not know. Respondents were asked what things are "holding back the development of solar energy." Mentioned most frequently were: (1) inadequate funding for technological development (41 percent); (2) the problem of rebuilding most of the houses and other buildings now is existence (38 percent); and (3) the technology is not yet developed (32 percent). Climate was mentioned as a barrier to solar use in four studies (Farhar et al. 1980). Other concerns about the effectiveness of solar energy, mentioned in seven studies, included: (1) solar energy is still experimental; (2) it would not produce enough heat; and (3) storage problems (Farhar et al. 1980).

Most of the public has heard about using solar energy for heating and generating electricity (Farhar et al. 1980). A frequently mentioned explanation for not considering a solar energy system purchase is that people have insufficient information about solar energy on which to base a decision (Farhar et al. 1980).

Motivations for using solar energy can be both economic and noneconomic. A 1979 national poll asked: "In order to develop solar energy for the future, energy prices would have to rise. How much more per month would you be willing to pay to replace our current sources of energy with solar and other renewable sources?" (AP-NBC News Poll 1979). Over half indicated either that they would pay nothing more (24 percent) or were unsure (31 percent). The remaining 45 percent indicated that they would be willing to pay more per month; 19 percent from $1 to $10, 14 percent from $11 to $25, and 6 percent over $50 a month more. The reasons people might be willing to pay more for solar energy have been suggested in studies on its perceived advantages. These are: conserving resources, desire for a cleaner environment, availability of sunlight, reduced dependence on utility companies, and patriotism (Farhar et al. 1980). In addition, many see solar energy as an investment against rising energy costs in the future (Rogers et al. 1979). Several of these advantages were mentioned by respondents in field studies conducted by SERI and in limited surveys of those who presently use solar energy (Unseld and Crews 1979). In addition, solar energy is also perceived as having economic advantages. Mentioned in studies and field work have been reduction of utility bills, low cost, and low maintenance costs (Farhar et al. 1980).

The disadvantages of solar energy mentioned in several studies were: (1) initial cost too high; (2) too experimental; (3) perceived infeasibility of retrofit in present home; and (4) appearance (Farhar et al. 1980). The most important factors affecting a decision to purchase solar energy in a 1977 study of homeowners who had not done so were identified as initial price of the system and reduction of utility bills (Scott 1976). This study was conducted prior to the federal income tax credit for solar energy.

Solar ownership in the United States is quite limited. Roper (1979a) found that fewer than 0.5 percent of a national sample reported that they owned a solar heating system, with no measurable change since 1978. A 1977 study in four areas of the country estimated that 0.01 percent of current homes have solar energy systems. Another estimate, based on national figures for housing units, is 0.05 percent solar homes, or about 1 solar home for every 2,000 homes (Farhar et al. 1979b). It has been reported that 34,800 homeowners received solar tax credits from the Internal Revenue Service for 1977 and 1978 (The Jordan Energy Line 1979). SERI researchers have observed that solar homes tend to be geographically clustered in special solar developments or in communities where local leaders have had a decisive effect on solar adoption patterns, for example, in Davis, California (Farhar et al. 1979c). This clustering phenomenon may account, in part, for low percentages of solar users found in national surveys. The sampling for these surveys is generally done in clusters, and unless solar clusters are included, which is not highly probable, solar owners would tend to be somewhat underrepresented.

A few differences in attitudes toward solar energy were identified by sociodemographic characteristics. Greater favorability toward solar energy was found among younger age groups, higher-income levels, professional/managerial workers, and whites. No differences were evident by other sociodemographic characteristics. Data on regional differences suggest that those in the West may be more favorable toward solar energy than the rest of the nation and that those in the South may be less favorable (Farhar et al. 1979b).

In identifying research gaps on solar energy, there are insufficient national data on perceived effectiveness, relative advantage, knowledge and information sources, behavioral intention, and action. Almost no data exist on policy preferences (Farhar et al. 1980).

STUDIES OF SOLAR OWNERS

This review contains an analysis of the results of 15 empirical studies of solar adopters, all the studies known to exist as of 1979. Since the universe of solar adopters in the United States is not known, it is impossible to assess the generalizability of the results of these 15 studies to a larger population of the United States (Unseld and Crews 1979). Many of the studies varied in sophistication of experimental design and the reporting of results and conclusions. None of the studies used random sampling or control groups, and few had a firm theoretical basis on which to interpret research findings.

The results of these studies indicated that significant numbers of early solar adopters have experienced problems relating to the design, installation, and operation of their solar energy systems. However, these problems were not readily apparent to some solar owners. Fortunately, many of these problems could be resolved without major technological breakthroughs by training programs for those who design and install solar energy systems. Accelerated commercialization of solar technologies will require improvement in all elements of the technology delivery system, ranging from system design to installation, financing, and consumer education in the maintenance of solar energy systems.

Despite the problems experienced by these innovators, their overall experiences with solar energy were positive, as determined by self-reports. The majority of adopters reported few or no problems with their solar energy systems. In a 1978 survey by the House Subcommittee on Oversight and Investigations, 52 percent of the solar adopters reported that they were "very satisfied" with their systems, while 34 percent said they were "satisfied" and 6 percent indicated that were "unsatisfied" or "very unsatisfied" (Unseld and Crews 1979).

However, another study also suggested that owner reports of system performance or satisfaction may not correlate highly with actual performance, as determined by site inspection conducted by solar technicians. The Florida Solar Energy Center found, in a mail survey of 800 Florida solar adopters, that they were generally satisfied with their systems (Yarosh et al. 1978). Center personnel later made 60 site visits to inspect a sample of the systems reported in the earlier study. The inspection team categorized 27 percent of the systems as performing acceptably, while 73 percent were classified as unacceptable from an engineering standpoint. However, 68 percent of the owners of "unacceptable" systems characterized their experiences as satisfactory.

One explanation for the apparent incongruity is that systems had problems evident to technicians but not perceived by system owners. Such problems included poor roof penetrations, inadequate collector mountings, lack of ability to drain the system to protect from freezing, and corrosion. In two systems, a backup heater was providing all of the hot water used in the residences, but because there was a sufficient supply of hot water, the owners assumed the solar systems were functioning satisfactorily.

In some instances, solar adopters may have believed that their decisions to adopt were good and that their experiences with a solar energy system were satisfactory because to believe otherwise would lead to "cognitive dissonance." They had already invested in solar energy and may have wanted their decisions to be "right." Another explanation lies in the "unsatisfactory" classification used by the technicians. Many of the problems they discovered were minor and were satisfactorily repaired. Thus, for the system to have a problem, as defined by technicians, did not necessarily indicate that owners had an unsatisfactory experience, at least up to the time of the inspection. The Florida study underscores the need for sound solar installation practices.

Current users of solar energy systems are sometimes classified as the "wheat germ and granola set," indicating that they are perceived as different from the average members of society. Solar innovations, like many others, are value laden. In fact, consumer attitudes and values may be even more important in the diffusion of solar energy systems than in other innovations that have been investigated, such as CB radios, microwave ovens, and 35mm cameras (Leonard-Barton 1978).

In summary, in their review of existing studies of solar owners, none of which are generalizable to any larger population of solar owners, Unseld and Crews (1979) found that important motivations to adopt solar energy were economic considerations, concern for the environment, and the perceived need for energy conservation. Other decision factors were interest in solar technology, desire to be innovative, desire for self-reliance, and fears of possible future energy shortages. Most present solar owners (from 70 to 90 percent) report that they are satisfied with their experiences with their solar energy systems. Very high proportions (over 85 percent) indicate that they would adopt solar energy if they had it to do over again. The limited data available suggest that solar owners come from several walks of life; ownership does not appear to be limited to any one social class.

APPENDIX B:
HOMEOWNER INTERVIEW SCHEDULE

SURVEY CONDUCTED FOR SERI
HOMEOWNERS ON SOLAR ENERGY

The Gallup Organization, Inc.
53 Bank Street
Princeton, New Jersey

OMB # CS038S8009

XM-9-8202-1

Time started: _______

INTERVIEWER'S NAME: _______________

Time finished: _______

Length: ___________

Date: ___________
Spot Number: _______
GO 80138

SUGGESTED INTRODUCTION: Hello, I am working on a nation-wide survey for The Gallup Organization in Princeton, New Jersey. This survey is funded by the Solar Energy Research Institute with Department of Energy sponsorship under the Solar Energy Research, Development, and Demonstration Act. We would like your opinion on some topics of interest. Your household is one of the many households chosen to participate in this survey. Your participation is voluntary, but we hope you will help provide information that will be used to develop policies about the use of solar energy. Once this questionnaire is returned to our office, it is validated to make sure it is complete, then all names are destroyed. Your answers then are tabulated by computer along with those provided by the other 3,000 people who cooperated in this project.

The first question I'd like to ask is . . .

1. Do you own or rent this home?

   1[ ] Own
   2[ ] Rent (TERMINATE - THANK RESPONDENT AND LEAVE)
   3[ ] Other arrangement (PLEASE SPECIFY) __________

2. INTERVIEWER CHECK OFF:

1[ ] Attached house (duplex, townhouse, etc.)
2[ ] One-family house
3[ ] Mobile home
4[ ] Other (PLEASE SPECIFY) ______________________

IF THEY OWN:

3. Do you have control over decisions to change the external appearance of this house without going to a board or committee for approval?

1[ ] Yes
2[ ] No (TERMINATE - THANK RESPONDENT AND LEAVE)

4. (HAND CARD 1) Here is a picture of a ladder. Let's suppose the top of the ladder represents the best possible energy situation for our country; and the bottom, the worst possible energy situation for our country. Please show me on which step of the ladder you think the United States is at the present time? (CIRCLE ANSWER)

10 9 8 7 6 5 4 3 2 1 0 DK

5. (RETAIN CARD 1) Just as your best guess, if things go pretty much as you now expect, where do you think the U.S. will be on the ladder, let us say, about five years from now? (CIRCLE ANSWER)

10 9 8 7 6 5 4 3 2 1 0 DK

6. (HAND CARD 2) In different parts of the Nation, people have reported a variety of effects of the energy situation on their daily lives. How has the energy situation affected your family (household)? Which statement on this card best describes your household? Just call off the number. (CHECK ONE)

1[ ] It really has had no effect on us.
2[ ] We have had to make a few adjustments, but our life-style has not been affected.
3[ ] Our life has been less comfortable and convenient, but it is not serious.
4[ ] We have had to make serious changes in our daily habits.
5[ ] DK

7a. (HAND CARD 3) Here is a list of several energy sources available to us. Please rank the top three energy sources you would prefer to see developed to meet our future energy needs. (1 = most preferred, 2 = next most preferred, 3 = third most preferred.) Which do you prefer most? Just call off the letter. Which second? Which third? INTERVIEWER: RECORD NUMBER, DO NOT USE X.

A____Coal
B____Nuclear energy
C____Energy conservation (reducing energy needs by adding more home insulation, driving a fuel-efficient car, etc.)
D____Water power from dams or falls (hydropower)
E____Solar energy (includes solar, wind, wood)
F____Oil and natural gas
G____Synfuels (for example, industrial plant converts oil shale or coal to a liquid or gas)
H____Don't know

7b. (RETAIN CARD 3) Now, please indicate the source you least prefer. (CHECK ONE)

1[ ] A. Coal
2[ ] B. Nuclear energy
3[ ] C. Energy conservation
4[ ] D. Water power from dams or falls
5[ ] E. Solar energy
6[ ] F. Oil and natural gas
7[ ] G. Synfuels
8[ ] H. DK

This part of the interview contains a number of statements. You may agree with some and disagree with others. For each statement, we will ask you to choose which of the following answers best fits your feelings: strongly agree, agree, unsure, disagree, or strongly disagree. (GET AFFIRMATIVE RESPONSE) O.K., here's the first statement:

8. (HAND CARD 4) People like me have a responsibility to help resolve our country's energy problems by cutting back on consumption, even if this means making some sacrifices in the way I live.

1[ ] Strongly agree
2[ ] Agree
3[ ] Unsure
4[ ] Disagree
5[ ] Strongly disagree

9. (RETAIN CARD 4) People like me should trust the federal government to find a solution to the energy crisis.

1[ ] Strongly agree
2[ ] Agree
3[ ] Unsure
4[ ] Disagree
5[ ] Strongly disagree

10. (RETAIN CARD 4) Our energy problems will be resolved when energy companies have a free hand to find the energy we need.

1[ ] Strongly agree
2[ ] Agree
3[ ] Unsure
4[ ] Disagree
5[ ] Strongly disagree

11. (RETAIN CARD 4) We should pay more attention to the particular energy needs of each city or town—and to meeting those needs through local resources whenever possible.

1[ ] Strongly agree
2[ ] Agree
3[ ] Unsure
4[ ] Disagree
5[ ] Strongly disagree

12. How well informed would you say you are about solar energy? Would you say you are very well informed, somewhat informed, slightly informed, or not at all informed?

1[ ] Very well informed
2[ ] Somewhat informed
3[ ] Slightly informed
4[ ] Not at all informed
5[ ] DK

13. Solar energy is a very broad term that includes many ways of using the sun to produce energy. (HAND CARD 5) Here is a list of solar features that can be used by individual homeowners to produce energy. Please tell me for each solar feature whether you have heard of it, have seen it, or own one of these features. Let's start with solar panels. Have you heard of, seen, or do you own solar panels? (CHECK ALL THAT APPLY)

(INTERVIEWER READ EACH ITEM BELOW)	Have Heard Of	Have Seen	Own	DK
Solar panels (or collectors) . . .	1[ ]	2[ ]	3[ ]	4[ ]
Solar panels (or collectors) to heat water for homes	1[ ]	2[ ]	3[ ]	4[ ]
Solar panels (or collectors) to heat homes	1[ ]	2[ ]	3[ ]	4[ ]
Wood stoves to heat homes. . . .	1[ ]	2[ ]	3[ ]	4[ ]
Solar swimming pool, hot tub, or spa heaters	1[ ]	2[ ]	3[ ]	4[ ]
Attached solar greenhouses to heat homes	1[ ]	2[ ]	3[ ]	4[ ]
Solar cells (photovoltaics) to produce electricity	1[ ]	2[ ]	3[ ]	4[ ]
Small wind machines (or windmills) to produce electricity or pump water .	1[ ]	2[ ]	3[ ]	4[ ]
Bioconversion (such as gasohol and using farm wastes to produce gas for fuel)	1[ ]	2[ ]	3[ ]	4[ ]
Passive solar design (such as using many south-facing windows to get heat from the sun, with few windows on the north; using water containers, walls, floors, or ceilings to collect and store the sun's heat, etc.).	1[ ]	2[ ]	3[ ]	4[ ]
None	1[ ]	2[ ]	3[ ]	4[ ]

(RESPONDENT RETAINS CARD) READ TO RESPONDENT: WHEN WE DISCUSS SOLAR ENERGY IN THIS INTERVIEW, WE MEAN TO INCLUDE ALL TYPES OF SOLAR ENERGY ON THIS LIST EXCEPT WOOD STOVES.

14. About how many solar homes and buildings have you noticed in your own neighborhood?

_______________ (# of homes and buildings)

15. (HAND CARD 6) Do you feel you presently know enough about solar energy to make a decision about using it in your home (or buying a solar home)? (CHECK ONE)

1[ ] Yes, definitely
2[ ] Yes, I think so
3[ ] Unsure
4[ ] No, probably not
5[ ] No, definitely not

16. Would you like to get more information regarding solar energy systems? (CHECK ONE)

1[ ] No - GO TO Q. 18
2[ ] Yes

17. What information would you like to have? (PROBE FOR COMPLETE RESPONSE)

____
____
____
____
____
____

18. (HAND CARD 7) Which of the following statements, if any, would you say accurately describes people who have solar energy systems? Just call off the numbers. (CHECK ALL THAT APPLY)

1[ ] Upper income people
2[ ] Do-it-yourselfers
3[ ] Environmentalists
4[ ] Economy-minded people
5[ ] People who enjoy being different
6[ ] Back-to-the-land types
7[ ] Not very different from my friends
8[ ] Socially-responsible people
9[ ] Political radicals
10[ ] Retired peoples
11[ ] Scientific types
12[ ] Young people
13[ ] Other (PLEASE SPECIFY) ____________________
14[ ] None of these
15[ ] DK

19. Now, as I read you the following, please tell me whether each is favorable to, or opposed to, using solar energy for homes? (VOLUNTEERED)

	Favorable	Mixed/ Neutral	Opposed	Don't Know
a. The federal government	1[ ]	2[ ]	3[ ]	4[ ]
b. Your state government	1[ ]	2[ ]	3[ ]	4[ ]
c. Your local government	1[ ]	2[ ]	3[ ]	4[ ]
d. Your neighbors and friends	1[ ]	2[ ]	3[ ]	4[ ]
e. People you work with	1[ ]	2[ ]	3[ ]	4[ ]
f. Your spouse/housemate	1[ ]	2[ ]	3[ ]	4[ ]

20. Based on your understanding of solar energy for homes, how do you feel about it—do you strongly favor, favor, oppose, or strongly oppose its use?

1[ ] Strongly favor
2[ ] Favor
3[ ] Unsure/neutral
4[ ] Oppose
5[ ] Strongly oppose

21. Given what you know about solar energy right now, do you strongly favor, favor, oppose, or strongly oppose the idea of using it on <u>your</u> home?

1[ ] Strongly favor
2[ ] Favor
3[ ] Unsure/neutral
4[ ] Oppose
5[ ] Strongly oppose

Please tell me whether you strongly agree, agree, disagree, or strongly disagree with the following statements about active and passive solar energy for homes.

22. (HAND CARD 8) In general, solar energy systems are technically and economically practical for homes.

1[ ] Strongly agree
2[ ] Agree
3[ ] Unsure
4[ ] Disagree
5[ ] Strongly disagree

23. (RETAIN CARD 8) If they are practical at all, solar energy systems are only practical for homes in warm, sunny climates.

1[ ] Strongly agree
2[ ] Agree
3[ ] Unsure
4[ ] Disagree
5[ ] Strongly disagree

24. (RETAIN CARD 8) Solar energy systems in general will perform satisfactorily over a long period of time. (INTERVIEWER: "A LONG PERIOD OF TIME" MEANS ABOUT 15-20 YEARS)

1[ ] Strongly agree
2[ ] Agree
3[ ] Unsure
4[ ] Disagree
5[ ] Strongly disagree

25. (HAND CARD 9) I will read you a list of things that might be done to promote the use of solar energy in homes. For each item I read to you, please tell me which of the groups on the card, if any, should be doing these things. Just call off the numbers, please. (READ EACH STATEMENT BELOW AND CIRCLE THE NUMBERS FOR EACH ANSWER. CIRCLE ALL THAT APPLY)

a) Provide financial incentives, such as grants, low-cost loans, or other kinds of financial assistance.

1 2 3 4 5 6 7 8 9

b) Educate and inform consumers.

1 2 3 4 5 6 7 8 9

c) Conduct home energy audits, in which an inspector would look over a house and make recommendations on how energy can be saved and what kinds of solar energy could be used.

1 2 3 4 5 6 7 8 9

d) Cover guarantees or warranties for solar energy systems if companies go out of business.

1 2 3 4 5 6 7 8 9

e) Require by law or ordinance passive solar design and/or solar domestic hot water in new home construction.

1 2 3 4 5 6 7 8 9

f) Identify and make a list available of qualified solar installers.

1 2 3 4 5 6 7 8 9

g) Inspect systems to see how they are working.

1 2 3 4 5 6 7 8 9

26. Do you happen to know if there is a federal credit available on your income tax for those who purchase solar energy systems for the home?

1[ ] Yes, aware of federal tax credit
2[ ] No, unaware of federal tax credit
3[ ] Don't know/unsure

27. As far as you know, does your state have an income tax credit for the purchase of solar energy systems for the home? (CHECK ONE)

1[ ] Yes
2[ ] No
3[ ] State does not have income tax
4[ ] Don't know/unsure

28. (HAND CARD 10) A number of alternatives for federal or state programs have been proposed for aiding homeowners in financing solar energy systems. I would like you to look at each of these options and tell me which one you would prefer, <u>assuming that each provided the same financial benefit</u>. Just call off the number. (CHECK ONE)

1[ ] Income tax credits (the amount of credit is subtracted from what you owe the government)
2[ ] Low-interest loans (apply to a bank for a loan guaranteed by the government)
3[ ] Direct rebate for a portion of the system cost (apply to the government after the system is purchased)
4[ ] Direct grant for a portion of the system cost (apply to the government before the system is purchased)
5[ ] Other (PLEASE SPECIFY) ______________________

6[ ] Prefer no financial assistance from government sources
7[ ] Makes no difference
8[ ] Don't know

29. (HAND CARD 11) Planning is underway to make home energy audits available to all homeowners desiring them starting early in 1981. The audits will cover both energy conservation measures (such as weatherstripping) and solar energy measures (such as solar panels). How interested would you say you are in having such an audit done in your home? (CHECK ONE)

1[ ] Very interested
2[ ] Interested
3[ ] Unsure
} GO TO Q. 30

4[ ] Not very interested
5[ ] Not at all interested
6[ ] Have already done it
} GO TO Q. 31

IF VERY INTERESTED, INTERESTED, OR UNSURE IN Q. 29, ASK Q. 30:

30. (HAND CARD 12) What is the most you would be willing to pay for such an audit? Just call off the number of your answer, please. (CHECK ONE)

1[ ] Nothing at all
2[ ] Under $25
3[ ] $25-$49
4[ ] $50-$74
5[ ] $75-$99
6[ ] $100-$149
7[ ] $150-$199
8[ ] $200 or more
9[ ] DK

ASK EVERYONE:

31. I am going to read a list of some things you might see as problems with solar energy. For each item on the list, please tell me how important a concern it would be for you, if you were thinking about using solar energy in your home. (HAND CARD 13 TO RESPONDENTS WITH RESPONSE CATEGORIES)

(INTERVIEWER READ EACH ITEM BELOW)	Not at all important	Slightly important	Somewhat important	Important	Very important	DK
a. Expense of maintaining solar energy systems	1[ ]	2[ ]	3[ ]	4[ ]	5[ ]	6[ ]
b. Effects of solar energy systems on attractiveness of house	1[ ]	2[ ]	3[ ]	4[ ]	5[ ]	6[ ]
c. Worry over whether getting a solar energy system would be time-consuming and difficult	1[ ]	2[ ]	3[ ]	4[ ]	5[ ]	6[ ]
d. Initial cost of a solar home or of solar energy systems	1[ ]	2[ ]	3[ ]	4[ ]	5[ ]	6[ ]
e. Might interfere with life-style	1[ ]	2[ ]	3[ ]	4[ ]	5[ ]	6[ ]
f. Possible problems with the utility company	1[ ]	2[ ]	3[ ]	4[ ]	5[ ]	6[ ]
g. Possibility that solar energy systems now on the market will soon be obsolete and better systems will be available later	1[ ]	2[ ]	3[ ]	4[ ]	5[ ]	6[ ]
h. What friends and neighbors would say	1[ ]	2[ ]	3[ ]	4[ ]	5[ ]	6[ ]
i. Dependability and reputability of solar firms	1[ ]	2[ ]	3[ ]	4[ ]	5[ ]	6[ ]
j. Problems of obtaining a loan to finance the solar energy system	1[ ]	2[ ]	3[ ]	4[ ]	5[ ]	6[ ]
k. Warranty coverage for solar energy system	1[ ]	2[ ]	3[ ]	4[ ]	5[ ]	6[ ]
l. Cost and difficulty of getting solar energy system covered by homeowner's insurance	1[ ]	2[ ]	3[ ]	4[ ]	5[ ]	6[ ]
m. Safety of solar energy systems	1[ ]	2[ ]	3[ ]	4[ ]	5[ ]	6[ ]
n. Operating reliability of solar energy systems (need for frequent repair and maintenance)	1[ ]	2[ ]	3[ ]	4[ ]	5[ ]	6[ ]
o. Possible damage to solar energy systems by dirt, vandalism, storms, corrosion, etc.	1[ ]	2[ ]	3[ ]	4[ ]	5[ ]	6[ ]
p. Codes or covenants might prohibit it	1[ ]	2[ ]	3[ ]	4[ ]	5[ ]	6[ ]
q. Possibility of lowering resale value of house	1[ ]	2[ ]	3[ ]	4[ ]	5[ ]	6[ ]
r. Climate too cold or cloudy or not enough wind	1[ ]	2[ ]	3[ ]	4[ ]	5[ ]	6[ ]
s. Is there anything else you consider a problem? Other (PLEASE SPECIFY) ____________ (How would you rate this?)	1[ ]	2[ ]	3[ ]	4[ ]	5[ ]	6[ ]

32. About how much do you think it would cost to buy a solar water heater for your house including the installation cost? (CHECK ONE) (INTERVIEWER: WE ARE REFERRING TO A HOT WATER HEATER TO HEAT WATER FOR DISH WASHING, BATHING, ETC.)

1[ ] Less than $500
2[ ] $500-$999
3[ ] $1000-$1499
4[ ] $1500-$2499
5[ ] $2500-$3499
6[ ] $3500-$4999
7[ ] $5000 or over
8[ ] Don't know

33. About how much do you think it would cost to buy a solar energy system to heat your home, including installation cost? (CHECK ONE)

1[ ] Less than $2500
2[ ] $2500-$3999
3[ ] $4000-$5499
4[ ] $5500-$6999
5[ ] $7000-$8499
6[ ] $8500-$9999
7[ ] $10,000 or over
8[ ] Don't know

34. (HAND CARD 14) At present costs, do you think that owning a solar energy system will result in sufficient utility savings to make it a good investment? (CHECK ONE)

1[ ] Yes, definitely
2[ ] Yes, I think so
3[ ] Maybe, not sure
4[ ] No, probably not
5[ ] No, definitely not

35. If you were thinking of investing in a solar system, how would you pay for it—through cash, finance through a loan, or both? (CHECK ONE)

1[ ] Pay cash
2[ ] Finance through a loan
3[ ] Both (part cash, part finance)
4[ ] Not interested in a solar energy system—GO TO Q. 37
5[ ] Don't know

IF PAY CASH, FINANCE, OR BOTH IN Q. 35, ASK Q. 36

36. (HAND CARD 15) Which financing arrangement would you prefer? Just call off the number. (CHECK ONE)

1[ ] Take out regular home improvement loan
2[ ] Receive an extension on your mortgage with the same monthly payments
3[ ] Pay increased monthly mortgage payments, but with no extension
4[ ] Have an automatic deduction from your paycheck
5[ ] Increase monthly payments to your utility company until the system is paid for

6[ ] Have your utility company install and maintain the system and lease it to you for a monthly rate
7[ ] Rent the system from a solar contractor at a monthly rate with an option to buy
8[ ] Have your utility pay for the system; you would repay the total cost when you sell your house
9[ ] Makes no difference
10[ ] Don't know

ASK EVERYONE:

37. During this past winter—say during December, January, and February—what were your average monthly energy costs for electricity, heating, and hot water?

$ ____________

(INTERVIEWER: 1[ ] Used bill or checkbook; 2[ ] Estimated)

38. During this past summer—say during June, July, and August—what were your average monthly energy costs for electricity, heating, hot water, and cooling?

$ ____________

(INTERVIEWER: 1[ ] Used bill or checkbook; 2[ ] Estimated)

39. (HAND CARD 16) How much more (if anything) than your present total monthly bill for utilities (heat, hot water, and electricity) would you be willing to pay per month for your own solar energy system in your home?

1[ ] Nothing more per month
2[ ] $8 per month
3[ ] $25 per month
4[ ] $40 per month
5[ ] $85 per month
6[ ] Even more
7[ ] Don't want solar energy system in my home
8[ ] Don't know

40. (HAND CARD 17) Which of the following do you know for certain are available in this area? Just call off the numbers, please. (CHECK ALL THAT APPLY)

1[ ] New solar homes
2[ ] Solar energy systems to purchase

3[ ] Reliable solar distributors and installers
4[ ] Small wind machines (windmills)
5[ ] Architects who can do solar design
6[ ] Contractors who can build or convert homes to solar energy
7[ ] Materials and products for solar energy systems
8[ ] Wood stoves
9[ ] None
10[ ] Don't know

41. I am going to read a list of possible advantages of using solar energy that could enter into your decision about using solar energy in your home. For each item on the list, please tell me how important it would be for you in making such a decision. (HAND CARD 18)

(INTERVIEWER READ EACH ITEM BELOW)	Not at all important	Slightly important	Somewhat important	Important	Very important	DK
a. Easing the energy shortage	1[ ]	2[ ]	3[ ]	4[ ]	5[ ]	6[ ]
b. Protecting against rising costs	1[ ]	2[ ]	3[ ]	4[ ]	5[ ]	6[ ]
c. Reducing need for more large power plants	1[ ]	2[ ]	3[ ]	4[ ]	5[ ]	6[ ]
d. Conserving natural resources; protecting the environment	1[ ]	2[ ]	3[ ]	4[ ]	5[ ]	6[ ]
e. Increasing overall self-reliance	1[ ]	2[ ]	3[ ]	4[ ]	5[ ]	6[ ]
f. Having a more reliable supply of energy	1[ ]	2[ ]	3[ ]	4[ ]	5[ ]	6[ ]
g. Saving money over the long term	1[ ]	2[ ]	3[ ]	4[ ]	5[ ]	6[ ]
h. Increasing status, prestige, and self-esteem	1[ ]	2[ ]	3[ ]	4[ ]	5[ ]	6[ ]
i. Increasing resale value of home	1[ ]	2[ ]	3[ ]	4[ ]	5[ ]	6[ ]
j. Increasing independence from utility company	1[ ]	2[ ]	3[ ]	4[ ]	5[ ]	6[ ]
k. Increasing independence from federal government policies	1[ ]	2[ ]	3[ ]	4[ ]	5[ ]	6[ ]
l. Reducing utility bills now	1[ ]	2[ ]	3[ ]	4[ ]	5[ ]	6[ ]
m. Availability of income tax credits	1[ ]	2[ ]	3[ ]	4[ ]	5[ ]	6[ ]
n. Getting to innovate or experiment	1[ ]	2[ ]	3[ ]	4[ ]	5[ ]	6[ ]
o. Increasing comfort of home	1[ ]	2[ ]	3[ ]	4[ ]	5[ ]	6[ ]
p. Is there anything else you consider an advantage? Other (PLEASE SPECIFY) ____ (HOW WOULD YOU RATE THIS?)	1[ ]	2[ ]	3[ ]	4[ ]	5[ ]	6[ ]

42. Based on what you know about using solar energy (domestic hot water, solar greenhouses, home heating systems), would you say your house could easily use solar energy, could use it with some difficulty, could only use it with major renovation, or could not use it? (CHECK ONE)

1[ ] Could use easily
2[ ] Could use with some difficulty
3[ ] Could use but only with major renovation
4[ ] Could not use
5[ ] Don't know

43. (HAND CARD 19) To what extent, if any, have you considered investing in a solar energy system of any kind for your house in the next 2 to 3 years?

1[ ] Have considered, definitely plan to invest (i.e., have obtained cost estimates)
2[ ] Have considered, may invest
3[ ] Have considered, and <u>will not</u> invest } GO TO Q. 45
4[ ] Have not considered } GO TO Q. 45
5[ ] DK

IF DEFINITELY PLAN TO INVEST OR MAY INVEST IN Q. 43, ASK Q. 44:

44, What kind of solar energy system have you thought of investing in? (CHECK ALL THAT APPLY)

1[ ] Solar panels (or collectors) for hot water
2[ ] Solar panels (or collectors) for home heating
3[ ] Swimming pool, hot tub, or spa heater
4[ ] Solar greenhouse
5[ ] Solar cell (photovoltaic) system
6[ ] Small wind energy conversion system
7[ ] Passive solar home or features
8[ ] Other (PLEASE SPECIFY) ____________________
9[ ] DK

ASK EVERYONE:

45. (HAND CARD 20) Listed on this card are several activities or measures that conserve energy. Please tell me which, if any, you have done since you've lived here. Just call off the letters, please.)

(FOR <u>EACH</u> ITEM NOT MENTIONED BY RESPONDENT, FOLLOW UP BY SAYING "YOU DID NOT MENTION ITEM (LETTER), IS THIS SOMETHING YOU HAVEN'T DONE OR DOESN'T THIS APPLY TO YOU?")

	Have done	Have not done	Not applicable	DK
A. Set thermostat at 65 degrees or lower in winter	1[ ]	2[ ]	3[ ]	4[ ]
B. Set thermostat at 78 degrees or higher in summer	1[ ]	2[ ]	3[ ]	4[ ]
C. Carpool	1[ ]	2[ ]	3[ ]	4[ ]
D. Drive a small car	1[ ]	2[ ]	3[ ]	4[ ]
E. Install shutters or shades on windows	1[ ]	2[ ]	3[ ]	4[ ]
F. Install weatherstripping or caulk around doors & windows	1[ ]	2[ ]	3[ ]	4[ ]
G. Install storm or doublepane windows	1[ ]	2[ ]	3[ ]	4[ ]
H. Lower temperature setting on hot water heater	1[ ]	2[ ]	3[ ]	4[ ]
I. Install glass doors, heatalator, or other energy-saving devices on fireplace	1[ ]	2[ ]	3[ ]	4[ ]
J. Buy energy-efficient appliances	1[ ]	2[ ]	3[ ]	4[ ]
K. Add insulation	1[ ]	2[ ]	3[ ]	4[ ]
L. Install a heat pump	1[ ]	2[ ]	3[ ]	4[ ]
M. Install a timer on heating system or furnace	1[ ]	2[ ]	3[ ]	4[ ]
N. Install wood stove	1[ ]	2[ ]	3[ ]	4[ ]

46. (HAND CARD 21) There is a lot of talk these days about what the aims of this country should be for the next ten years. Here are some of the goals which different people would give top priority. Would you please indicate which of these you personally consider the most important? And which would be the next most important? Just call off the number, please.

Most important ________ (FILL IN NUMBER)   Next most important ________ (FILL IN NUMBER)

47. (HAND CARD 22) If you had to choose, which one item in the following list would you say is most importnat? What would be your second choice?

Most important ________ (FILL IN NUMBER)   Next most important ________ (FILL IN NUMBER)

48. (HAND CARD 23) In your opinion, which one of these is most important? Which is next most important?

Most important ________ (FILL IN NUMBER)   Next most important ________ (FILL IN NUMBER)

49. (HAND CARD 24) I am going to read you a list of possible sources of information about solar energy. I would like you to indicate how likely it is that you would use (or go to) each source to get information about solar energy, assuming these were available to you.

(INTERVIEWER READ EACH ITEM BELOW)	Very Unlikely	Unlikely	Unsure	Likely	Very Likely
a. Friends, relatives, neighbors, and acquaintances	1[ ]	2[ ]	3[ ]	4[ ]	5[ ]
b. People who have solar energy systems	1[ ]	2[ ]	3[ ]	4[ ]	5[ ]
c. Demonstration buildings or model solar homes	1[ ]	2[ ]	3[ ]	4[ ]	5[ ]
d. Books, journals, and reports about solar energy	1[ ]	2[ ]	3[ ]	4[ ]	5[ ]
e. Magazines and newspapers	1[ ]	2[ ]	3[ ]	4[ ]	5[ ]
f. Television and radio programs	1[ ]	2[ ]	3[ ]	4[ ]	5[ ]
g. Seminars/workshops	1[ ]	2[ ]	3[ ]	4[ ]	5[ ]
h. Adult education classes	1[ ]	2[ ]	3[ ]	4[ ]	5[ ]
i. Local contractors (heating, plumbing, etc.)	1[ ]	2[ ]	3[ ]	4[ ]	5[ ]
j. Energy fairs, exhibits, home shows	1[ ]	2[ ]	3[ ]	4[ ]	5[ ]
k. Home builders, architects	1[ ]	2[ ]	3[ ]	4[ ]	5[ ]
l. Home energy audits	1[ ]	2[ ]	3[ ]	4[ ]	5[ ]
m. Is there any other information source you would use? (PLEASE SPECIFY) ________ (How would you rate this source?)	1[ ]	2[ ]	3[ ]	4[ ]	5[ ]

n. [ ] None of these

50. (RETAIN CARD 24) Now as I read another list of organizations that have information about solar energy, please indicate for each organization how likely it is that you would contact each source for information, assuming these were available to you.

(INTERVIEWER READ EACH ITEM BELOW)	Very Unlikely	Unlikely	Unsure	Likely	Very Likely
a. Solar companies	1[ ]	2[ ]	3[ ]	4[ ]	5[ ]
b. Utility companies	1[ ]	2[ ]	3[ ]	4[ ]	5[ ]
c. Consumer groups (e.g., Consumers Union, etc.)	1[ ]	2[ ]	3[ ]	4[ ]	5[ ]
d. State energy or solar office	1[ ]	2[ ]	3[ ]	4[ ]	5[ ]
e. Solar or alternative technology clubs/associations	1[ ]	2[ ]	3[ ]	4[ ]	5[ ]
f. Federal energy agencies or laboratories (e.g., Department of Energy, Solar Energy Research Institute, Regional Solar Energy Centers, etc.)	1[ ]	2[ ]	3[ ]	4[ ]	5[ ]
g. Federal information organizations (e.g., Government Printing Office, National Solar Heating & Cooling Information Center, etc.)	1[ ]	2[ ]	3[ ]	4[ ]	5[ ]
h. State or county agricultural extension service agents	1[ ]	2[ ]	3[ ]	4[ ]	5[ ]
i. Environmental organizations (e.g., Sierra Club, Wilderness Society, etc.)	1[ ]	2[ ]	3[ ]	4[ ]	5[ ]
j. Public library	1[ ]	2[ ]	3[ ]	4[ ]	5[ ]
k. Local government	1[ ]	2[ ]	3[ ]	4[ ]	5[ ]
l. Lobbying organizations (e.g., Solar Lobby, etc.	1[ ]	2[ ]	3[ ]	4[ ]	5[ ]
m. Trade and professional organizations (e.g., Solar Energy Industries Assoc.)	1[ ]	2[ ]	3[ ]	4[ ]	5[ ]
n. Other (SPECIFY) ______________ (How would you rate?	1[ ]	2[ ]	3[ ]	4[ ]	5[ ]

o. [ ] None of these

I am going to read you a set of statements about activities that some people engage in. I would like you to tell me whether and how much you and other members of your household engage in these activities. Just call off the number, please.

51. (HAND CARD 25) Use a bicycle for transportation or recreation. (CHECK ONE)

1[ ] Regularly
2[ ] Usually
3[ ] Frequently
4[ ] Sometimes
5[ ] Never
6[ ] DK

52. (RETAIN CARD 25) Change oil in the family car(s) at home. (CHECK ONE)

1[ ] Regularly
2[ ] Usually
3[ ] Frequently
4[ ] Sometimes
5[ ] Never
6[ ] DK

53. (RETAIN CARD 25) Developing and using skills to increase self-reliance, such as in carpentry, car repair, food preservation. (CHECK ONE)

1[ ] Regularly
2[ ] Usually
3[ ] Frequently
4[ ] Sometimes
5[ ] Never
6[ ] DK

54. (RETAIN CARD 25) Engage in exercise for physical fitness (e.g., running, swimming, calisthenics, etc.) (CHECK ONE)

1[ ] Regularly
2[ ] Usually
3[ ] Frequently
4[ ] Sometimes
5[ ] Never
6[ ] DK

55. (HAND CARD 26) Recycle the newspapers, glass, or cans used at home. (CHECK ONE)

1[ ] Recycle all of this material
2[ ] Most of this material
3[ ] About half of this material
4[ ] Some of this material
5[ ] Never recycle
6[ ] DK

56. (HAND CARD 27) Buying clothing at a garage sale or second-hand stores. (CHECK ONE)

1[ ] All of the household's clothing
2[ ] Most items
3[ ] About half of the household's clothing
4[ ] A few items
5[ ] None of the household's clothing
6[ ] DK

57. (HAND CARD 28) Contribute to ecologically oriented organizations (such as the Sierra Club, etc.) (CHECK ONE)

1[ ] Contribute regularly to 2 or more organizations
2[ ] Contribute regularly to 1 organization
3[ ] Occasionally contribute
4[ ] Used to contribute, but no longer do
5[ ] Never have contributed
6[ ] DK

Now, a few questions about you and your household.

58. Does anyone in your household own any of the following? (CHECK ALL THAT APPLY)

1[ ] Video game
2[ ] Shower flow restrictor
3[ ] Microwave oven
4[ ] SOLAR toys and gadgets (calculators, wristwatches, etc.)
5[ ] None
6[ ] DK

59. (HAND CARD 29) The following list represents different kinds of magazines widely read today. Please tell me the number of each one which is frequently read in your household. (CHECK ALL THAT APPLY) (INTERVIEWER: "FREQUENTLY" MEANS 3 OR 4 OUT OF 4 ISSUES.)

1[ ] National Geographic
2[ ] Time
3[ ] Solar Age
4[ ] Parents
5[ ] Better Homes and Gardens
6[ ] Rain
7[ ] Redbook
8[ ] Newsweek
9[ ] Field and Stream
10[ ] Mother Earth News

11[ ] People
12[ ] Popular Mechanics
13[ ] Psychology Today
14[ ] Reader's Digest
15[ ] U.S. News and World Report
16[ ] Woman's Day
17[ ] Cosmopolitan
18[ ] Family Circle
19[ ] Playboy
20[ ] Sunset

21[ ] Consumer Reports
22[ ] Retirement Living
23[ ] Popular Science
24[ ] Southern Living
25[ ] None of these
26[ ] DK

60. How likely is it that you will move within the next three years? Would you say it is very likely, likely, unlikely, or very unlikely? (CHECK ONE)

1[ ] Very likely
2[ ] Likely
3[ ] Unsure
4[ ] Unlikely
5[ ] Very unlikely

61. What is the main fuel used to heat your home? (CHECK ONE)

1[ ] Natural gas
2[ ] Fuel oil
3[ ] Electricity
4[ ] Propane
5[ ] Solar
6[ ] Wood
7[ ] Other (PLEASE SPECIFY) ____________________
8[ ] DK

62. (HAND CARD 30) Which of the following activities have you engaged in during the last year? (CHECK ALL THAT APPLY) Just call off the number, please.

1[ ] Worked with others to help solve community problems
2[ ] Contacted political officials or politicians
3[ ] Ran for political office
4[ ] Wrote a letter to the editor
5[ ] Made a speech
6[ ] Worked on a political campaign
7[ ] Wrote an article
8[ ] Was an officer of a community organization
9[ ] Signed a petition
10[ ] Have often given friends and neighbors advice
11[ ] None of these
12[ ] DK

63. Regardless of your political party identification, in general would you call yourself . . . ? (READ LIST, CHECK ONE)

1[ ] A strong conservative
2[ ] A moderate conservative
3[ ] Middle of the road
4[ ] A moderate liberal

5[ ] A strong liberal
6[ ] A radical
7[ ] DK

64. (HAND CARD 31) Please tell me which of the categories of this card most nearly describes the kind of work the chief wage earner in your immediate family does. Just call off the number, please. (IF CHIEF WAGE EARNER IS UNEMPLOYED, ASK WHAT TYPE OF WORK HE/SHE WOULD DO IF EMPLOYED.)

1[ ]
2[ ]
3[ ]
4[ ]
5[ ]
6[ ]
7[ ]
8[ ]
9[ ]
10[ ]
11[ ]
12[ ]
13[ ]
14[ ]
15[ ]
16[ ] Other (SPECIFY) __________
17[ ] DK

65. What is the highest level of education you have completed? (CHECK ONE)

1[ ] Less than high school
2[ ] High school graduate
3[ ] Trade or technical school
4[ ] Some college
5[ ] College graduate
6[ ] Graduate work or more
7[ ] DK

66. We would like to know a little about your household—the number of people who live here and their ages. Please start with yourself—what is your age, please? What are the names of all other people living or staying here? (RECORD NAMES FOR PURPOSES OF ASKING THE FOLLOWING FOR EACH PERSON NAMED: WHAT IS _____'S RELATIONSHIP TO YOU, AND WHAT IS HIS/HER AGE?)

Codes:

D = daughter	U = uncle
S = son	G = grandparent
W = wife	SIB = sibling (brother, sister)
H = husband	OR = other related by blood or marriage
R = roommate or housemate	ON = other nonrelated
A = aunt	

Respondent: Age _____

Name	Relationship	Age
____________	_____	_____
____________	_____	_____
____________	_____	_____
____________	_____	_____
____________	_____	_____
____________	_____	_____
____________	_____	_____
____________	_____	_____

67. (HAND CARD 32) Which category on this card best represents the total annual income, before taxes, of all persons living in your household? Just call off the number (CHECK ONE)

1[ ] Under $5,000 a year (or under $96/week)
2[ ] $5000-$6999 a year (or $96-$134.50/week)
3[ ] $7000-$9999 a year (or $135-$192.50/week)
4[ ] $10,000-$11,999 a year (or $193-$229.50/week)
5[ ] $12,000-$14,999 a year (or $230-$288.50/week)
6[ ] $15,000-$19,999 a year (or $289-$383.50/week)
7[ ] $20,000-$24,999 a year (or $384-$484.50/week)
8[ ] $25,000-$34,999 a year (or $481-$672.50/week)
9[ ] $35,000-$44,999 a year (or $673-$864.50/week)
10[ ] $45,000-$54,999 a year (or $865-$1056.50/week)
11[ ] Over $55,000 a year (or over $1057/week)
12[ ] DK

68. (HAND CARD 33) Which of the following best describes your racial or ethnic identification? Just call off the letter from the card, please. (CHECK ONE) (INTERVIEWER: IF RESPONDENT ASKS, THIS IS THE FEDERAL STANDARD FOR ASKING ABOUT RACE. IT IS THE SAME MEASURE USED BY THE GOVERNMENT FOR THEIR SURVEYS, E.G., THE CENSUS.)

1[ ] American Indian or Alaskan Native
2[ ] Asian or Pacific Islander
3[ ] Black, not of Hispanic origin
4[ ] Hispanic (Chicano)
5[ ] White, not of Hispanic origin
6[ ] Refused to answer
7[ ] DK

(INTERVIEWER: IF RESPONDENT REFUSES TO ANSWER, PLEASE CHECK THE CATEGORY BELOW WHICH YOU FEEL BEST DESCRIBES THE RESPONDENT AFTER YOU LEAVE THE HOUSEHOLD.)

1[ ] American Indian or Alaskan Native
2[ ] Asian or Pacific Islander
3[ ] Black, not of Hispanic origin
4[ ] Hispanic (including Chicano)
5[ ] White, not of Hispanic origin

HAND CARDS

Q. 4,5

GO 80138/80142
1-0-1

CARD 1

LADDER SCALE
10
9
8
7
6
5
4
3
2
1
0

Q.6

GO 80138/80142
1-0-2

CARD 2

1. It really has had no effect on us.
2. We have had to make a few adjustments, but our lifestyle has not been affected.
3. Our life has been less comfortable and convenient, but it is not serious.
4. We have had to make serious changes in our daily habits.

Qs. 7a, 7b

GO 80138/80142
2-0-2

CARD 3

G. Synthetic fuels (for example, industrial plant converts oil shale or coal to a liquid or gas)

F. Oil and natural gas

E. Solar energy (includes solar, wind, wood)

D. Water power from dams or falls (hydropower)

C. Energy conservation (reducing energy needs by additing more home insulation, driving a fuel-efficient car, etc.)

B. Nuclear energy

A. Coal

Q. 8-11

GO 80138/80142
1-0-1

CARD 4

1. Strongly agree
2. Agree
3. Unsure
4. Disagree
5. Strongly disagree

Q. 13

GO 80138/80142
1-0-1

CARD 5

Solar panels (or collectors)

Solar panels (or collectors) to heat water for homes

Solar panels (or collectors) to heat homes

Wood stoves to heat homes

Solar swimming pool, hot tub, or spa heaters

Attached solar greenhouses to heat homes

Solar cells (photovoltaics) to produce electricity

Small wind machines (or windmills) to produce electricity or pump water

Bioconversion (such as gasohol and using farm wastes to produce gas or fuel)

Passive solar design (such as using many south-facing windows to get heat from the sun, with few windows on the north; using water containers, walls, floors, or ceilings to collect and store the sun's heat, etc.)

None

Q. 15

GO 80138/80142
1-0-1

CARD 6

1. Yes, definitely
2. Yes, I think so
3. Unsure
4. No, probably not
5. No, definitely not

Q. 18

GO 80138/80142
1-0-2

CARD 7

1. Upper income people
2. Do-it-yourselfers
3. Environmentalists
4. Economy-minded people
5. People who enjoy being different
6. Back-to-the-land types
7. Not very different from my friends
8. Socially responsible people
9. Political radicals
10. Retired people
11. Scientific types
12. Young people

Qs. 22-24

GO 80138/80142
1-0-1

CARD 8

1. Strongly agree
2. Agree
3. Unsure
4. Disagree
5. Strongly disagree

Q. 25

GO 80138/80142
2-0-2

CARD 9

6. Banks
5. Private industry
4. Utility companies
3. Local government
2. State government
1. The federal government
7. It doesn't matter
8. Don't know
9. Should not be done.

Q. 28

GO 80138/80142
1-0-2

CARD 10

1. Income tax credits (the amount of credit is subtracted from what you owe the government)
2. Low-interest loans (apply to a bank for a loan guaranteed by the government)
3. Direct rebate for a portion of the system cost (apply to the government after the system is purchased)
4. Direct grant for a portion of the system purchase (apply to the government before the system is purchased)
5. Other (please specify)
6. Prefer no financial assistance from government sources
7. Makes no difference

Q. 29

GO 80138/80142
1-0-1

CARD 11

1. Very interested
2. Interested
3. Unsure
4. Not very interested
5. Not at all interested
6. Have already had it done

Q. 30

GO 80138/80142
2-0-2

CARD 12

8. $200 or more
7. $150-$199
6. $100-$149
5. $75-$99
4. $50-$74
3. $25-$49
2. Under $25
1. Nothing at all

Q. 31

GO 80138/80142
1-0-1

CARD 13

1. Not at all important
2. Slightly important
3. Somewhat important
4. Important
5. Very important

Q. 34

GO 80138/80142
1-0-1

CARD 14

1. Yes, definitely
2. Yes, I think so
3. Unsure
4. No, probably not
5. No, definitely not

Q.36 GO 80138/80142
1-0-2

CARD 15

1. Take out regular home improvement loan
2. Receive an extension on your mortgage with the same monthly payments
3. Pay an increased monthly mortgage payment, but with no extension
4. Have an automatic deduction from your paycheck
5. Increase monthly payments to your utility company until the system is paid for
6. Have your utility company install and maintain the system and lease it to you for a monthly rate
7. Rent the system from a solar contractor at a monthly rate with an option to buy
8. Have your utility pay for the system; you would repay the total cost when you sell the house
9. Makes no difference

Q.39 GO 80138/80142
1-0-1

CARD 16

1. Nothing more per month
2. $8 per month
3. $25 per month
4. $40 per month
5. $85 per month
6. Even more
7. Don't want solar energy system in my home

Q.40 GO 80138/80142
2-0-2

CARD 17

8. Wood stoves
7. Materials and products for solar energy systems
6. Contractors who can build or convert homes to solar energy
5. Architects who can do solar design
4. Small wind machines (windmills)
3. Reliable solar distributors and installers
2. Solar energy systems to purchase
1. New solar homes

Q. 41

GO 80138/80142
1-0-1

CARD 18

1. Not at all important
2. Slightly important
3. Somewhat important
4. Important
5. Very important

Q. 43

GO 80138/80142
1-0-1

CARD 19

1. Have considered, definitely plan to invest (i.e., have obtained information about systems and/or cost estimates)
2. Have considered, may invest
3. Have considered, and will not invest
4. Have not considered

Q. 45

GO 80138/80142
2-0-2

CARD 20

N. Install wood stove
M. Install timer on heating system or furnace
L. Install a heat pump
K. Add insulation
J. Buy energy-efficient appliances
I. Install glass doors, heatalator, or other energy saving device on fireplace
H. Lower temperature setting of hot water heater
G. Install storm or doublepane windows
F. Install weatherstripping or caulk around doors or windows
E. Install shutters or shades on windows
D. Drive a small car
C. Carpool
B. Set thermostat to 78 degrees or higher in summer
A. Set thermostat to 65 degrees or lower in winter

Q. 46 GO 80138/80142

CARD 21 1-0-2

1. Maintaining a high level of economic growth
2. Making sure that this country has strong defense forces
3. See that people have more say about how things are done at their jobs and in their communities
4. Protecting nature from being spoiled or polluted

Q. 47 GO 80138/80142

CARD 22 2-0-2

4. Protecting freedom of speech
3. Fighting rising prices
2. Giving the people more to say in important government decisions
1. Maintaining order in the nation

Q. 48 GO 80138/80142

CARD 23 1-0-2

1. A stable economy
2. Progress toward a less impersonal and more humane society
3. Progress toward a society in which ideas count more than money
4. The fight against crime

Q. 49, 50 GO 80138/80142

CARD 24 1-0-1

1. Very unlikely
2. Unlikely
3. Unsure
4. Likely
5. Very likely

Q. 51-54 GO 30138/80142

CARD 25 1-0-1

1. Regularly
2. Usually
3. Frequently
4. Sometimes
5. Never

Q. 55 GO 80138/80142

CARD 26 1-0-1

1. Recycle all of this material
2. Most of this material
3. About half of this material
4. Some of this material
5. Never recycle

Q. 56 GO 80138/80142

CARD 27 1-0-1

1. All of the household's clothing
2. Most items
3. About half of the household's clothing
4. A few items
5. None of the household's clothing

Q. 57 GO 80138/80142

CARD 28 1-0-1

1. Contribute regularly to 2 or more organizations
2. Contribute regularly to 1 organization
3. Occasionally contribute
4. Used to contribute, but no longer do
5. Never have contributed

Q. 59 GO 80138/80142

CARD 29 2-0-2

24. Southern Living
23. Popular Science
22. Retirement Living
21. Consumer Reports
20. Sunset
19. Playboy
18. Family Circle
17. Cosmopolitan
16. Woman's Day
15. U.S. News and World Report
14. Reader's Digest
13. Psychology Today
12. Popular Mechanics
11. People
10. Mother Earth News
9. Field and Stream
8. Newsweek
7. Redbook
6. Rain
5. Better Homes and Gardens
4. Parents
3. Solar Age
2. Time
1. National Geographic
25. None of these

Q. 62 GO 80138/80142

CARD 30 1-0-2

1. Worked with others to help solve community problems
2. Contacted political officials or politicians
3. Ran for political office
4. Wrote a letter to the editor
5. Made a speech
6. Worked on a political campaign
7. Wrote an article
8. Was an officer of a community organization
9. Signed a petition
10. Have often given friends and neighbors advice
11. None of these

Q. 64 GO 80138/80142

CARD 31 2-0-2

12. Farm laborer, farm helper, or farm foreman
11. Laboring worker (other than farm): e.g., plumber's assistant, construction laborer, longshoreman, or other physical work
10. Service worker who performs services: e.g., policeman, fireman, waiter, maid, or barber
9. Manufacturer's representation: e.g., outside salesperson, sales representative
8. Sales worker: e.g., a clerk in a store or a door-to-door salesman
7. Clerical or office worker, in business, government agency, or other type of organization: e.g., typist, secretary, postal clerk, or telephone operator
6. Farm owner, farm manager
5. Runs own business with one or more employees: e.g., store, factory, plumbing contractor, etc.
4. Manager, executive, or official: in business, government agency, or other organization
3. Semi-skilled worker: e.g., operates a machine in a factory, is an assembly-line worker in a factory, drives a truck, taxi
2. Works as a skilled trade or craft: e.g., printer, baker, tailor, RR engineer, plumber, or does mechanical work such as garage mechanic, carpenter
1. Professional worker: e.g., lawyer, doctor, scientist, teacher, systems analyst, musician, etc.
13. Retired
14. Full-time student
15. Housewife
16. Other (please specify)

Q. 67

GO 30138/80142
1-0-1

CARD 32

1. Under $5000 a year (or under $96/week)
2. $5000-$6999 a year (or $96-$134.50/week)
3. $7000-$9999 a year (or $135-$192.50/week)
4. $10,000-$11,999 a year (or $193-$229.50/week)
5. $12,000-$14,999 a year (or $230-$288.50/week)
6. $15,000-$19,999 a year (or $289-$383.50/week)
7. $20,000-$24,999 a year (or $384-$480.50/week)
8. $25,000-$34,999 a year (or $481-$672.50/week)
9. $35,000-$44,999 a year (or $673-$864.50/week)
10. $45,000-$54,999 a year (or $865-$1056.50/week)
11. Over $55,000 a year (or over $1057/week)

Q. 68

GO 80138/80142
1-0-1

CARD 33

1. American Indian or Alaskan native
2. Asian or Pacific Islander
3. Black, not of Hispanic origin
4. Hispanic
5. White, not of Hispanic origin

APPENDIX C:
SOLAR HOMEOWNER QUESTIONNAIRE

National Study of the Residential Solar Consumer

Solar Energy Research Institute
A Division of Midwest Research Institute
1617 Cole Boulevard
Golden, Colorado 80401

Operated for the
U.S. Department of Energy
under Contract No. EG-77-C-01-4042

1. Do you presently own (or did you ever own) a solar home or a house with solar features? ("Solar" includes: collectors or panels; passive features such as many south-facing windows and skylights installed specifically for solar benefit, specially constructed walls or floors for heat storage, solar greenhouses; photovoltaic or solar cells; small wind energy conversion systems or windmills. Solar energy does not include wood stoves for the purposes of this question.) (CHECK ONE)

1[ ]Yes, presently own
2[ ]Yes, once owned but no longer do
3[ ]No

IF NO, PLEASE STOP HERE AND RETURN QUESTIONNAIRE IN THE ENCLOSED ENVELOPE.

2. When did you first become a solar owner?

_____, _____
Month Year

3. When did you discontinue solar use?

_____, _____ [ ]Not applicable
Month Year

4. Solar energy is a very broad term that includes many ways of using the sun to produce energy. Here is a list of solar features that can be used by individual homeowners to produce energy. Please indicate those you have heard of, those you have actually seen, and those you actually own. (CHECK ALL THAT APPLY)

	Have heard of	Have seen	Own
Solar panels (or collectors) to heat water for homes	1[ ]	2[ ]	3[ ]
Solar panels (or collectors) to heat homes	1[ ]	2[ ]	3[ ]
Wood stoves to heat homes	1[ ]	2[ ]	3[ ]
Solar swimming pool, hot tub, or spa heaters	1[ ]	2[ ]	3[ ]
Attached solar greenhouses to heat homes	1[ ]	2[ ]	3[ ]
Solar cells (photovoltaics) to produce electricity	1[ ]	2[ ]	3[ ]
Small wind machines (or windmills) to produce electricity or pump water	1[ ]	2[ ]	3[ ]
Bioconversion (such as gasohol and using farm wastes to produce gas for fuel)	1[ ]	2[ ]	3[ ]
Passive solar design (such as using many south-facing windows to get heat from sun, with few windows on the north; using water containers, walls, floors, or ceilings to collect and store the sun's heat, etc.)	1[ ]	2[ ]	3[ ]
Other (PLEASE SPECIFY) __________	1[ ]	2[ ]	3[ ]

WHEN WE MENTION SOLAR ENERGY IN THIS QUESTIONNAIRE, WE MEAN TO INCLUDE ALL TYPES OF SOLAR ENERGY ON THIS LIST EXCEPT WOOD STOVES. (LATER THERE WILL BE A FEW QUESTIONS IN THE QUESTIONNAIRE ON WOOD STOVES FOR THOSE OWNING THEM ALONG WITH OTHER SOLAR ENERGY SYSTEMS.)

5. What other kinds of heating or cooling systems do you use with your solar energy system(s)? **(CHECK ALL THAT APPLY)**

1[ ]Electricity for hot water or home heat
2[ ]Natural gas for hot water or home heat
3[ ]Heating oil
4[ ]Wood stoves/fireplaces
5[ ]Propane or other fuel for hot water or home heat
6[ ]Heat pump
7[ ]Air conditioner for cooling system
8[ ]None
9[ ]Don't know

6. Does your solar energy system have storage capability? **(CHECK ONE)**

1[ ]Yes (GO TO Q. 7)
2[ ]No (SKIP TO Q. 9)

7. What type of storage? **(CHECK ALL THAT APPLY)**

1[ ]Rock
2[ ]Water
3[ ]Batteries
4[ ]Phase change materials (e.g., eutectic salts)
5[ ]Mass wall, floor, ceiling
6[ ]Other (PLEASE SPECIFY)____________________

8. Approximately what is the storage capability of the system in days?

____________(# of days)

9. Altogether, how satisfied are you with your experiences as a solar owner? **(CHECK ONE)**

1[ ]Very satisfied
2[ ]Satisfied
3[ ]Unsure, no opinion yet
4[ ]Dissatisfied
5[ ]Very dissatisfied

10. Has living in a solar home changed your lifestyle in terms of family routines or living patterns, such as times for showers, use of rooms, etc? Would you say your family routines changed a great deal, changed somewhat, or did not change? **(CHECK ONE)**

1[ ]Yes, changed a great deal (GO TO Q. 11)
2[ ]Yes, changed somewhat (GO TO Q. 11)
3[ ]No, unchanged (SKIP TO Q. 13)

11. How have your routines changed? ____________________

____

____

____

____

12. In general, how do you feel about these changes? **(CHECK ONE)**

1[ ]Positive
2[ ]Neutral/mixed
3[ ]Negative

13. On the basis of your experience, have you recommended solar energy to others or not? **(CHECK ONE)**

1[ ]Yes, have recommended it highly (GO TO Q. 14)
2[ ]Yes, with some reservations (GO TO Q. 14)
3[ ]Have not recommended for or against it (SKIP TO INSTRUCTIONS BEFORE Q. 15)
4[ ]No, have recommended against it (SKIP TO INSTRUCTIONS BEFORE Q. 15)

14. About how many people to whom you have recommended solar energy are now using it?

_____________(# of people)
[ ]Don't know

THIS SECTION OF THE QUESTIONNAIRE IS FOR THOSE OWNING AN ACTIVE SOLAR ENERGY SYSTEM OR A PASSIVE SYSTEM EMPLOYING COLLECTORS (SUCH AS THERMOSYPHON OR BREADBOX SYSTEMS). IF YOU DO NOT OWN SUCH A SYSTEM, SKIP TO INSTRUCTIONS BEFORE QUESTION 29.

Please answer the following questions (Q.'s 15-28) if you own an active or hybrid system using solar collectors. If you have more than one such system, please answer each question for your first two systems (System 1 and System 2) using the order indicated in Question 15. For example, if you have a solar heating system and a solar swimming pool heater, System 1 is the solar heating system and System 2 is the swimming pool heater.

15. What is the function of your solar energy system(s)? **(CHECK ALL THAT APPLY)**

System 1	System 2	
1[ ]	1[ ]	Combined domestic water heating/home heating
2[ ]	2[ ]	Home heating (separate system)
3[ ]	3[ ]	Domestic water heating
4[ ]	4[ ]	Swimming pool heating
5[ ]	5[ ]	Other (PLEASE SPECIFY) ______________________________

16. In what year was your system installed?

System 1: ________(Year)
[ ]Don't know

System 2: ________(Year)
[ ]Don't know

17. Was the system(s) ......? **(CHECK ONE)**

System 1	System 2	
1[ ]	1[ ]	Built into the house when originally constructed
2[ ]	2[ ]	Added to an existing house
3[ ]	3[ ]	Part of a room or wing addition to an existing house

18. Was your solar energy system ......? **(CHECK ONE)**

System 1	System 2	
1[ ]	1[ ]	Built on site
2[ ]	2[ ]	Bought as components and assembled on site
3[ ]	3[ ]	Bought as a complete "packaged" system and assembled on site
4[ ]	4[ ]	Other (PLEASE SPECIFY) ________________

19. Who installed your solar energy system? **(CHECK ONE)**

System 1	System 2	
1[ ]	1[ ]	Yourself
2[ ]	2[ ]	A contractor
3[ ]	3[ ]	Other (PLEASE SPECIFY) ________________

20. What type of solar panels or collectors do you own? **(CHECK ONE)**

System 1	System 2	
1[ ]	1[ ]	Flat plate collectors
2[ ]	2[ ]	Evacuated tube collectors
3[ ]	3[ ]	"Bread box" or thermosyphon collectors
4[ ]	4[ ]	Other (PLEASE SPECIFY) ________________

21. What circulates through the collectors or panels? **(CHECK ONE)**

System 1	System 2	
1[ ]	1[ ]	Air
2[ ]	2[ ]	Water
3[ ]	3[ ]	Water and anti-freeze
4[ ]	4[ ]	Chemical or non-water solution
5[ ]	5[ ]	Don't know
6[ ]	6[ ]	Not applicable

22. Many solar energy systems require owners to engage in activities to operate or maintain them properly. For each of the following activities you engage in, please indicate how often you do so. **(CHECK ONE RESPONSE FOR EACH ACTIVITY)**

Activity	Daily	Weekly	Monthly	Occasionally	Seasonally	Yearly	Never	Not Applicable
Drain down collectors or other freeze protection activities	1[ ]	2[ ]	3[ ]	4[ ]	5[ ]	6[ ]	7[ ]	8[ ]
Clean collectors	1[ ]	2[ ]	3[ ]	4[ ]	5[ ]	6[ ]	7[ ]	8[ ]
Maintenance of pumps, etc.	1[ ]	2[ ]	3[ ]	4[ ]	5[ ]	6[ ]	7[ ]	8[ ]
Change liquids in system	1[ ]	2[ ]	3[ ]	4[ ]	5[ ]	6[ ]	7[ ]	8[ ]
Turn auxiliary system on/off	1[ ]	2[ ]	3[ ]	4[ ]	5[ ]	6[ ]	7[ ]	8[ ]

23. How do you feel about these activities? Are they ......? **(CHECK ONE)**

1[ ]Enjoyable
2[ ]Not at all inconvenient
3[ ]Somewhat inconvenient
4[ ]Very inconvenient
5[ ]Not applicable (engage in none of these activities)

24. What overall condition describes your solar energy system(s) right now? **(CHECK ONE)**

System 1	System 2	
1[ ]	1[ ]	Excellent condition, never were any problems
2[ ]	2[ ]	Very good, the few problems have been solved
3[ ]	3[ ]	Good, some minor problems remain
4[ ]	4[ ]	Fair, system problems difficult to resolve
5[ ]	5[ ]	Poor, system has chronic problems

IF YOU PURCHASED YOUR SYSTEM, OR PARTS AND MATERIALS FOR IT, PLEASE ANSWER QUESTIONS 25-28. IF NOT, PLEASE SKIP TO INSTRUCTIONS BEFORE QUESTION 29.

25. From whom did you purchase your system(s) or parts and materials for it? **(CHECK ALL THAT APPLY)**

System 1	System 2	
1[ ]	1[ ]	Mail order from a non-local solar manufacturer
2[ ]	2[ ]	Local solar company
3[ ]	3[ ]	Local plumbing, heating, ventilating, or air conditioning store
4[ ]	4[ ]	Local outlet of a chain store (e.g., Sears, etc.)
5[ ]	5[ ]	Local building contractor
6[ ]	6[ ]	Local architect
7[ ]	7[ ]	Local hardware, lumber, etc. supply house
8[ ]	8[ ]	Other (PLEASE SPECIFY) ______________________________

26. About how many firms did you contact before purchasing your system(s) or components?

System 1: ___________ (# of firms)

System 2: ___________ (# of firms)

27. Why did you choose the firm(s) you did?

____

____

____

____

____

28. What kind of firm manufactured your solar energy system(s) or components? **(CHECK ONE)**

System 1	System 2	
1[ ]	1[ ]	Small local firm
2[ ]	2[ ]	Small national firm
3[ ]	3[ ]	Large national firm
4[ ]	4[ ]	Foreign firm
5[ ]	5[ ]	Other (PLEASE SPECIFY) ______________________
6[ ]	6[ ]	Don't know

THIS SECTION OF THE QUESTIONNAIRE IS FOR THOSE OWNING A PASSIVE OR HYBRID (COMBINED ACTIVE AND PASSIVE) SOLAR ENERGY SYSTEM OR HOME. IF YOU DO NOT OWN PASSIVE FEATURES AS DEFINED IN QUESTION 29, SKIP TO INSTRUCTIONS BEFORE QUESION 42.

Please answer the following questions (Q.'s 29-41) if you own a passive or hybrid system or home. If you have more than one passive system, please answer each question for your first two systems (System 1 and System 2) using the priority indicated in Question 29. For example, if you have a solar greenhouse and a Trombe wall, System 1 is the greenhouse and System 2 is the Trombe wall.

29. What kind of passive system(s) or home do you own **(CHECK ALL THAT APPLY)**

System 1	System 2	
1[ ]	1[ ]	House with major passive features (e.g., extra south glazing, thermal mass materials in floors, walls, ceilings—brick, concrete, water containers)
2[ ]	2[ ]	House with limited passive features (e.g., broad side facing south, extra south glazing but no significant increase in thermal mass materials over conventional house)
3[ ]	3[ ]	Attached solar greenhouse
4[ ]	4[ ]	Thermal storage wall (Trombe wall)
5[ ]	5[ ]	Earth-bermed house with good southern exposure
6[ ]	6[ ]	Envelope house
7[ ]	7[ ]	Other (PLEASE SPECIFY) ______________________

30. When was your passive energy system(s) constructed?

System 1: __________, ________ (Month, Year)
[ ]Don't know

System 2: __________, ________ (Month, Year)
[ ]Don't know

31. Was the system(s)......? **(CHECK ONE)**

System 1	System 2	
1[ ]	1[ ]	Built into the house when originally constructed
2[ ]	2[ ]	Added to an existing house
3[ ]	3[ ]	Part of a room or wing addition to an existing house

32. Was your passive system(s)......? **(CHECK ONE)**

System 1	System 2	
1[ ]	1[ ]	Built on site
2[ ]	2[ ]	Bought as components and assembled on site
3[ ]	3[ ]	Bought as a complete "packaged" system and assembled on site
4[ ]	4[ ]	Other (PLEASE SPECIFY) ______________________

33. Who designed your passive system(s)? **(CHECK ONE)**

System 1	System 2	
1[ ]	1[ ]	Yourself
2[ ]	2[ ]	Acquired plans
3[ ]	3[ ]	A contractor
4[ ]	4[ ]	An architect
5[ ]	5[ ]	Other (PLEASE SPECIFY) ______________________

34. Who installed your passive system(s)? **(CHECK ONE)**

System 1	System 2	
1[ ]	1[ ]	Yourself
2[ ]	2[ ]	A contractor
3[ ]	3[ ]	Other (PLEASE SPECIFY) ______________________

35. Many solar energy systems require owners to engage in activities to operate or maintain them properly. For each of the following activities you engage in, please indicate how often you do so. **(CHECK ONE RESPONSE FOR EACH ACTIVITY)**

Activity	Daily	Weekly	Monthly	Occasion-ally	Season-ally	Yearly	Never	Not Applicable
Open/close drapes or shutters on passive house	1[ ]	2[ ]	3[ ]	4[ ]	5[ ]	6[ ]	7[ ]	8[ ]
Open/close vents/skylights	1[ ]	2[ ]	3[ ]	4[ ]	5[ ]	6[ ]	7[ ]	8[ ]
Open/close flues or dampers	1[ ]	2[ ]	3[ ]	4[ ]	5[ ]	6[ ]	7[ ]	8[ ]
Control insects in greenhouse	1[ ]	2[ ]	3[ ]	4[ ]	5[ ]	6[ ]	7[ ]	8[ ]

36. How do you feel about these activities? Are they.....? **(CHECK ONE)**

1[ ]Enjoyable
2[ ]Not at all inconvenient
3[ ]Somewhat inconvenient
4[ ]Very inconvenient
5[ ]Not applicable (engage in none of these activities)

37. What overall condition best describes your solar energy system(s) right now? **(CHECK ONE)**

System 1	System 2	
1[ ]	1[ ]	Excellent condition, never were any problems
2[ ]	2[ ]	Very good, the few problems have been solved
3[ ]	3[ ]	Good, some minor problems remain
4[ ]	4[ ]	Fair, system problems difficult to resolve
5[ ]	5[ ]	Poor, system has chronic problems

IF YOU PURCHASED YOUR SYSTEM OR MAJOR COMPONENTS AND MATERIALS FOR IT, PLEASE ANSWER QUESTIONS 38-41. IF NOT, PLEASE SKIP TO INSTRUCTIONS BEFORE QUESTION 42.

38. From whom did you purchase your system(s) or parts and materials for it? **(CHECK ALL THAT APPLY)**

System 1	System 2	
1[ ]	1[ ]	Mail order from a non-local solar manufacturer
2[ ]	2[ ]	Local solar company
3[ ]	3[ ]	Local heating, plumbing, ventilating, or air conditioning company
4[ ]	4[ ]	Local outlet of a chain store (e.g., Sears, etc.)
5[ ]	5[ ]	Local building contractor
6[ ]	6[ ]	Local architect
7[ ]	7[ ]	Local hardware, lumber, etc. supply house
8[ ]	8[ ]	Other (PLEASE SPECIFY) ______________________

39. How many firms or architects, etc. did you contact before purchasing your system(s)?

System 1: __________ (# of firms)

System 2: __________ (# of firms)

40. Why did you choose the firm you did?

____

____

____

____

____

41. What kind of firm built your system(s)?

System 1	System 2	
1[ ]	1[ ]	General building contractor
2[ ]	2[ ]	Solar building company
3[ ]	3[ ]	Other (PLEASE SPECIFY) ______________________
4[ ]	4[ ]	Don't know

THE FOLLOWING QUESTIONS ARE ABOUT THE ENERGY SITUATION IN GENERAL.

42. Here is a picture of a ladder. Suppose the top of the ladder represents the **best** possible energy situation for our country; and the bottom, the **worst** possible energy situation for our country. Please indicate on which step of the ladder you think the United States is at the present time?

__________ (# of step)

43. Just as your best guess, if things go pretty much as you now expect, where do you think the U.S. will be on the ladder **about five** years from now?

__________ (# of step)

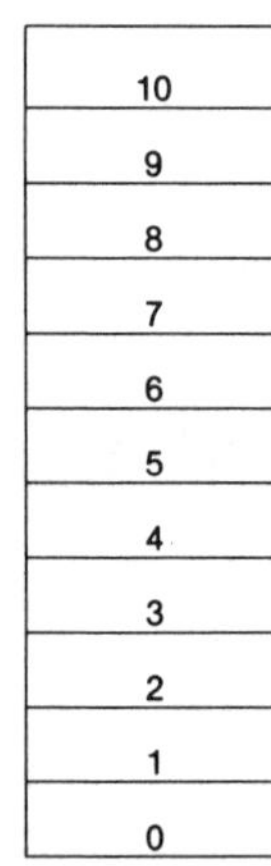

44. In different parts of the nation, people have reported a variety of effects of the energy situation on their daily lives. How has the energy situation affected your family (household)? Would you say ......? **(CHECK ONE)**

1[ ]It really has had no effect on us.
2[ ]We have had to make a few adjustments, but our lifestyle has not been affected.
3[ ]Our life has been less comfortable and convenient, but it is not serious.
4[ ]We have had to make serious changes in our daily habits.

THIS PART OF THE QUESTIONNAIRE CONTAINS A NUMBER OF STATEMENTS. YOU MAY AGREE WITH SOME AND DISAGREE WITH OTHERS. FOR EACH STATEMENT, PLEASE INDICATE, BY CHECKING YOUR RESPONSE, WHETHER YOU (1) STRONGLY AGREE, (2) AGREE, (3) ARE UNSURE, (4) DISAGREE, OR (5) STRONGLY DISAGREE.

45. People like me have a responsibility to help resolve our country's energy problems by cutting back on consumption, even if this means making some sacrifices in the way I live. **(CHECK ONE)**

1[ ]Strongly agree
2[ ]Agree
3[ ]Unsure
4[ ]Disagree
5[ ]Strongly disagree

46. People like me should trust the federal government to find a solution to the energy crisis. **(CHECK ONE)**

1[ ]Strongly agree
2[ ]Agree
3[ ]Unsure
4[ ]Disagree
5[ ]Strongly disagree

47. Our energy problems will be resolved when energy companies have a free hand to find the energy we need. **(CHECK ONE)**

1[ ]Strongly agree
2[ ]Agree
3[ ]Unsure
4[ ]Disagree
5[ ]Strongly disagree

48. We should pay more attention to the particular energy needs of each city or town—and to meeting those needs through local resources wherever possible. **(CHECK ONE)**

1[ ]Strongly agree
2[ ]Agree
3[ ]Unsure
4[ ]Disagree
5[ ]Strongly disagree

49. How well informed would say you are about solar energy? **(CHECK ONE)**

1[ ]Very well informed
2[ ]Somewhat informed
3[ ]Slightly informed
4[ ]Not at all informed

50. Which of the following statements, if any, would you say accurately describe people who have solar energy systems? **(CHECK ALL THAT APPLY)**

1[ ]Upper income people
2[ ]Do-it-yourselfers
3[ ]Environmentalists
4[ ]Economy-minded people
5[ ]People who enjoy being different
6[ ]Back-to-the-land types
7[ ]Not very different from my friends
8[ ]Socially-responsible people
9[ ]Political radicals
10[ ]Retired people
11[ ]Scientific types
12[ ]Young people
13[ ]Other (PLEASE SPECIFY)________________________________
14[ ]None of these
15[ ]Don't know

51. Please indicate whether each of the following is favorable to, or opposed to, using solar energy for homes? **(CHECK ONE FOR EACH LETTER)**

	Favorable	Mixed/ Neutral	Opposed	Don't Know
a. The federal government	1[ ]	2[ ]	3[ ]	4[ ]
b. Your state government	1[ ]	2[ ]	3[ ]	4[ ]
c. Your local government	1[ ]	2[ ]	3[ ]	4[ ]
d. Your neighbors and friends	1[ ]	2[ ]	3[ ]	4[ ]
e. People you work with	1[ ]	2[ ]	3[ ]	4[ ]
f. Your spouse/housemate	1[ ]	2[ ]	3[ ]	4[ ]

PLEASE INDICATE, BY CHECKING YOUR RESPONSE, WHETHER YOU (1) STRONGLY AGREE, (2) AGREE, (3) ARE UNSURE, (4) DISAGREE, OR (5) STRONGLY DISAGREE WITH THE FOLLOWING STATEMENTS ABOUT ACTIVE AND PASSIVE SOLAR ENERGY SYSTEMS FOR HOMES.

52. In general solar energy systems are technically and economically practical today for homes. **(CHECK ONE)**

1[ ]Strongly agree
2[ ]Agree
3[ ]Unsure
4[ ]Disagree
5[ ]Strongly disagree

53. If they are practical at all, solar energy systems are only practical for homes in warm, sunny climates. **(CHECK ONE)**

1[ ]Strongly agree
2[ ]Agree
3[ ]Unsure
4[ ]Disagree
5[ ]Strongly disagree

54. Solar energy systems in general will perform satisfactorily over a long period of time. **(CHECK ONE)**

1[ ]Strongly agree
2[ ]Agree
3[ ]Unsure
4[ ]Disagree
5[ ]Strongly disagree

55. Here is a list of things that might be done to promote the use of solar energy in homes. For each item, please indicate which of the groups listed below, if any, should be doing these things.

Groups	Write letter(s) of your response(s)		Measure
A. The federal government	______	1.	Provide financial incentives
B. State government	______	2.	Educate and inform consumers
C. Local government	______	3.	Conduct home energy audits
D. Utility companies	______	4.	Cover guarantees or warranties for solar energy systems if companies go out of business
E. Private industry			
F. Banks	______	5.	Require by law or ordinance passive solar design and/or solar domestic hot water in new home construction
G. Don't know			
H. It doesn't matter			
I. Should not be done	______	6.	Identify and make available a list of qualified solar installers
	______	7.	Inspect systems to see how they are working

56. What was the total installed cost of your solar energy system(s)? (NOTE: If you own a solar home, give cost of solar portion.)

Total cost: $______________________

57. Did you receive an income tax credit for your solar energy system or home? **(CHECK ONE)**

1[ ]Yes (GO TO Q. 58)
2[ ]No (GO TO Q. 60)
3[ ]Not applicable, no credits existed when you acquired system (SKIP TO Q. 60)

58. Which tax credits? **(CHECK ALL THAT APPLY)**

1[ ]Federal
2[ ]State

59. What proportion of the cost of the solar energy system(s) did your tax credit(s) cover?

______% of the total solar energy system(s) cost.

60. Aside from income tax credits, have you received any government or industry aid in connection with your solar energy system(s) or home? **(CHECK ONE)**

1[ ]Yes (GO TO Q. 61)
2[ ]No (SKIP TO Q. 62)
3[ ]Don't know (SKIP TO Q. 62)

61. What kind of aid have you received? **(CHECK ALL THAT APPLY)**

1[ ]HUD Solar Heating and Cooling Demonstration Program Grant
2[ ]HUD Solar Hot Water Demonstration Program Grant
3[ ]State sales tax credit or exemption
4[ ]Local sales tax credit or exemption
5[ ]State property tax exemption
6[ ]Local property tax exemption
7[ ]State low-income loan
8[ ]Rebate from industry
9[ ]Other (PLEASE SPECIFY) ______________________________

THE NEXT QUESTION REFERS TO THE SITUATION YOU WERE IN BEFORE YOU ACQUIRED YOUR SOLAR ENERGY SYSTEM(S) OR HOME.

62. Here is a list of some things you might see as problems with solar energy. Please indicate which of these were concerns to you **when you were thinking about using solar energy in your home.** For each item on the list, please enter the number on the scale that shows how important that concern was to you.

Not at all Important	Slightly Important	Somewhat Important	Important	Very Important
1	2	3	4	5

_____Expense of maintaining solar energy systems
_____Effects of solar energy systems on attractiveness of house/property
_____Worry over whether getting a solar energy system would be time-consuming and difficult
_____Initial cost of a solar home or of solar energy systems
_____Might interfere with lifestyle
_____Possible problems with the utility company
_____Possibility that solar energy systems now on the market will soon be obsolete and better systems will be available later
_____What friends and neighbors would say
_____Dependability and reputability of solar firms
_____Problems of obtaining a loan to finance the solar energy system
_____Warranty coverage for solar energy systems
_____Cost and difficulty of getting solar energy systems covered by homeowner's insurance
_____Safety of solar energy systems
_____Operating reliability of solar energy systems (need for frequent repair and maintenance)
_____Possible damage to solar energy systems by dirt, vandalism, storms, corrosion, etc.
_____Codes or covenants might prohibit it
_____Possibility of lowering resale value of house
_____Climate too cold or cloudy or not enough wind

_____Other (PLEASE SPECIFY) ______________________________

63. When you bought or built your solar energy system(s), approximately what percent of your total monthly utility costs did you expect to save? **(CHECK ONE)**

1[ ]I did not expect to save at all
2[ ]Between one and 10%
3[ ]Between 10 and 20%
4[ ]Between 20 and 30%
5[ ]Between 30 and 40%
6[ ]Between 40 and 50%
7[ ]Between 50 and 60%
8[ ]Between 60 and 70%
9[ ]Between 70 and 80%
10[ ]Over 80%
11[ ]Don't know

64. On what source of information did you base your estimate of savings? **(CHECK ALL THAT APPLY)**

1[ ]Installer/builder
2[ ]Own calculations
3[ ]Advertisements
4[ ]Books, magazines
5[ ]Architect/engineer
6[ ]Seller of system/home
7[ ]Other (PLEASE SPECIFY) ______________________________

65. Since you acquired your solar energy system, what percent of your total monthly utility costs do you think you have **actually** saved? **(CHECK ONE)**

1[ ]I have not saved at all
2[ ]Between one and 10%
3[ ]Between 10 and 20%
4[ ]Between 20 and 30%
5[ ]Between 30 and 40%
6[ ]Between 40 and 50%
7[ ]Between 50 and 60%
8[ ]Between 60 and 70%
9[ ]Between 70 and 80%
10[ ]Over 80%
11[ ]Don't know

66. Do you think that your solar energy system(s) will eventually pay for itself in savings on utility bills? **(CHECK ONE)**

1[ ]Yes (GO TO Q. 67)
2[ ]Unsure (SKIP TO Q. 68)
3[ ]No (SKIP TO Q. 68)

67. How many years from the time of installation do you think that it will take to pay for itself?

__________ (# of years)

68. During this past winter—say during December, January, and February—what were your average monthly energy costs for electricity, heating, and hot water?

$__________

69. During this past summer—say during June, July, and August—what were your average monthly energy costs for electricity, heating, hot water, and cooling?

$__________

70. Were your figures for Q 68 and 69 based on a bill or checkbook, or an estimate? **(CHECK ONE)**

1[ ]Bill or checkbook
2[ ]Estimate

THE NEXT QUESTION REFERS TO THE SITUATION YOU WERE IN BEFORE YOU ACQUIRED YOUR SOLAR ENERGY SYSTEM OR HOME.

71. Here is a list of factors that might have influenced your decision to adopt solar energy. For each item on the list, please enter the number on the scale that shows how important this advantage **actually was** to you in making your decision to use solar energy.

Not at all Important	Slightly Important	Somewhat Important	Important	Very Important
1	2	3	4	5

______Easing the energy shortage
______Protecting against rising energy costs
______Reducing need for more large power plants
______Conserving natural resources; protecting the environment
______Increasing overall self-reliance

______Having a more reliable supply of energy
______Saving money over the long term
______Increasing status, prestige, and self-esteem
______Increasing resale value of home
______Increasing independence from utility company

______Increasing independence from federal government policies
______Reducing utility bills now
______Availability of federal income tax credits
______Availability of state tax credits

______Getting to innovate or experiment
______Increasing comfort of home
______Increasing awareness of the natural environment, for example, weather patterns
______Increasing awareness of household energy use

72. Please identify the 5 most important advantages that have actually occurred since you became a solar owner. (Feel free to use the above list and your own experiences.)

____

____

____

____

____

73. Based on your understanding of solar energy for homes, how do you feel about it? **(CHECK ONE)**

1[ ]Strongly favor
2[ ]Favor
3[ ]Unsure/neutral
4[ ]Oppose
5[ ]Strongly oppose

74. How likely is it that you will add other solar applications to your present home? **(CHECK ONE)**

1[ ]Very likely (GO TO Q. 75)
2[ ]Likely (GO TO Q. 75)
3[ ]Unsure (GO TO Q. 75)
4[ ]Unlikely (SKIP TO Q. 76)
5[ ]Very unlikely (SKIP TO Q. 76)

75. What kind(s) of solar energy systems(s) might you invest in? **(CHECK ALL THAT APPLY)**

1[ ]Solar panels (or collectors) for hot water
2[ ]Solar panels (or collectors) for home heating
3[ ]Swimming pool, hot tub, or spa heater
4[ ]Solar greenhouse
5[ ]Solar cell (photovoltaic) system
6[ ]Small wind energy conversion system
7[ ]Passive solar home or features
8[ ]Other (PLEASE SPECIFY) ____________________

76. Listed below are several activities or measures that conserve energy. Please indicate by checking the appropriate column which you have not done, which you did **before** acquiring your solar energy system or house, and which you have done **since** acquiring your solar energy system or house.

	Have not done	Did before using solar energy	Did after using solar energy	Not applicable
Set thermostat to 65 degrees or lower in winter	1[ ]	2[ ]	3[ ]	4[ ]
Set thermostat to 78 degrees or higher in summer	1[ ]	2[ ]	3[ ]	4[ ]
Carpool	1[ ]	2[ ]	3[ ]	4[ ]
Drive a small car	1[ ]	2[ ]	3[ ]	4[ ]
Install shutters or shades on windows	1[ ]	2[ ]	3[ ]	4[ ]
Install weatherstripping or caulk around doors or windows	1[ ]	2[ ]	3[ ]	4[ ]
Install storm or double pane windows	1[ ]	2[ ]	3[ ]	4[ ]
Lower temperature setting on hot water heater	1[ ]	2[ ]	3[ ]	4[ ]
Install glass doors, heatalator, or other energy-saving device on fireplace	1[ ]	2[ ]	3[ ]	4[ ]
Buy energy-efficient appliances	1[ ]	2[ ]	3[ ]	4[ ]
Add insulation	1[ ]	2[ ]	3[ ]	4[ ]
Install heat pump	1[ ]	2[ ]	3[ ]	4[ ]
Install timer on heating system or furnace	1[ ]	2[ ]	3[ ]	4[ ]
Install wood stove	1[ ]	2[ ]	3[ ]	4[ ]

77. Did you receive written operation and maintenance instructions for your solar energy system(s)? **(CHECK ONE)**

1[ ]Yes (GO TO Q. 78)
2[ ]No (SKIP TO Q. 79)
3[ ]Don't know (SKIP TO Q. 79)
4[ ]Not applicable for my system (SKIP TO Q. 79)

78. How helpful were the instructions to you? **(CHECK ONE)**

1[ ]Extremely helpful
2[ ]Helpful
3[ ]Don't know/didn't use
4[ ]Not helpful
5[ ]Harmed instead of helped

79. Did you receive a written warranty for your solar energy system(s)? **(CHECK ONE)**

1[ ]Yes (GO TO Q. 80)
2[ ]No (SKIP TO Q. 82)
3[ ]Don't know (SKIP TO Q. 82)
4[ ]Not applicable to my system (SKIP TO Q. 82)

80. Does your warranty cover......? **(CHECK ALL THAT APPLY)**

1[ ]Service and maintenance calls
2[ ]On-site repairs and replacement of parts
3[ ]Don't know

81. How satisfied are you with your warranty coverage? **(CHECK ONE)**

1[ ]Very satisfied
2[ ]Satisfied
3[ ]Don't know/didn't use
4[ ]Dissatisfied
5[ ]Very dissatisfied

82. Relative to the repair service for conventional heating or hot water systems, how would you rate the quality of your solar repair service? **(CHECK ONE)**

1[ ]Better
2[ ]About the same
3[ ]Less reliable, but adequate
4[ ]Inadequate
5[ ]No opinion/don't know
6[ ]Never needed repair
7[ ]Not applicable to my system
8[ ]I do my own repairs

83. As a solar owner, which of the following problems have you actually experienced? **(CHECK ALL THAT APPLY)**

1[ ]Difficulty in obtaining financing
2[ ]Utility raised rates for solar owners
3[ ]Lack of clear, reliable information about solar energy
4[ ]Conflicts with local building codes
5[ ]Vandalism to the solar energy system
6[ ]Maintenance problems
7[ ]Installation problems
8[ ]Difficulty in finding parts and equipment for initial purchase or repair
9[ ]Difficulty in evaluating and choosing among competing systems
10[ ]Overheating
11[ ]Increased property taxes
12[ ]Opposition of neighbors
13[ ]Obstruction of your solar energy system by vegetation, new buildings, etc.
14[ ]Difficulty obtaining solar access rights
15[ ]House has become less comfortable
16[ ]Resale value of house has decreased
17[ ]Finding a qualified contractor, architect, or builder
18[ ]Difficulties arising from improper system design
19[ ]None of these

THIS SECTION OF THE QUESTIONNAIRE (Q.'s 84-95) IS FOR THOSE OWNING AN ACTIVE SOLAR ENERGY SYSTEM OR A PASSIVE SYSTEM EMPLOYING COLLECTORS (SUCH AS THERMOSYPHON OR BREADBOX SYSTEMS). IF YOU DO NOT OWN SUCH A SYSTEM, SKIP TO INSTRUCTIONS BEFORE QUESTION 96.

84. Have you ever had any of the following problems with your **pumps** or **fans**?**(CHECK ALL THAT APPLY)**

System 1	System 2	
1[ ]	1[ ]	Does not apply (system has no pumps or fans) (GO TO Q. 85)
2[ ]	2[ ]	No problems with pumps or fans (GO TO Q. 85)
3[ ]	3[ ]	Pump operates on warm days when the sun is not shining
4[ ]	4[ ]	Pump or fan operates noisily
5[ ]	5[ ]	Pump or fan operates all the time
6[ ]	6[ ]	Pump or fan seldom operates
7[ ]	7[ ]	Pump or fan never operates
8[ ]	8[ ]	Pump or fan has been repaired
9[ ]	9[ ]	Pump or fan has been replaced
10[ ]	10[ ]	Other (PLEASE SPECIFY) ______________________
11[ ]	11[ ]	Don't know

85. Have you ever had any of the following problems with your **controller**? **(CHECK ALL THAT APPLY)**

System 1	System 2	
1[ ]	1[ ]	Does not apply (system has no controller) (GO TO Q. 86)
2[ ]	2[ ]	No problems with controller (GO TO Q. 86)
3[ ]	3[ ]	Controller does not activate the pump or fan to provide efficient operation of the system
4[ ]	4[ ]	Controller fails to initiate freeze protection for the collector
5[ ]	5[ ]	Controller has been repaired
6[ ]	6[ ]	Controller has been replaced
7[ ]	7[ ]	Other (PLEASE SPECIFY) ______________________
8[ ]	8[ ]	Don't know

86. Have you ever had any of the following problems with your **collectors**? **(CHECK ALL THAT APPLY)**

System 1	System 2	
1[ ]	1[ ]	No problems with collectors (GO TO Q. 87)
2[ ]	2[ ]	Water has condensed inside the collector
3[ ]	3[ ]	Collector has corroded
4[ ]	4[ ]	Collector was mounted loosely
5[ ]	5[ ]	Transparent part of collector has corroded
6[ ]	6[ ]	Collector glass or cover has separated from the collector box
7[ ]	7[ ]	Collector has frozen or ruptured
8[ ]	8[ ]	Other (PLEASE SPECIFY) ______________________
9[ ]	9[ ]	Don't know

87. Have you ever had any of the following problems? **(CHECK ALL THAT APPLY)**

System 1	System 2	
1[ ]	1[ ]	Insufficient insulation anywhere in system
2[ ]	2[ ]	Storage tank has leaked
3[ ]	3[ ]	Storage tank was undersized
4[ ]	4[ ]	House was damaged during installation of the solar energy system
5[ ]	5[ ]	Piping or pipe joints have corroded
6[ ]	6[ ]	Pipes were mounted loosely
7[ ]	7[ ]	Panel mounting failed
8[ ]	8[ ]	Roof penetration leaked
9[ ]	9[ ]	Health hazards from anti-freeze mixture or chemicals in system
10[ ]	10[ ]	Other (PLEASE SPECIFY) ______________________________
11[ ]	11[ ]	None of these
12[ ]	12[ ]	Don't know

88. Approximately how many incidents of failure or "down time" have you had with your solar energy system(s)?

System 1	System 2	
______	______	(# of incidents) (GO TO Q. 89)
1[ ]	1[ ]	None (SKIP TO Q. 92)
2[ ]	2[ ]	Not applicable to my system (SKIP TO Q. 92)

89. How many occurred......?

System 1	System 2	
______	______	# Within three months of installation?
______	______	# Within 3-6 months of installation?
______	______	# Within 6 months to 1 year of installation?
______	______	# Within 1-2 years of installation?
______	______	# After 2 years of installation?

90. Taken together, how much have these problems cost you so far?

System 1: $__________

System 2: $__________

91. What percent of the total costs involved were paid by each of the following sources? (Indicate actual percentage)

System 1	System 2	
______%	______ %	Warranty
______%	______ %	Insurance
______%	______ %	Selves

92. In light of your experience since your purchase, in general, how accurate do you think the claims regarding your solar energy system(s) were? **(CHECK ONE)**

System 1	System 2	
1[ ]	1[ ]	System performed better than claimed
2[ ]	2[ ]	System performed about as claimed
3[ ]	3[ ]	System performed worse than claimed
4[ ]	4[ ]	Don't know
5[ ]	5[ ]	Not applicable, no claims were made

93. Do you have a way of knowing how well your solar energy system(s) is working? **(CHECK ONE)**

System 1	System 2	
1[ ]	1[ ]	Yes (GO TO Q. 94)
2[ ]	2[ ]	Unsure (SKIP TO Q. 95)
3[ ]	3[ ]	No (SKIP TO Q. 95)

94. How can you tell? **(CHECK ALL THAT APPLY)**

System 1	System 2	
1[ ]	1[ ]	From reduced utility bills/energy consumption
2[ ]	2[ ]	From meters or instruments
3[ ]	3[ ]	From your inspection of the equipment
4[ ]	4[ ]	From professional inspection
5[ ]	5[ ]	From personal comfort
6[ ]	6[ ]	Other (PLEASE SPECIFY) ____________________

95. How long do you expect your solar energy system(s) to last?

System 1: ____________ (# of years)

System 2: ____________ (# of years)

THIS SECTION OF THE QUESTIONNAIRE (Q.'s 96-104) IS FOR THOSE OWNING A PASSIVE OR HYBRID (COMBINED ACTIVE AND PASSIVE) SOLAR ENERGY SYSTEM. IF YOU DO NOT OWN SUCH A PASSIVE SYSTEM, SKIP TO INSTRUCTIONS BEFORE QUESTION 105.

96. Have you ever had any of the following problems? **(CHECK ALL THAT APPLY)**

1[ ]Excessive condensation
2[ ]Mechanical failure of movable insulation or shutters
3[ ]Mechanical failure of fans or motorized dampers
4[ ]Overheating
5[ ]Excessive glare
6[ ]Unevenness of temperature throughout the house
7[ ]None of these

97. Approximately how many incidents of failure or "down time" have you had with your solar energy system(s)?

System 1	System 2	
_______	_______	(# of incidents) (GO TO Q. 98)
1[ ]	1[ ]	None (SKIP TO Q. 101)
2[ ]	2[ ]	Not applicable to my system (SKIP TO Q. 101)

98. How many occurred......?

System 1	System 2	
_______	_______	# Within three months of installation?
_______	_______	# Within 3-6 months of installation?
_______	_______	# Within 6 months to a year of installation?
_______	_______	# Within 1-2 years of installation?
_______	_______	# After 2 years of installation?

99. Taken together, how much have these problems cost you so far?

System 1: $__________

System 2: $__________

100. What percent of the total costs involved were paid by each of the following sources? (Indicate actual percentage)

System 1	System 2	
_______%	_______%	Warranty
_______%	_______%	Insurance
_______%	_______%	Selves

101. In light of your experience since your purchase, in general, how accurate do you think the claims regarding your solar energy system(s) were?

System 1	System 2	
1[ ]	1[ ]	System performed better than claimed
2[ ]	2[ ]	System performed about as claimed
3[ ]	3[ ]	System performed worse than claimed
4[ ]	4[ ]	Don't know
5[ ]	5[ ]	Not applicable, no claims were made

102. Do you have a way of knowing how well your solar energy system(s) is working? **(CHECK ONE)**

System 1	System 2	
1[ ]	1[ ]	Yes (GO TO Q. 103)
2[ ]	2[ ]	Unsure (SKIP TO Q. 104)
3[ ]	3[ ]	No (SKIP TO Q. 104)

103. How can you tell? **(CHECK ALL THAT APPLY)**

System 1	System 2	
1[ ]	1[ ]	From reduced utility bills/energy consumption
2[ ]	2[ ]	From meters or instruments
3[ ]	3[ ]	From your inspection of the equipment
4[ ]	4[ ]	From professional inspection
5[ ]	5[ ]	From personal comfort
6[ ]	6[ ]	Other (PLEASE SPECIFY) ______________________________

104. How long do you expect your solar energy system(s) to last?

System 1: __________ (# of years)

System 2: __________ (# of years)

IF YOU OWN A WOOD STOVE, PLEASE ANSWER QUESTION 105. IF NOT, SKIP TO INSTRUCTIONS BEFORE QUESTION 106.

105. Have any of the following ever occurred? **(CHECK ALL THAT APPLY)**

1[ ]Had difficulty obtaining wood
2[ ]Experienced wood price increases of over 50% since installation
3[ ]Problems with installation of wood stove
4[ ]Problems with indoor air quality or ventilation
5[ ]Increase in cost of homeowners insurance
6[ ]Had difficulty in cleaning or maintaining system
7[ ]Local inspection required
8[ ]Chimney fire
9[ ]Other (PLEASE SPECIFY) ______________________________
10[ ]None of these
11[ ]Don't know

IF YOU OWN A SMALL WIND ENERGY CONVERSION SYSTEM, PLEASE ANSWER QUESTION 106. IF NOT, SKIP TO QUESTION 107.

106. Have any of the following ever occurred? **(CHECK ALL THAT APPLY)**

1[ ]Blade throw
2[ ]Excessive vibration
3[ ]Excessive noise
4[ ]Lack of sufficient wind
5[ ]Tower collapse
6[ ]Leakage of battery acid
7[ ]"Attractive nuisance" problems with people
8[ ]Signal interference with TV
9[ ]Birds flying into the system
10[ ]Hazard to utility repairmen
11[ ]Problems hooking up with utility grid
12[ ]Hazards in servicing the machine
13[ ]Increase in homeowners' insurance
14[ ]Other (PLEASE SPECIFY) ______________________________
15[ ]None of these
16[ ]Don't know

107. Please describe what you consider to be the most negative experiences associated with owning a solar energy system.

____

____

____

____

____

____

____

____

____

____

____

____

____

THAT FINISHES THE SERIES OF QUESTIONS ABOUT SYSTEMS. THE FOLLOWING THREE QUESTIONS ARE CONCERNED WITH YOUR IDEAS ON SOME NATIONAL ISSUES.

108. There is a lot of talk these days about what the aims of this country should be for the next 10 years. Listed below are some of the goals which different people would give top priority. Would you please indicate which of these you personally consider the most important? And which would be the next most important? **(PLACE LETTERS ON LINES BELOW)**

a. Maintaining a high level of economic growth

b. Making sure that this country has strong defense forces

c. Seeing that people have more say about how things are done at their jobs and in their communities

d. Protecting nature from being spoiled and polluted

__________Most important

__________Next most important

109. If you had to choose, which one item in the following list would you say is most desirable? What would be your second choice? **(PLACE LETTERS ON LINES BELOW)**

e. Maintaining order in the nation

f. Giving the people more say in important government decisions

g. Fighting rising prices

h. Protecting freedom of speech

__________Most important

__________Next most important

110. In your opinion, which one of these is most important? Which is next most important? **(PLACE LETTERS ON LINES BELOW)**

i. A stable economy

j. Progress toward a less impersonal and more humane society

k. Progress toward a society in which ideas count more than money

l. The fight against crime

__________Most important

__________Next most important

THE NEXT TWO QUESTIONS DEAL WITH SOURCES OF INFORMATION ABOUT SOLAR ENERGY.

111. Here is a list of possible sources of information about solar energy. Please check the **three best** information sources you found.

1[ ]Friends, relatives, neighbors, and acquaintances
2[ ]People who have solar energy systems
3[ ]Demonstration buildings or model solar homes
4[ ]Books, journals, and reports about solar energy
5[ ]Magazines and newspapers
6[ ]Television and radio programs
7[ ]Seminars/workshops
8[ ]Adult education classes
9[ ]Local contractors (heating, plumbing, etc.)
10[ ]Energy fairs, exhibits, home shows
11[ ]Homebuilders, architects
12[ ]Home energy audits
13[ ]Other (PLEASE SPECIFY) ____________________________
14[ ]None of these

112. Here is a list of organizations that have information about solar energy. Please check the **three best** information sources you found.

1[ ]Solar companies
2[ ]Utility companies
3[ ]Consumer groups (e.g., Consumers Union, etc.)
4[ ]State Energy or Solar Office
5[ ]Solar or alternative technology clubs/associations
6[ ]Federal energy agencies or laboratories (e.g., Department of Energy, Solar Energy Research Institute, Regional Solar Energy Centers, etc.)
7[ ]Federal information organizations (e.g., Government Printing Office, National Solar Heating and Cooling Information Center, etc.)
8[ ]State or county agricultural extension service agents
9[ ]Environmental organizations (e.g., Sierra Club, Wilderness Society, etc.)
10[ ]Public library
11[ ]Local government
12[ ]Lobbying organizations (e.g., Solar Lobby, etc.)
13[ ]Trade or professional organizations (e.g., Solar Energy Industries Association)
14[ ]Other (PLEASE SPECIFY) ____________________
15[ ]None of these

HERE IS A SET OF STATEMENTS ABOUT ACTIVITIES THAT SOME PEOPLE ENGAGE IN. PLEASE INDICATE WHETHER AND HOW MUCH YOU OR MEMBERS OF YOUR HOUSEHOLD ENGAGE IN THESE ACTIVITIES.

113. Use a bicycle for transportation or recreation. **(CHECK ONE)**

1[ ]Regularly
2[ ]Usually
3[ ]Frequently
4[ ]Sometimes
5[ ]Never

114. Recycle the newspapers, glass, or cans used at home. **(CHECK ONE)**

1[ ]Recycle all of this material
2[ ]Most of this material
3[ ]About half of this material
4[ ]Some of this material
5[ ]Never recycle

115. Change oil in the family car(s) at home. **(CHECK ONE)**

1[ ]Regularly
2[ ]Usually
3[ ]Frequently
4[ ]Sometimes
5[ ]Never

116. Developing and using skills to increase self-reliance, such as in carpentry, car repair, food preservation. **(CHECK ONE)**

1[ ]Regularly
2[ ]Usually
3[ ]Frequently
4[ ]Sometimes
5[ ]Never

117. Buying clothing at a garage sale or second-hand store. **(CHECK ONE)**

1[ ]All of the household's clothing
2[ ]Most items
3[ ]About half of the household's clothing
4[ ]A few items
5[ ]None of the household's clothing

118. Contribute to ecologically-oriented organizations (such as the Sierra Club, etc.) **(CHECK ONE)**

1[ ]Contribute regularly to 2 or more organizations
2[ ]Contribute regularly to 1 organization
3[ ]Occasionally contribute
4[ ]Used to contribute, but no longer do
5[ ]Never have contributed

119. Engage in exercise for physical fitness (e.g., running, swimming, calisthenics, etc.).

1[ ]Regularly
2[ ]Usually
3[ ]Frequently
4[ ]Sometimes
5[ ]Never

FOLLOWING ARE A FEW QUESTIONS ABOUT YOU AND YOUR HOUSEHOLD.

120. Does anyone in your household own any of the following? **(CHECK ALL THAT APPLY)**

1[ ]Video game
2[ ]Shower flow restrictor
3[ ]Microwave oven
4[ ]Solar toys and gadgets (calculators, wristwatches, etc.)

121. The following list represents different kinds of magazines widely read today. Please indicate which of these are frequently read in your household. **(CHECK ALL THAT APPLY)**

1[ ]National Geographic
2[ ]Time
3[ ]Solar Age
4[ ]Parents
5[ ]Better Homes and Gardens
6[ ]Rain
7[ ]Redbook
8[ ]Newsweek
9[ ]Field and Stream
10[ ]Mother Earth News
11[ ]People
12[ ]Popular Mechanics
13[ ]Psychology Today
14[ ]Reader's Digest
15[ ]U.S. News and World Report
16[ ]Woman's Day
17[ ]Cosmopolitan
18[ ]Family Circle
19[ ]Playboy
20[ ]Sunset
21[ ]Consumer Report
22[ ]Retirement Living
23[ ]Popular Science
24[ ]Southern Living
25[ ]None of these

122. What type of structure do you live in? **(CHECK ONE)**

1[ ]One-family house
2[ ]Attached house (duplex, townhouse, etc.)
3[ ]Mobile home
4[ ]Other (PLEASE SPECIFY) ______________________________

123. How likely is it that you will move within the next three years? **(CHECK ONE)**

1[ ]Very likely
2[ ]Likely
3[ ]Unsure
4[ ]Unlikely
5[ ]Very unlikely

124. What is the main fuel used to heat your home? **(CHECK ONE)**

1[ ]Natural gas
2[ ]Fuel oil
3[ ]Electricity
4[ ]Propane
5[ ]Solar energy
6[ ]Wood
7[ ]Other (PLEASE SPECIFY) ____________________

125. Which of the following activities have you engaged in during the last year? **(CHECK ALL THAT APPLY)**

1[ ]Worked with others to help solve community problems
2[ ]Contacted political officials or politicians
3[ ]Ran for political office
4[ ]Wrote a letter to the editor
5[ ]Made a speech
6[ ]Worked on a political campaign
7[ ]Wrote an article
8[ ]Was an officer of a community organization
9[ ]Signed a petition
10[ ]Often gave friends and neighbors advice
11[ ]None of these

126. Regardless of your political party identification, in general would you call yourself......?

1[ ]A strong conservative
2[ ]A moderate conservative
3[ ]Middle of the road
4[ ]A moderate liberal
5[ ]A strong liberal
6[ ]A radical

127. Please check off which one of the following categories **most nearly** describes the kind of work the chief wage earner in your immediate family does. **(IF CHIEF WAGE EARNER IS UNEMPLOYED, CHECK OFF WHAT TYPE OF WORK HE/SHE WOULD DO IF EMPLOYED.) (CHECK ONE)**

1[ ]**Professional worker:** e.g., lawyer, doctor, scientist, teacher, systems analyst, musician, etc.

2[ ]**Works at a skilled trade or craft:** e.g., printer, baker, tailor, RR engineer, plumber, or does mechanical work such as garage mechanic, carpenter

3[ ]**Semi-skilled worker:** e.g., operates a machine in a factory, is an assembly-line worker in a factory, drives a truck, taxi

4[ ]**Manager, executive, or official:** in business, government agency, or other organization

5[ ]**Runs own business with one or more employees:** e.g., store, factory, plumbing, contractor, etc.

6[ ]**Farm owner, farm manager**

7[ ]**Clerical or office worker:** in business, government agency, or other type of organization, e.g., typist, secretary, postal clerk, or telephone operator

8[ ]**Sales worker:** e.g., a clerk in a store or a door-to-door salesperson

9[ ]**Manufacturer's representative:** e.g., outside salesperson, sales representative

10[ ]**Service worker who performs services:** e.g., policeman, fireman, waiter, maid, or barber

11[ ]**Laboring worker** (other than farm): e.g., plumber's assistant, construction laborer, longshore man, or other physical work

12[ ]**Farm laborer, farm helper, or farm foreman**

13[ ]**Retired**

14[ ]**Full-time student**

15[ ]**Housewife**

16[ ]**Other** (PLEASE SPECIFY) ____________________

128. Are you professionally involved with solar energy? **(CHECK ONE)**

1[ ]Yes (GO TO Q. 129)

2[ ]No (SKIP TO Q. 130)

129. How are you involved? **(CHECK ALL THAT APPLY)**

1[ ]Installer of systems
2[ ]Designer of systems
3[ ]Architect
4[ ]Sales of systems
5[ ]Teacher of solar courses
6[ ]Other (PLEASE SPECIFY) ____________________

130. Are you a member of a solar energy group, such as a local solar energy association or a chapter of the International Solar Energy Society? **(CHECK ONE)**

1[ ]Yes
2[ ]No

131. What is the highest level of education you have completed? **(CHECK ONE)**

1[ ]Less than high school
2[ ]High school graduate
3[ ]Trade or technical school
4[ ]Some college
5[ ]College graduate
6[ ]Graduate work or more

132. We would like to know a little about your household—the number of people who live here and their ages. Please start with yourself and indicate your age. Then, please fill in a code letter for each other person living here to indicate that person's relationship to you, and indicate that person's age.

	Relationship	Age
You yourself		______
	______	______
	______	______
	______	______
	______	______
	______	______
	______	______
	______	______
	______	______
	______	______

Codes:
- D= daughter
- S= son
- W= wife
- H= husband
- R= roommate or housemate
- A= aunt
- U= uncle
- G= grandparent
- SIB= sibling (brother, sister)
- OR= other related by blood or marriage
- ON= other nonrelated

133. Please check off which category best represents the total annual income, before taxes, of your immediate family living in your household?

1[ ]Under $5000 a year (or under $96/wk.)
2[ ]$5000 - $6999 a year (or $96 - $134.50/wk.)
3[ ]$7000 - $9999 a year (or $135 - $192.50/wk.)
4[ ]$10,000 - $11,999 a year (or $193 - $229.50/wk.)
5[ ]$12,000 - $14,999 a year (or $230 - $288.50/wk.)
6[ ]$15,000 - $19,999 a year (or $289 - $383.50/wk.)
7[ ]$20,000 - $24,999 a year (or $384 - $480.50/wk.)
8[ ]$25,000 - $34,999 a year (or $481 - $672.50/wk.)
9[ ]$35,000 - $44,999 a year (or $673 - $864.50/wk.)
10[ ]$45,000 - $54,999 a year (or $865 - $1056.50/wk.)
11[ ]Over $55,000 a year (or over $1057/wk.)

134. Which of the following best describes your racial or ethnic identification? **(CHECK ONE)**

1[ ]American Indian or Alaskan Native
2[ ]Asian or Pacific Islander
3[ ]Black, not of Hispanic origin
4[ ]Hispanic
5[ ]White, not of Hispanic origin

135. What is your zip code?

☐ ☐ ☐ ☐ ☐

Thank you very much for participating in this survey!

Just put completed questionnaire in the postage paid return envelope and drop in the mail.

☆ U.S. GOVERNMENT PRINTING OFFICE: 1980—677-175/259

APPENDIX D: DEMOGRAPHICS, VALUES, AND LIFE-STYLE CHARACTERISTICS OF THE NATIONAL HOMEOWNERS SAMPLE

Gallup determined that the homeowners data were sufficiently representative of the population of U.S. homeowners that no weighting of data was necessary. Where available, somewhat comparable U.S. Census data are included in this discussion with the sample data tables. However, it should be noted that census data represent the total population of all households (including renters and homeowners);

TABLE D1. Home Ownership

Do you own or rent this home?

	%	N
Own	100	(2013)
Other arrangement	—	(10)
	100	(2023)
Interviewer check-off on type of housing.		
One-family detached house	90	(1819)
Attached house	6	(136)
Mobile home	3	(58)
Other	1	(10)
	100	(2023)

Note: All respondents indicated that they "have control over decisions to change the external appearance of this house without going to a board or committee for approval."

All occupied housing units, national data	%*
Owner-occupied	65
Renter-occupied	35
	100

*Figures from Statistical Abstracts of the U.S., 1980.

they are, thus, not directly comparable to the homeowners sample data. Nevertheless, they give some approximation of how representative the national homeowners sample is on key demographic characteristics. The homeowners sample appears to be representative on all variables for which there were data.

As indicated in Table D1, virtually all respondents were homeowners with full control over decisions to change the external appearance of their homes without going to a board or committee for approval. Thus, row house owners with full decision control are included in the same, while single-family residence owners who must get approval for external changes are excluded. The rationale for this is that the decision process for an individual or family innovation de-

TABLE D2. Age

What is your age, please?

Response	%	N
24 or less	4	(76)
25 to 34	17	(345)
35 to 44	20	(403)
45 to 54	21	(414)
55 to 64	18	(364)
65 to 74	13	(270)
75 and older	6	(127)
	99	(1999)
	Missing data	(24)
	Total N	(2023)

All adults, national data	%*
22-24	8.6
25-34	24.8
35-44	17.8
45-54	16.3
55-64	14.9
65 and over	17.6
	100.0

*Figures based on Statistical Abstracts of the U.S., 1980. Younger adults are underrepresented in this homeowners sample, as would be expected.

TABLE D3. Educational Level

What is the highest level of education you have completed?

Response	%	N
Less than high school	22	(452)
High school graduate	32	(639)
Trade or technical school	6	(119)
Some college	18	(369)
College graduate	13	(252)
Graduate work or more	9	(172)
Don't know	—	(3)
	100	(2006)
	Missing data	(17)
	Total N	(2023)

All adults 25 and over, national data	%*
Less than high school	32.3
High school graduate	36.6
Some college	14.7
College graduate or more	16.4
	100.0

*Figures based on Statistical Abstracts of the U.S., 1980.

cision is theoretically distinct from the decision process in organizations and communities. Table D1 shows that 90 percent of the sample live in single-family residences, 6 percent in attached residences, and 3 percent in mobile homes. Homeowners represent about 65 percent of all occupied housing units in the country.

DEMOGRAPHICS

The sample is 55 percent male and 45 percent female. The age distribution of the sample is presented in Table D2. About a fifth of the sample is younger than 34, while 19 percent are over 65. In the middle adult years between 35 and 64 are 59 percent of the sample. Younger persons are somewhat underrepresented in the sample, as would be expected among homeowners.

The educational level of the sample is presented in Table D3. Over a fifth have less than a high school education; about a third are high school graduates; 6 percent went to trade or technical school. About 40 percent have at least some college. College graduates are 13 percent; 9 percent went on to graduate work. Those with lower educational achievement are slightly underrepresented in the sample, probably as a result of the link between educational attainment and income, which would in turn be related to home ownership.

Data on occupations are presented in Table D4. The modal occupational category is "retired," at 19 percent. Occurring next frequently are skilled workers (17 percent) and professionals (16 percent); the distribution across categories is diverse. There are very few full-time students, sales workers, and farm laborers or foremen in the sample. The small proportion of housewives, 2 percent, underscores the demise of this role in society but also suggests that housewives may have been more likely to refuse the interview.

TABLE D4. Occupation of Chief Wage Earner

Please tell me which of the categories on this card most nearly describes the kind of work the chief wage earner in your immediate family does.*

Response	%	N
Professional worker	16	(307)
Works at skilled trade or craft	17	(333)
Semi-skilled worker	10	(196)
Manager, executive or official	9	(186)
Runs own business with one or more employees	6	(113)
Farm owner, farm manager	4	(70)
Clerical or office worker	5	(99)
Sales worker	1	(24)
Manufacturers representative	2	(34)
Service worker who performs services	4	(88)
Laboring worker (other than farm)	4	(87)
Farm laborer, helper or foreman	—	(11)
Retired	19	(383)
Full time student	—	(7)
Housewife	2	(45)
Other	—	(1)
	99	(1984)
Missing data		(39)
Total N		(2023)

*The U.S. census data employ categories sufficiently different to make any meaningful comparisons with the sample difficult to draw.

TABLE D5. Gross Annual Income

Which category best represents the total annual gross income of all persons living in your household?

Response	%	N
Under $5,000	6	(117)
5,000 - 6,999	5	(100)
7,000 - 9,999	7	(140)
10,000 - 11,999	6	(125)
12,000 - 14,999	9	(169)
15,000 - 19,999	14	(273)
20,000 - 24,999	15	(287)
25,000 - 34,999	16	(302)
35,000 - 44,999	10	(186)
45,000 - 54,999	4	(75)
Over $55,000	4	(74)
Don't know	5	(97)
	101	(1945)
	Missing data	(78)
	Total N	(2023)

All households, national data	%*
Under $5,000	14.7
5,000 - 6,999	7.6
7,000 - 9,999	10.8
10,000 - 11,999	7.1
12,000 - 14,999	9.6
15,000 - 24,999	27.5
25,000 and over	22.7
	100.0

*Figures based on Statistical Abstracts of the U.S., 1980. Lower income groups are under-represented and higher income groups are over-represented in this homeowners sample, as would be expected.

Income levels of the sample are as varied as national data. Over half of households earn between $15,000 and $45,000 annually; a third of the households live on under $15,000 a year. Data on income

TABLE D6. Political Orientation

Regardless of your political party identification, in general would you call yourself...

Response	%	N
A strong conservative	9	(180)
A moderate conservative	34	(694)
A middle of the roader	35	(699)
A moderate liberal	13	(260)
A strong liberal	2	(43)
A radical	1	(16)
Don't know	6	(124)
	100	(2016)
Missing data		(7)
Total N		(2023)

TABLE D7. Race

Which of the following best describes your racial or ethnic identification?

Response[a]	%	N
White, not of Hispanic origin	88	(1767)
Black, not of Hispanic origin	8	(167)
Hispanic/Chicano	2	(44)
Asian/Pacific Islander	1	(16)
American Indian or Alaskan native	1	(11)
	100	(2005)
Missing data		(18)
Total N		(2023)

[a]Standard federal statistical categories.

Total population, national data	%*
White	86
Black	12
Other	2

*Figures from Statistical Abstracts of the U.S., 1980.

TABLE D8. Family Type

We would like to know a little about your household—the number of people who live here and their ages. What are the names of all other people living or staying here? What is his/her age/relationship to you?

Response	%		N
Two adults with children, all under 13	19	45	(358)
Two adults with children age 13 or over	16		(312)
Two adults with children, both over and under age 13	10		(182)
Married couple without children	30		(569)
One adult	17		(326)
Single parent family	5		(86)
Extended family	4		(82)
	101		(1915)
		Missing data	(108)
		Total data	(2023)

Number of Children

Response	%	N
None	45	(913)
One	19	(389)
Two	20	(409)
Three	11	(215)
Four	3	(68)
Five or more	1	(29)
	99	(2023)
$\bar{X}$ 1.14		
Range 0-9		

All households, national data	%*
None	47.5
One	20.6
Two	19.0
Three	8.5
Four or more	4.5
	100.1

*Figures from Statistical Abstracts of the U.S., 1980.

are presented in Table D5. Lower-income persons are somewhat underrepresented in the sample, as would be expected among homeowners.

Reflecting the conservative victory in the November 1980 election, 78 percent of the sample characterize themselves as middle-of-the-road to strongly conservative in political orientation; 15 percent say they are liberals, and 1 percent say they are radicals. These data are shown in Table D6.

TABLE D9. City Size

Objective Coding	%	N
Open country	30	(598)
Under 2,500	5	(107)
2,500 to 9,999	4	(77)
10,000 to 24,999	10	(181)
25,000 to 49,999	4	(90)
50,000 to 99,999	5	(97)
100,000 to 249,999 (including urban fringe)	10	(193)
250,000 to 499,999 (including urban fringe)	8	(168)
500,000 to 999,999 (including urban fringe)	10	(193)
1,000,000 and over (including urban fringe)	16	(319)
	102	(2023)

Total population, national data	%*
Other rural	21
Under 2,500	5
2,500 to 10,000	10
10,000 to 25,000	11
25,000 to 50,000	9
50,000 to 100,000	8
100,000 to 250,000	7
250,000 to 500,000	5
500,000 to 1,000,000	6
1,000,000 and over	9
Unincorporated parts of urbanized areas	7
	98

*Figures based on Statistical Abstracts of the U.S., 1980.

The sample is primarily white (88 percent), with 8 percent blacks, 2 percent Hispanic or Chicano, and 1 percent each Asian and American Indian. Table D7 presents the sample's racial distribution.

Over half the homeowners (55 percent) live in families with children, while 30 percent are married couples without children (see Table D8). Single adults comprise 17 percent of these households. Two-parent families are 45 percent of households, single parent families are 5 percent, and extended families with three generations are 4 percent. The families are small, however; the mean number of children is 1.14, with the range from 0 (45 percent) to 9 children.

As Table D9 shows, about half the sample lives in cities of 50,000 or more, with 16 percent in the largest SMSAs; 30 percent are rural in residence, while almost one in four live in smaller towns (under 50,000 in population).

The majority of respondents heat their homes with natural gas, as shown in Table D10. Used next most frequently for home heating is fuel oil, by 19 percent. Electricity is used by 13 percent, wood as a primary heat source by 8 percent, propane by 4 percent, and other sources by 1 percent.

Respondents were queried about their average energy costs for electricity, heating, and hot water. Most respondents simply esti-

TABLE D10. Main Fuel Used for Space Heating

What is the main fuel used to heat your home?

Response	%		N
Natural gas	56		(1101)
Fuel oil	19		(368)
Electricity	13		(249)
Wood	8		(161)
Propane	4		(72)
Other	1		(17)
Don't know	—		(5)
	101		(1973)
		Missing data	(50)
		Total N	(2023)

All households, national data	%*
Natural gas	55
Fuel oil, kerosene	22
Electricity	16
Liquid petroleum gas	4
Wood	2
Other	1
	100

*Figures from Statistical Abstracts of the U.S., 1980.

TABLE D11. Average Monthly Utility Costs in Winter

During this past winter (December, January, and February) what were your average energy costs for electricity, heating, and hot water?

Response	%		N
Under $50	10		(190)
$50 - 99	33		(661)
$100 - 149	27		(546)
$150 - 199	12		(249)
$200 or more	16		(315)
Don't know	2		(33)
	100		(1994)
		Missing data	(29)
		Total N	(2023)
$\bar{X}$ $129.39			
Range $4 - 1300			
SD 107.51			

Average Monthly Utility Costs in Summer

During this past summer (June, July, August) what were your average monthly energy costs for electricity, heating, hot water, and cooling?

Response	%		N
Under $25	7		(137)
$25 - 49	28		(555)
$50 - 74	25		(506)
$75 - 99	14		(275)
$100 - 124	10		(202)
$125 and over	15		(303)
Don't know	1		(25)
	100		(2003)
		Missing data	(20)
		Total N	(2023)
$\bar{X}$ $76.59			
Range $4 - 900			
SD 61.23			

mated these costs from memory; a few looked them up in their records. Data are summarized in Table D11. Average monthly costs in winter vary widely; however, most people report paying between $50 and $150 per month. Almost 30 percent pay over $150 a month. The average winter utility bill is $129.30 a month. Average monthly costs in summer are also highly variable; most people report paying between $25 and $75 per month. Almost 40 percent pay over $75 a month in summer. The average summer utility bill is $76.59 a month.

The sample was asked how likely it was that they would move within the next three years, as a measure of mobility/stability. Most people said they are unlikely to move (77 percent); 16 percent said

TABLE D12. Household Stability

How likely is it that you will move within the next three years? Would you say it is very likely, likely, unlikely, or very unlikely?

Response	%		N
Very likely	10	16	(197)
Likely	6		(124)
Unsure	7		(142)
Unlikely	20	77	(397)
Very unlikely	57		(1157)
	100		(2107)
		Missing data	(6)
		Total N	(2023)

TABLE D13. Magazine Readership

The following list represents different kinds of magazines widely read today. Please tell me the number of each one which is frequently read in your household. (CHECK ALL THAT APPLY)

Magazine	%	N
Reader's Digest	50	(1019)
Better Homes and Gardens	38	(758)
Family Circle	26	(531)
National Geographic	26	(524)
Woman's Day	25	(509)
Time	19	(388)
Newsweek	18	(356)
Popular Mechanics	14	(290)
Consumer Report	14	(289)
Field and Stream	12	(237)
Redbook	11	(227)
U.S. News and World Report	11	(217)
Popular Science	10	(206)
People	10	(201)
Southern Living	8	(159)
Playboy	8	(155)
Cosmopolitan	7	(147)
Parents	6	(111)
Psychology Today	4	(84)
Mother Earth News	3	(69)
Retirement Living	3	(64)
Sunset	3	(62)
Solar Age	—	(13)
Rain	—	(1)
None of these	15	(305)
Don't know	—	(11)
N	(2015)	
Missing data	(8)	
Total N	(2023)	

TABLE D14. Opinion Leadership

Which of the following activities have you engaged in during the last year? (CHECK ALL THAT APPLY)

Response	%	N
Have often given friends/neighhbors advice	40	(816)
Signed a petition	39	(789)
Worked with others to help solve community problems	23	(464)
Contacted political officials or politicians	16	(316)
Was an officer of a community organization	11	(229)
Made a speech	10	(196)
Wrote a letter to the editor	7	(146)
Worked on a political campaign	7	(132)
Wrote an article	6	(117)
Ran for political office	1	(18)
None of these	36	(723)
Don't know	—	(11)

N (2017)
Missing data (6)
Total N (2023)

they were likely to move soon, and 7 percent were unsure. Data are presented in Table D12.

Table D13 presents data on the magazines read by sample households. The five magazines most frequently mentioned are: Reader's Digest (50 percent), Better Homes and Gardens (38 percent), Family Circle (26 percent), National Geographic (26 percent), and Woman's Day (25 percent).

To measure opinion leadership, an item was included on leadership and community activities during the past year. Data on frequency of response are presented in Table D14. Those engaging in a high number of these activities (except for signing petitions) would score high on the opinion leadership index used in the study's analyses.

TABLE D15. Innovativeness

Does anyone in your household own any of the following (CHECK ALL THAT APPLY)

Response	%	N
Shower flow restrictor	19	(383)
Microwave oven	19	(379)
Video game	16	(319)
Solar toys or gadgets (calculators, wristwatches, etc.)	14	(285)
Don't know	—	(4)
None of these items	55	(1115)

N (2013)
Missing data (10)
Total N (2023)

The item used to measure innovativeness asked whether anyone in the household owned any of the following items: (1) shower flow restrictor, (2) microwave oven, (3) video game, and (4) solar toys or gadgets, including calculators, wristwatches, etc. Inexpensive items were included so that the item would not be a measure of income. Data are presented in Table D15. The more of these items a respondent reports having, the higher his or her score on the innovativeness index used in the study's data analyses.

Over half of the sample reported owning none of these items. Almost one in five had a shower flow restrictor and a similar proportion had a microwave oven; 16 percent reported owning a video game and 14 percent had solar toys and gadgets.

VALUES

A values scale used for 20 years in cross-cultural research was replicated in this survey. The scale is presented in three separate items, which present four statements each and ask respondents to choose the most important and next most important goals of the country from the four listed.

In all, 12 statements are presented; half of the statements are intended to represent "materialist" values and half "postmaterialist" values (Inglehart 1977). Materialist values were represented by the following statements:

- Maintaining a high level of economic growth
- Making sure this country has strong defense forces
- Maintaining order in the nation
- Fighting rising prices
- A stable economy
- The fight against crime

Postmaterialist values were represented by the following statements:

- Protecting nature from being spoiled or polluted
- Seeing that people have more say about how things are done at their jobs and in their communities
- Protecting freedom of speech
- Progress toward a less impersonal and more humane society
- Giving people say in important government decisions
- Progress toward a society in which ideas count more than money

Each of the three items presented two materialist and two postmaterialist statements.

TABLE D16. Values

There is a lot of talk these days about what the aims of this country should be for the next ten years. Here are some goals which different people would give top priority. Would you please indicate which of these you consider the most important? And which would be the next most important?

	Most Important	Next Most Important	Total
Making sure that this country has strong defense forces.	37%	30	67
Maintaining a high level of economic growth.	33	28	61
Protecting nature from being spoiled or polluted.	15	22	37
Seeing that people have more say about how things are done at their jobs and in their communities.	14	20	34
Don't know.	1	—	
	100	100	
N	(2008)	(1991)	
Missing data	(15)	(32)	
Total N	(2023)	(2023)	

If you had to choose, which one item in the following list would you say is most desirable? What would be your second choice?

	Most Important	Next Most Important	Total
Fighting rising prices.	45%	30	75
Maintaining order in the nation.	23	26	49
Giving people say in important government decisions.	18	30	48
Protecting freedom of speech.	13	15	28
Don't know.	—	—	
	99	101	
N	(2006)	(1985)	
Missing data	(17)	(38)	
Total N	(2023)	(2023)	

In your opinion, which one of these is most important? Which is the next most important?

	Most Important	Next Most Important	Total
A stable economy.	60%	25	85
The fight against crime.	24	40	64
Progress toward a less impersonal and more humane society.	8	16	24
Progress toward a society in which ideas count more than money.	8	19	27
Don't know.	—	—	
	100	100	
N	(2009)	(1987)	
Missing data	(14)	(36)	
Total N	(2023)	(2023)	

Data on the frequency of response for the three items are presented in Table D16. In general, the statements representing materialist values are designated most important and next most important by a majority of the sample. In the second item, 49 percent defined "maintaining order in the nation" as important, and 48 percent defined "giving people say in important government decisions" as important. The latter was the only statement of postmaterialist values to achieve such a high level of importance. The other postmaterialist value statements were designated as important by sample minorities ranging from 24 to 37 percent. Clearly, homeowners

TABLE D17. Voluntary Simplicity Life-style

I am going to read statements about activities that some people engage in. I would like you to tell me whether and how much you and other members of your household engage in these activities.

1. Use a bicycle for transportation or recreation.

Response	%	N
Regularly	9	(173)
Usually	3	(53)
Frequently	10	(203)
Sometimes	26	(524)
Never	53	(1063)
Don't know	—	(3)
	101	(2019)
	Missing data	(4)
	Total N	(2023)

2. Recycle the newspapers, glass, or cans used at home.

Response	%	N
Recycle all this material	12	(236)
Most of this material	11	(212)
About half of this material	9	(179)
Some of this material	24	(478)
Never recycle	45	(908)
Don't know	—	(6)
	101	(2019)
	Missing data	(4)
	Total N	(2023)

3. Change oil in the family car(s) at home.

Response	%	N
Regularly	43	(866)
Usually	7	(135)
Frequently	10	(194)
Sometimes	10	(202)
Never	30	(611)
Don't know	—	(7)
	100	(2015)
	Missing data	(8)
	Total N	(2023)

(continued)

TABLE D17 (continued)

4. Developing and using skills to increase self-reliance, such as in carpentry, car repair, and food preservation.

Response	%	N
Regularly	42	(849)
Usually	12	(235)
Frequently	15	(299)
Sometimes	18	(360)
Never	13	(270)
Don't know	—	(4)
	100	(2017)
	Missing data	(4)
	Total N	(2021)

5. Buying clothing at a garage sale or second hand store.

Response	%	N
All of the household's clothing	1	(22)
Most times	2	(42)
About half of the household's clothing	6	(117)
A few items	33	(661)
None of the household's clothing	58	(1164)
Don't know	—	(11)
	100	(2017)
	Missing data	(6)
	Total N	(2023)

6. Contribute to ecologically-oriented organizations (such as the Sierra Club, etc.)

Response	%	N
Contribute regularly to two or more organizations	3	(51)
Contribute regularly to one organization	3	(65)
Occasionally contribute	13	(263)
Used to contribute, but no longer do	3	(64)
Never have contributed	77	(1552)
Don't know	1	(21)
	100	(2016)
	Missing data	(7)
	Total N	(2023)

(continued)

TABLE D17 (continued)

7. Engage in exercise for physical fitness (e.g., running, swimming, calisthenics, etc.)

Response	%	N
Regularly	23	(464)
Usually	10	(193)
Frequently	15	(312)
Sometimes	27	(540)
Never	25	(508)
Don't know	—	(1)
	100	(2018)
		Missing data (5)
		Total N (2023)

express a preference for materialist values in their definition of important national goals.

VOLUNTARY SIMPLICITY LIFE-STYLE

Some of the items used in previous research to measure preference for voluntary simplicity were selected for inclusion in this survey (Leonard-Barton 1978). Each item included represented a factor or dimension of voluntary simplicity that had been defined in earlier research.

Seven statements on activities associated with voluntary simplicity were included; respondents were asked how frequently they engage in these behaviors. Data are presented in Table D17.

The seven statements are:

- Use a bicycle for transportation or recreation
- Recycle the newspapers, glass, or cans used at home
- Change oil in the family car(s) at home
- Developing and using skills to increase self-reliance, such as in carpentry, car repair, and food preservation
- Buying clothing at a garage sale or secondhand store
- Contribute to ecologically oriented organizations (such as the Sierra Club, etc.)
- Engage in exercise for physical fitness (e.g., running, swimming, calisthenics, etc.)

The majority of the sample engage in activities to increase skills for self-reliance; exercising for physical fitness; changing oil in cars at home; and recycling newspapers, glass, or cans used at home. A majority does not bicycle, buy used clothing, or contribute to environmental organizations.

APPENDIX E: STATES WITH SOLAR TAX CREDITS

States with solar tax credits are:

Alaska
Arizona
Arkansas
California
Colorado
Delaware
Hawaii
Idaho
Kansas
Massachusetts
Michigan
Minnesota
Montana
Nebraska
New Mexico
North Carolina
North Dakota
Oklahoma
Oregon
Utah
Vermont
Wisconsin

Sources: *State Solar Energy Incentives Primer* (1979); Anderson (1980).

APPENDIX F: DEMOGRAPHICS, VALUES, AND LIFE-STYLE CHARACTERISTICS OF THE NATIONAL SOLAR HOMEOWNERS SAMPLE

As reported earlier, the sample of solar owners is not necessarily representative of solar homeowners in the country. In addition, diffusion theory would lead us to expect that present solar homeowners would not be typical of the nation's homeowners on key descriptive variables. In this appendix, the solar homeowner sample is described and compared with national data in cases where the latter are available.

As indicated in Table F4, virtually all respondents (95 percent) are solar owners living in single-family detached houses; 3 percent live in attached houses and 1 percent live in mobile homes.

Well over half (59 percent) of the sample became solar owners in 1978 or subsequent years. An additional 19 percent adopted in 1977; 17 percent adopted in the years 1973-76, with just 4 percent having adopted prior to 1973 (see Table F1). The mean length of system ownership was three years.

Tables F2 and F3 show year of installation of active and passive systems, respectively. Passive system owners tended to be earlier

TABLE F1. Date of Solar Ownership

Response	%	N
Year, prior to 1973	4	(143)
1973	1	(37)
1974	2	(74)
1975	5	(185)
1976	9	(329)
1977	19	(705)
1978	25	(916)
1979	26	(975)
1980	9	(351)
	100	(3715)
		Missing data (94)
		Total N (3809)

$\bar{X}$ years = 3.00
Range of years = 1926-1980
Range in years = 0-54

TABLE F2. Year of Installation of Passive System

When was your passive energy system(s) constructed?

Response	System 1 %	System 2 %	Total %
Prior to 1973	10	5	9
1973	2	1	2
1974	2	1	2
1975	4	4	4
1976	8	7	8
1977	15	10	14
1978	22	22	22
1979	22	25	22
1980	11	22	13
Don't know	4	3	4
	100	100	100
N	(1088)	(271)	(1359)
Missing data	(11)	(20)	(31)
Total N	(1099)	(291)	(1390)

System 1

$\bar{X}$ = 3.96 years
Range = 1-68 years

System 2

$\bar{X}$ = 2.96 years
Range = 1-64 years

TABLE F3. Year of Installation of Active System

In what year was your (active) system(s) installed?

Response	System 1 %	System 2 %	Total %
Prior to 1973	2	3	2
1973	1	1	1
1974	1	3	2
1975	5	6	5
1976	9	10	9
1977	19	16	19
1978	25	22	24
1979	27	19	26
1980	10	18	11
Don't know	1	2	1
	100	100	100
N	(3210)	(496)	(3706)
Missing data	(—)	(2)	(2)
Total N	(3210)	(498)	(3708)

System 1

$\bar{X}$ = 2.96 years
Range = 0–78 years

System 2

$\bar{X}$ = 2.68 years
Range = 0–71 years

TABLE F4. Housing Type

What type of structure do you live in?

Response	%	(N)
One family house	95	(3565)
Attached house (duplex, townhouse, etc.)	3	(98)
Other	1	(47)
Mobile home	1	(33)
	100	(3743)
	Missing data	(66)
	Total N	(3809)

adopters, with the mean close to four years of ownership; 57 percent of passive systems were installed in 1978 or later years.

The active system sample, with a mean of about three years of ownership, consisted of owners of systems, 61 percent of which were purchased in 1978 or later.

DEMOGRAPHICS

The age distribution of the sample is presented in Table F5. About 17 percent of the sample are younger than 34, while 16 percent are over 65. In the middle adult years between 35 and 64 are 68 percent of the sample. Middle adults are overrepresented in the solar owners sample, compared with the homeowners sample and with national data.

The educational level of the sample is presented in Table F6. The solar owners sample is highly educated. Over 80 percent have

TABLE F5. Age

What is your age, please?

Response	%	(N)
24 or less	—	(14)
25 to 34	17	(608)
35 to 44	26	(970)
45 to 54	22	(799)
55 to 64	20	(716)
65 to 74	13	(467)
75 or older	3	(94)
	101	(3668)
	Missing data	(141)
	Total N	(3809)

TABLE F6. Educational Level

What is the highest level of education you have completed?

Response	%	(N)
Less than high school	3	(97)
High school graduate	10	(360)
Trade or technical school	6	(232)
Some college	19	(688)
College graduate	24	(890)
Graduate work or more	39	(1451)
	101	(3718)
	Missing data	(91)
	Total N	(3809)

some college or above. About one in four are college graduates, and about 40 percent have graduate work or more. This is extremely different from both homeowners nationally, 9 percent of whom have graduate work, and from national educational levels, where 16.4 percent are college graduates and above.

The modal occupational category in the solar owners sample is "professional workers," at 41 percent (see Table F7). Occurring next most frequently are "manager, executive, or official" (15 percent) and "retired" (14 percent). Skilled workers comprise 10 percent

TABLE F7. Occupation of Chief Wage Earner

Please check off which of the following categories most nearly describes the kind of work the chief wage earner in your immediate family does.

Response	%	(N)
Professional worker	41	(1507)
Manager, executive or official	15	(557)
Retired	14	(523)
Works at skilled trade or craft	10	(382)
Runs own business with one or more employees	7	(254)
Service worker who performs services	2	(63)
Manufacturers representative	2	(59)
Clerical or office worker	2	(59)
Semi-skilled worker	1	(46)
Farm owner, farm manager	1	(41)
Laboring worker (other than farm)	—	(18)
Housewife	—	(17)
Sales worker	—	(17)
Farm laborer, helper or foreman	—	(3)
Full time student	—	(2)
Other	4	(156)
	99	(3704)
	Missing data	(105)
	Total N	(3809)

TABLE F8. Professional Involvement with Solar Energy

Are you professionally involved with solar energy?

Response	%		N
Yes	17		(633)
No	83		(3071)
	100		(3704)
		Missing data	(54)
		Total N	(3758)

How are you involved? (CHECK ALL THAT APPLY)

Response	%	N
Designer of systems	52	(273)
Installer of systems	39	(208)
Sales of system	34	(182)
Teacher of solar courses	31	(164)
Architect	16	(82)
Other	41	(214)
N (528)		
Missing data (105)		
Total N (633)		

of the sample. There are very few other occupational groups with significant representation in the sample. Clearly, the sample disproportionately represents professionals, as compared with homeowners nationally.

TABLE F9. Solar Voluntary Association Membership

Are you a member of a solar energy group, such as a local solar energy association or a chapter of the International Solar Energy Society?

Response	%		N
Yes	15		(555)
No	85		(3108)
	100		(3663)
		Missing data	(146)
		Total N	(3809)

TABLE F10. Gross Annual Income

Please check off which category best represents the total annual income, before taxes, of your immediate family living in your household.

Response	%	(N)
Under $5,000	1	(51)
5,000 - 6,999	1	(35)
7,000 - 9,999	2	(87)
10,000 - 11,999	3	(102)
12,000 - 14,999	5	(175)
15,000 - 19,999	10	(366)
20,000 - 24,999	14	(510)
25,000 - 34,999	24	(853)
35,000 - 44,999	17	(617)
45,000 - 54,000	10	(341)
Over $55,000	12	(423)
	99	(3560)
	Missing data	(249)
	Total N	(3809)

Most of the sample is not professionally involved with solar energy (83 percent). Of those who are professionally involved (17 percent), about half are system designers, 39 percent are system installers, 34 percent are in sales, and 31 percent in teaching. Data are presented in Table F8.

Neither are the sample's solar owners members of voluntary associations for solar energy, such as a local solar energy association or the International Solar Energy Society. Most (85 percent) report no such membership (see Table F9).

The solar owners in this sample are more affluent than either the homeowners sample or all households nationally; 12 percent have annual incomes of under $15,000, compared with 33 percent of the homeowners sample and 49.8 percent of all households. About one-

TABLE F11. Political Orientation

Regardless of your political party identification, in general would you call yourself...

Response	%	(N)
A strong conservative	8	(306)
A moderate conservative	34	(1228)
A middle of the roader	30	(1097)
A moderate liberal	21	(743)
A strong liberal	5	(167)
A radical	2	(65)
	100	(3606)
	Missing data	(203)
	Total N	(3809)

TABLE F12. Race

Which of the following best describes your racial or ethnic identification?

Response[a]	%	(N)
White, not of Hispanic origin	85	(3110)
Asian/Pacific Islander	13	(467)
Hispanic/Chicano	1	(40)
American Indian or Alaskan native	1	(22)
Black, not of Hispanic origin	—	(16)
	100	(3655)
	Missing data	(154)
	Total N	(3809)

[a]Standard federal statistical categories.

TABLE F13. Family Size

We would like to know a little about your household—the number of people who live here and their ages. What are the names of all other people living or staying here? What is his/her age/relationship to your? (Paraphrase)

Response	%		(N)
Two adults with children, all under age 13	22	51	(818)
Two adults with children, all 13 or over	18		(660)
Two adults with children, both 13 or over and under age 13	11		(387)
Two adults (married couple)	32		(1168)
One adult	5		(180)
Single parent family	1		(45)
Extended family (with children, grandparents, aunts, uncles, etc.)	4		(165)
Other	7		(257)
	100		(3680)
	Missing data		(129)
	Total N		(3809)

Number of Children

Response	%	(N)
None	41	(1521)
One	18	(664)
Two	24	(876)
Three	10	(389)
Four	4	(134)
Five or more	3	(100)
	100	(3684)
	Missing data	(125)
	Total N	(3809)

TABLE F14. Main Fuel Used for Space Heating

What is the main fuel used to heat your home?

Response	%		N
Electricity	32		(1181)
Solar energy	18		(650)
Fuel oil	16		(580)
Natural gas	12		(426)
Wood	10		(383)
Propane	2		(81)
Other	10		(365)
	100		(3666)
		Missing data	(143)
		Total N	(3809)

TABLE F15. Average Monthly Utility Costs

During this past winter—say during December, January, and February—what were your average monthly costs for electricity, heating, and hot water?

Response	%		N
Under $50	28		(960)
$50-99	37		(1276)
$100-149	15		(530)
$150-199	8		(257)
$200 or more	10		(338)
Don't know	2		(77)
	100		(3438)
		Missing data	(371)
		Total N	(3809)

$\bar{X}$ = $105.94
Range: $0-8000

During this past summer—say during June, July, and August—what were your average monthly costs for electricity, heating, hot water, and cooling.

Response	%		N
Under $25	11		(382)
$25-49	33		(1179)
$50-74	24		(836)
$75-99	12		(414)
$100-124	8		(280)
$125 and over	10		(371)
Don't know	2		(60)
	100		(3522)
		Missing data	(287)
		Total N	(3809)

$\bar{X}$ = $95.32
Range: $0-6,000

fifth of the solar owners sample have annual incomes above $45,000. Data are presented in Table F10.

Reflecting the conservative victory in the November 1980 elections, 72 percent of the solar owners sample characterize themselves as middle-of-the-road to strongly conservative in political orientation. About one in four say they are liberals, and 2 percent say they are radicals; these data are presented in Table F11. The solar owners sample is almost as conservative as the homeowners sample generally and is slightly more liberal in political orientation than the homeowners sample.

The solar owners sample is primarily white (85 percent), with 13 percent Asian/Pacific Islanders, 1 percent Chicano or Hispanic, and 1 percent American Indian or Alaskan. There are almost no blacks in the sample. Table F12 presents the sample's racial distribution. Compared with the U.S. total population, blacks are underrepresented in this sample, while other minorities are overrepresented. Compared with the homeowners sample generally, the same pattern applies.

Over half of the homeowners sample (51 percent) live in nuclear families with children, while 32 percent are married couples without children (see Table F13). This is comparable to the national homeowners sample. Single adults comprise 5 percent of the sample households, fewer than the 17 percent of homeowners. Single parent families are 1 percent of the sample and 5 percent of the homeowners sample, and extended families are 4 percent, identical to homeowners generally.

The plurality of solar owners in the sample heat their homes with electricity (32 percent), as shown in Table F14. This is double the proportion of homeowners and almost double the proportion of

TABLE F16. Household Stability

How likely is it that you will move within the next three years?

Response	%		N
Very likely	11	16	(414)
Likely	5		(189)
Unsure	11		(425)
Unlikely	28	73	(1042)
Very unlikely	45		(1672)
	100		(3742)
		Missing data	(67)
		Total N	(3809)

TABLE F17. Magazine Readership

The following list represents different kinds of magazines widely read today. Please indicate which of these are frequently read in your household. (CHECK ALL THAT APPLY)

Response	%	(N)
National Geographic	53	(1940)
Reader's Digest	43	(1575)
Better Homes and Gardens	41	(1485)
Time	28	(1023)
Consumer Report	27	(974)
Popular Science	23	(844)
Popular Mechanics	22	(803)
Newsweek	21	(770)
Family Circle	21	(753)
Woman's Day	17	(629)
U.S. News and World Report	17	(620)
Mother Earth News	16	(593)
Solar Age	14	(496)
Sunset	13	(464)
Playboy	11	(412)
Redbook	9	(341)
Southern Living	9	(328)
People	8	(281)
Psychology Today	7	(255)
Cosmopolitan	7	(254)
Parents	5	(184)
Retirement Living	5	(166)
None of these	4	(153)
Field and Stream	4	(151)
Rain	2	(81)
Don't know	—	(17)
N	(3650)	
Missing data	(159)	
Total N	(3809)	

TABLE F18. Opinion Leadership

Which of the following activities have you engaged in during the last year? (CHECK ALL THAT APPLY)

Response	%	(N)
Signed a petition	54	(1964)
Have often given friends/neighbors advice	51	(1860)
Worked with others to help solve community problems	41	(1477)
Contacted political officials or politicians	39	(1419)
Was an officer of a community organization	26	(956)
Made a speech	24	(864)
Wrote a letter to the editor	19	(689)
None of these	16	(594)
Wrote an article	16	(579)
Worked on a political campaign	11	(416)
Ran for political office	3	(96)
Don't know	—	(17)
N	(3625)	
Missing data	(184)	
Total N	(3809)	

households generally relying on electricity for their main source of heating. Solar energy is the main source of heating for 18 percent of the sample, far more than for the nation as a whole. Those who use fuel oil comprise 16 percent, which is a few percentage points under national data; 12 percent use natural gas, compared to 56 percent of the homeowners sample and 55 percent of households nationally. Clearly, homeowners in the sample overrepresent electricity users for household heating.

Solar homeowners were queried about their average energy costs for electricity, heating, and hot water. In answering these questions, 64 percent reported that they referred to their utility bills or checkbooks, while 35 percent estimated the costs from memory. The responses, summarized in Table F15, should, therefore, be reasonably accurate for most of the solar households in the sample. Average monthly costs in winter are $105.94, lower than the $129.39 monthly average reported by the homeowners sample. Summer monthly utility costs also vary; the majority of solar homeowners (57 percent) report paying between $25 and $74. Average monthly costs in summer are $95.32, higher than the average $76.59 reported by homeowners. On the basis of these differences, solar energy appears to be saving more energy in winter than in summer, but probably the situation is more complex than that, depending on such factors as climate, size of residence, existence of central air conditioning, and so on.

As shown in Table F16, about three out of four solar owners do not anticipate moving in the near future, while 11 percent are unsure. The pattern of findings on mobility/stability is similar to that for the homeowners sample.

Table F17 presents data on the magazines read by the solar households. The three magazines mentioned most frequently are National Geographic (53 percent), Reader's Digest (43 percent), and

TABLE F19. Innovativeness

Does anyone in your household own any of the following? (CHECK ALL THAT APPLY)

Response	%	(N)
Shower flow restrictor	60	(1740)
Microwave oven	54	(1558)
Solar toys and gadgets (calculators, wrist watches, etc.)	25	(724)
Video game	25	(715)
	N	(2903)
	Missing data	(906)
	Total N	(3809)

TABLE F20. Values

There is a lot of talk these days about what the aims of this country should be for the next 10 years. Listed below are some of the goals which different people would give top priority. Would you please indicate which of these you personally consider the most important? And which would be the next most important?

Response	Most Important %	Next Most Important %	Total %
Maintaining a high level of economic growth	37	23	60
Making sure that this country has strong defense forces	30	28	58
Protecting nature from being spoiled and polluted	21	29	50
Seeing that the people have more say about how things are done at their jobs and in their communities	12	19	31
	100	99	
N	(3559)	(3462)	
Missing data	(250)	(347)	
	(3809)	(3809)	

If you had to choose, which one item in the following list would you say is most desirable? What would be your second choice?

Response	Most Important %	Next Most Important %	Total %
Fighting rising prices	39	34	73
Maintaining order in the nation	27	26	53
Giving the people more say in government decisions	16	24	40
Protecting freedom of speech	19	17	36
	101	101	
N	(3530)	(3443)	
Missing data	(279)	(366)	
	(3809)	(3809)	

In your opinion, which one of these is most important? Which is next most important?

Response	Most Important %	Next Most Important %	Total %
A stable economy	64	20	84
The fight against crime	11	40	51
Progress towards a less impersonal and more humane society	16	22	38
Progress toward society in which ideas count more than money	10	18	28
	101	100	
N	(3511)	(3421)	
Missing data	(298)	(388)	
	(3809)	(3809)	

TABLE F21. Voluntary Simplicity Life-Style

Here is a set of statements about activities that some people engage in. Please indicate whether and how much you or members of your household engage in these activities.

1. Use a bicycle for transportation or recreation.

Response	%		N
Regularly	10		(382)
Usually	2		(79)
Frequently	14		(505)
Sometimes	43		(1584)
Never	32		(1174)
	101		(3724)
		Missing data	(85)
		Total N	(3809)

2. Recycle the newspapers, glass or cans used at home.

Response	%		N
Recycle all of this material	14		(537)
Most of this material	22		(831)
About half of this material	13		(470)
Some of this material	35		(1292)
Never recycle	16		(592)
	100		(3722)
		Missing data	(87)
		Total N	(3809)

3. Change oil in the family car(s) at home.

Response	%		N
Regularly	46		(1725)
Usually	8		(313)
Frequently	5		(179)
Sometimes	14		(527)
Never	26		(970)
	99		(3714)
		Missing data	(95)
		Total N	(3809)

4. Developing and using skills to increase self-reliance, such as in carpentry, car repair, food preservation.

Response	%		N
Regularly	50		(1867)
Usually	12		(444)
Frequently	14		(534)
Sometimes	19		(708)
Never	4		(152)
	99		(3705)
		Missing data	(104)
		Total N	(3809)

(continued)

TABLE F21 (continued)

5. Buying clothing at a garage sale or second-hand store.

Response	%	N
All of the household's clothing	1	(22)
Most items	3	(70)
About half of the household's clothing	5	(194)
A few items	36	(1335)
None of the household's clothing	56	(2080)
	101	(3701)
	Missing data	(108)
	Total N	(3809)

6. Contribute to ecologically-oriented organizations (such as Sierra Club, etc.)

Response	%	N
Contribute regularly to 2 or more organizations	14	(527)
Contribute regularly to 1 organization	7	(268)
Occasionally contribute	25	(934)
Used to contribute	6	(213)
Never have contributed	47	(1734)
	99	(3676)
	Missing data	(133)
	Total N	(3809)

7. Engage in exercise for physical fitness (e.g., running, swimming, calisthenics, etc.)

Response	%	N
Regularly	35	(1290)
Usually	11	(393)
Frequently	17	(645)
Sometimes	30	(1121)
Never	7	(249)
	100	(3698)
	Missing data	(111)
	Total N	(3809)

Better Homes and Gardens (41 percent). About twice as many solar owners read National Geographic as do homeowners generally.

The opinion leadership item is presented in Table F18. Those engaging in a high number of the listed activities would score high on the opinion leadership index, using the study's analyses.

The item used to measure innovativeness is reported in Table F19; 60 percent of solar owners report that they have shower flow restrictors, and half own microwave ovens. Of solar households, one in four owns solar toys and gadgets and video games. These proportions are considerably higher than those reported by homeowners generally.

VALUES

The items on materialist and postmaterialist values in the homeowners survey were also asked of solar homeowners; data are presented in Table F20. All statements expressing materialist values are ranked as most important and next most important by majorities of solar owners. The most important of these was "a stable economy," selected by 84 percent. Postmaterialist values are ranked as important by 28 to 50 percent of the sample, with "protecting nature from being spoiled and polluted" selected most frequently as important. In all cases, then, materialist values are the most important values expressed by this sample of solar owners.

VOLUNTARY SIMPLICITY LIFE-STYLE

Solar owners data on activities associated with voluntary simplicity as a life-style are presented in Table F21. The majority of solar owners in the sample engage in increasing self-reliance skills; exercising for physical fitness; recycling newspaper, glass, or cans used at home; changing oil in the family car; bicycling; and contributing to ecological organizations. Many more solar owners than homeowners report engaging in these activities.

APPENDIX G: THE RELEVANCE OF VALUES TO SOLAR ADOPTION

There could be an important relationship between the values a person holds and his willingness to adopt solar energy. For example, three "paradigms" or "representative views of social reality" that were thought to be related to preference for solar energy were described by Reuyl et al. (1977).

Paradigm A was held to be the view most prevalent in current U.S. society; it was characterized by preferences for a high rate of growth, increasing regulation, institutional centralization, the achievement of humane goals through technological progress, and rational economic decision making.

Paradigm B was thought to be characterized by environmental concern, desire for decentralization of control, emphasis on humane values accompanied by a deemphasis on materialism and economic growth, and increased participative decision making. Changes in society deemed necessary by those holding this world view would not necessarily imply radical social reordering. Change could be achieved within the existing system by reforming institutional structures and procedures.

Paradigm C was described as embracing values similar to those of Paradigm B, perhaps somewhat more emphatically, but was differentiated from Paradigm B in two ways: (1) a conviction that if the system continues in its present course, disaster or catastrophe will occur, and (2) a conviction that to achieve the social changes deemed necessary would mean large-scale change in the fundamental structure and values of society.

The idea was that preference for solar energy was most likely to be a characteristic of those favoring Paradigm C. However, an empirical attempt to link these paradigms to solar favorability resulted in negative findings (Farhar et al. 1979a). Furthermore, since solar energy is widely preferred over other energy supply sources, preference for it is obviously not related to a set of values held by a small minority. It was, therefore, decided not to use the paradigmatic approach in the present research.

Some pilot research using a Q-sort technique was done on energy issues (Brewer and Brunner 1978; Brunner 1978). Most of the Q-sort statements used in these studies related to what citizens, government, and industry should be doing about the energy situation or involved energy-environment trade-offs. Using two very small samples (N = 19; N = 27), the researchers found two "basically distinctive viewpoints" in one study and three in the other. Characteristics of these different viewpoints:

Study 1

1. • Conservation and renewable energy resources
   • Perceive energy crisis
   • Do not believe in market prices as a fair mechanism
   • Little confidence in the government
2. • Against government intervention
   • Express faith in market prices to increase energy production

Study 2

1. • Conservationist
   • Environmentalist-Activist
2. • Expect energy production to maintain living standards
   • Reject higher consumer prices
   • Alienation
3. • Reject government intervention
   • Express faith in market
   • Prices to increase production
   • Question wisdom of the public

While some political and environmental positions are expressed in these viewpoints, the statements used did not appear to measure fundamental values adequately. It was decided not to use them in the National Study.

Yet other researchers have also discovered differences in paradigms. Dunlap and Van Liere (1977) described two antithetical paradigms, based on surveys they had done using special group samples. The "dominant social paradigm" is characterized by preferences for growth, economic expansion, and mastery over nature. The "new environmental paradigm" is characterized by preferences for limits to growth, "steady state" economics, and emphasis on the balance of nature. The scale used in the research had been evaluated and found valid and reliable; however, it was felt that, conceptually, the scale was too closely limited to environmentalism to be used for the purposes of the National Study.

To lend further weight to the notion that values could be important in solar adoption, Shama and Jacobs (1979) found value differences between government policymakers and solar advocates. Policymakers put most emphasis on economic values, using "cost effectiveness" as the basis for policy decision making. "Policymakers tend to view economic values not only as ends in themselves, but also as means to achieve all other values" (Shama and Jacobs 1979, v). Solar advocates stress both economics and social values, including environmentalism, political decentralization, conservation of natural resources, and personal freedom and growth. The two groups were found to differ in the relative emphasis they placed on different values that they share.

NAME INDEX

SUBJECT INDEX

ABOUT THE AUTHOR

BARBARA FARHAR-PILGRIM has been studying technology/society interactions and diffusion of innovations for over 10 years. A Ph.D. from the University of Colorado in sociology and psychology, she has gained national recognition for her work on the social aspects of weather modification, both planned and inadvertent. She joined the Solar Energy Research Institute in 1977 as a consultant, and initiated and directed the National Study of the Residential Solar Consumer. She is expert in social impact assessment, technology assessment, and survey methodology. Dr. Farhar-Pilgrim has over 60 publications and papers, with monographs and articles on public opinion about energy alternatives and the relevance of behavioral science research to marketing and public policy. She is presently teaching and consulting in Boulder, Colorado

CHARLES T. UNSELD has a B.S. in engineering from Lehigh University and is a Ph.D. candidate in Environmental Psychology at the City University of New York. He has eleven years of experience in environmental research, specializing in social impact assessment. He worked for two years for the National Academy of Sciences' Committee on Nuclear and Alternative Energy Systems, where he edited and contributed to a major volume on the sociopolitical impacts of alternative energy futures, has worked as a consultant to the National Park Service and the National Bureau of Standards, headed an impact assessment project for the New York State Energy Office, and was at the Solar Energy Research Institute for over three years, where he worked on community energy management planning, in addition to the research for this book. He is now a partner in Black Hawk Associates, a research planning and management consulting firm.